Czechoslovak Academy of Sciences

CZECHOSLOVAK ACADEMY OF SCIENCES

Scientific Editor: Dr. Marta Vojtíšková

Scientific Adviser: Dr. Tomáš Hraba

Foreign Language Editor: Dr. Marta Vojtíšková

BLOOD GROUPS OF ANIMALS

PROCEEDINGS OF THE 9th EUROPEAN

ANIMAL BLOOD GROUP CONFERENCE

(FIRST CONFERENCE ARRANGED BY E.S.A.B.R.)

HELD IN PRAGUE, AUGUST 18-22, 1964

Edited by Josef Matoušek

Springer-Science+Business Media, B.V.

Additional material to this book can be downloaded from http://extras.springer.com.

ISBN 978-94-017-5834-5   ISBN 978-94-017-6289-2 (eBook)
DOI 10.1007/978-94-017-6289-2

PREFACE

An international conference on blood groups and other individual differences in animals, organized by the Laboratory of Physiology and Genetics of Animals, Czechoslovak Academy of Sciences, Liběchov, upon invitation of the European Society for Animal Blood Group Research, was held from 18 to 22 August 1964 in Prague.

This publication contains all papers presented and discussed in plenary and separate sessions. The order in which the lectures were actually delivered has been changed in the present volume, with regard to the subject and the methods employed. Accordingly, the discussions have had to be arranged.

It is a pleasant duty to thank all the participants for describing the latest results of their research and for discussing the various findings. To Prof. C. Stormont we owe many thanks for his very original and stimulating leadership of the discussion.

We hope that the meeting in Prague provided the opportunity to bring together the investigators working with animal blood group systems and discuss common problems and exchange ideas and points of view. It is my sincere wish that this book should remind us not only of the scientific results, but also of the cordial atmosphere — not less important and necessary for research and human relationships.

The Editor

Liběchov, September 1964

CONTENTS

Preface   5
List of participants   11
Opening addresses (J. Matoušek, M. Hašek)   19
Introduction (M. Braend)   21

Blood groups in cattle

Developments of blood group studies in cattle (J. Bouw)   25
Blood group studies on B-groups in Polish Red Cattle (J. Rapacz, L. Dola and J. Jakóbiec)   39
Blood group studies on Pinzgau-cattle (L. Erhard and D. O. Schmid)   43
Studies on blood groups in cattle, horses and pigs (M. Hesselholt, B. Larsen, P. B. Nielsen and B. Palludan)   49
Research work on the J system of insemination bulls in the German Democratic Republic (R. Ebertus)   63
The inheritance of blood groups in the blood group system C in cattle (G. E. Nasrat)   69
Bovine isohaemolysins seeming to have several specificities (P. Millot)   75
Studies on the S blood-group system in French cattle breeds (F. Grosclaude)   79
Aspects of relationships between genetically determined characters in cattle (G. J. Kraay)   87
Investigations on the blood groups of Wisents (Bison bonasus) and hybrids in comparison with the blood groups of cattle (J. Gasparski)   93
Fluctuations of the level of conglutinin in bovine sera (D. G. Ingram and D. A. Barnum)   99
Discussion   105

Blood groups of pigs

A study of blood groups in pigs (P. Imlah)   109
Investigations on producing test sera for determination of blood groups in pigs (S. Alexandrowicz, A. Kaczmarek and I. Wiatroszak)   123
Blood group studies in pigs (H. Buschmann)   129
A new approach to boar progeny testing (R. W. Widdowson and T. A. Newton)   137

Study of pig blood groups in Vojvodina (V. Jovanović and Z. Stojanović)   149
A contribution to the study of the blood group system A in pigs (J. Hojný and K. Hála)   155
Blood group system O in pigs (J. Hojný and K. Hála)   163
Discussion   169

Blood groups in chickens, ducks, rabbits, rats and mink

The effect of blood group genotypes of the B system on the performance of hybrid chickens (E. M. Mc Dermid)   173
Red cell antigenic polymorphism in a strain of the Wyandotte hen (M 11) (A. Perramon)   179
The blood groups of ducks (L. Podliachouk)   187
Blood groups in rabbits (M. Varga, M. Tolarová-Koutková, M. Tolar)   193
Erythrocyte B 1 antigen in inbred rat strains (B. Frenzl, R. Brdička, V. Křen and O. Štark)   197
Studies on erythrocytic factors in rats (M. Spiteri and A. Eyquem)   205
Blood group studies in the domestic mink (J. Rapacz, R. M. Shackelford and J. Jakóbiec)   211
Discussion   217

Blood groups and serum protein polymorphism in horses

Application of blood typing and protein tests in horses (C. Stormont, Y. Suzuki and J. Rendel)   221
The blood groups of equidae (L. Podliachouk)   229
Blood group studies in horses (D. O. Schmid)   237
Serum proteins in equidae: species, race and individual differences (M. Kaminski)   245
Hemoglobins, haptoglobins and albumins of horses (M. Braend and G. Efremov)   253
Discussion   261

Serum protein polymorphism in man, cattle, sheep, goats, pigs and canidae

Serum protein polymorphism in man and other primates (M. Harboe)   267
Studies on protein polymorphism in pigs, horses and cattle (M. A. Graetzer, M. Hesselholt, J. Moustgaard and M. Thymann)   279
A new haemoglobin variant in zebu cattle (S. N. Naik, L. D. Sanghvi)   295
Transferrin types in South African cattle breeds (D. R. Osterhoff, J. H. van Heerden)   301
New haemoglobin differentiation in cattle (D. R. Osterhoff, J. A. H. van Heerden)   309
Tf^G — a new transferrin allele in cattle (D. R. Osterhoff, J. A. H. van Heerden)   311
Haemoglobins, transferrins and albumins of sheep and goats (G. Efremov, M. Braend)   313
Genetic determination of the serum "thread proteins" and the slow α_2 globulin polymorphism in pigs (J. Schröffel)   321

Influence of some gonadotrophic and androgenic hormones on the male sexual fraction
in the serum of pigs (J. Matoušek, J. Schröffel) 331
Serum proteins in Canidae: Species, race and individual differences (M. Kaminski,
H. Balbierz) 337
Discussion 343

Protein polymorphism in some sexual gland fluids

Polymorphism of protein fractions in the fluids of accessory genital glands in bulls
and boars (M. Valenta, J. Matoušek, E. Petrovský, A. Stratil) 349
Antigenicity and polymorphism of the ovarian follicle fluids in cows (J. Matoušek) 359
The polarographic analysis of seminal fluids as a method for the study of their
antigenicity (R. Petrovská, E. Petrovský) 369
Protein polymorphism of the seminal vesicles of bulls and the sensitivity of bull
spermatozoa to cold shock (J. Fulka, H. Šulcová, M. Valenta) 381
Discussion 387

Immunological tolerance and transplantation antigens

Immunological tolerance and blood groups (M. Hašek) 391
The relationship between erythrocytes and transplantation antigens in chicks (F.
Knížetová, V. Hašková) 397
Relation of blood groups to transplantation antigens in rabbits (P. Iványi) 401
The ontogenetic development of H-2 antigens in vivo and in vitro (J. Klein) 405
Chimaerism in sheep (E. M. Tucker) 415
Discussion 423

Activity reports from laboratories

Working report of the Blood Group Laboratory of the Institute for Artificial Insemi-
nation at Schönow near Bernau, G. D. R. (J. Pilz) 429
Report on the present stage of investigations in the problem of cattle blood groups
in Rumania (V. Derlogea, I. Gavrilet, I. Granciu, Z. Sirbu, I. Soceanu, S. Rusu) 431
Report from Genetics Research Laboratory, Dairy Husbandry Division, National
Dairy Research Institute, Karnal, Punjab, India (P. G. Nair) 433
The activity report of the Blood Group Laboratory (A. Schindler) 435
Some results of blood group studies in animals (V. N. Tikhonov) 437
Report on some working methods developed and applied in our laboratory (R. Ebertus)439

Related questions

Comparison of complements in the haemolytic test (B. Busch) 445
Comparative tests for obtaining isoantibodies against cattle erythrocytes (V. Dikov) 447
Importance of antiglobulin reaction for detection of new factors in cattle (P. Millot) 449

Blood typing of cattle of two indigenous breeds: Podolic and Red Spotted breed of Vojvodina (Yugoslavia), and of the original Simmental breed (V. Jovanović, L. Končar)   453

Investigation of the subgroups in the blood group antigens P and J' of cattle (L. Erhard, D. O. Schmid)   457

Blood groups in chicken of Spanish strains and breeds (A. Jover, A. Rodero)   459

On the relation between blood group genes and a lethal gene for hairlessness and prolonged gestation (K. Maijala, G. Lindström)   461

Relationship between blood groups and beef production in Chiana breed cattle (A. Salerno)   463

A contribution to the problem of blood groups in ducks (V. Drobná, J. Hort, P. Iványi, J. Mardiak)   467

Investigations of the blood transfusion in cattle (B. Busch)   469

The serological analysis of the A blood group substance(s) (J. F. Borel)   471

Discussion   479

Report of the business meeting of the E. S. A. B. R.

(Chairman: M. Braend)

General business   485

Elections of officers   486

Technical aspects of the business meeting   487

Closing of the Conference (M. Braend)   489

PARTICIPANTS OF THE 9th EUROPEAN ANIMAL BLOOD GROUPS CONFERENCE

Borel J. F.	Schweizerisches rotes Kreuz, Blutspendedienst, Zentrallaboratorium (The Swiss Red Cross, Blood Donor Service, Central Laboratory), Universitätsstr. 2, Bern, Switzerland.
Bouquet Y. H.	Stichting Bloedgroepen Onderzoek (The Foundation for Blood Group Research), Casinoplein 21, Gent, Belgium.
Bouw J.	Stichting Bloedgroepen Onderzoek (The Foundation for Blood Group Research), Duivendaal 5, Wageningen, The Netherlands.
Böhm O.	Veterinarski zavod Slovenije (Veterinary Institute of Slovenia), Ljubljana, Yugoslavia.
Brdička R.	Ústav pro obecnou biologii, Lékařská fakulta (Institute of General Biology, Medical Faculty, Charles University), Albertov 4, Praha, Czechoslovakia.
Braend M.	Institut for Indremedisin, Norges Veterinaerhogskole (Institute of Internal Medicine, Norwegian Veterinary College), Ullevalsvn 72, Oslo, Norway.
Busch B.	Institut f. Tierzuchtsforschung, Deutsche Akademie der Landwirtschaftswissenschaften, Dummerstorf bei Rostock, G. D. R. (Institute of Animal Breeding Research, The German Agricultural Academy).
Buschmann H.	Institut für Blutgruppenforschung (Institute of Blood Group Research), Haydnstr. 11, München, G. F. R.
Čuta J.	Ústřední plemenářská stanice (Central Breeding Station), Liběchov, Czechoslovakia.
De Ligny W.	Rijksinstitut voor Visserijonderzoek (Institute of Fishery Research), Haringkade 1, Ijmuiden, The Netherlands.
De San Martin	Institut Pasteur (Pasteur Institute) 25, Rue du Docteur Roux, Paris, France.
Dikov V.	Institut biologii i patologii razmnoženija selskochozjastvennych životnych (Institute of Biology and Pathology of Reproduction of Farm Animals), Bul. Lenin 55, Sofia 13, Bulgaria.
Dola L.	Wyzsza Szkola Rolnicza (Agricultural College), Al. Mickiewicza 21, Kraków, Poland.
Dostál J.	Laboratoř fysiologie a genetiky živočichů, Československá akademie věd (Laboratory of Physiology and Genetics of Animals, The Czechoslovak Academy of Sciences), Liběchov, Czechoslovakia.

Drobná J.	Výzkumný ústav drůbeže (Poultry Research Institute), Ivanka pri Dunaji, Czechoslovakia.
Ebertus R.	Institut für künstliche Besamung (Institute of Artificia Insemination), Schönow bei Bernau, G. D. R.
Efremov G.	Institut for Indremedisin, Norges Veterinaerhogskole (Institute for Internal Medicine, Norwegian Veterinary College), Ullevalsvn 72, Oslo, Norway.
Erhard L.	Institut für Blutgruppenforschung (Institute of Blood Group Research), Haydnstr. 11, München, G. F. R.
Eyquem A.	Institut Pasteur (Pasteur Institute), 25 Rue du Docteur Roux, Paris, France.
Fábián G.	University of Agriculture, Gödöllo, Hungary.
Fésüs L.	Department of Animal Nutrition, University of Veterinary Medicine, VII Rottenbiller 23, Budapest, Hungary.
Fleischer J.	Krajská plemenářská správa (District Breeding Administration), Potštejn, p. Kostelec nad Orlicí, Czechoslovakia.
Frenzl B.	Ústav pro obecnou biologii, Lékařská fakulta Karlovy University (Institute of General Biology, Medical Faculty, Charles University), Albertov 4, Praha, Czechoslovakia.
Fulka J.	Laboratoř fysiologie a genetiky živočichů, Československá akademie věd (Laboratory of Physiology and Genetics of Animals, The Czechoslovak Academy of Sciences), Liběchov, Czechoslovakia.
Gahne B.	Lantbrukshögskolans, Institution för Husdjursförädling (Agricultural College, Institute of Animal Breeding), Uppsala, Sweden.
Gasparska J.	Zaklad Hodowli Doswiadczalnej Zwierzat, Polska Akademia Nauk (Institute of Experimental Breeding of Animals, The Polish Academy of Sciences), Nowy Swiat 72, Warszawa, Poland.
Gasparski J. M.	Zaklad Hodowli Doswiadczalnej Zwierzat (Institute of Experimental Breeding of Animals, The Polish Academy of Sciences), Nowy Swiat 72, Warszawa, Poland.
Gippert E.	Orszagos Mesterséges Termékenyítési Központ (State Centre of Artificial Insemination), Remény 42, Budapest, Hungary.
Granciu I.	Institutul de cercetari zootehnice (Institute of Zootechnical Research), Str. Dr. Staicovici 63, Bucuresti, Roumania.
Grosclaude F.	Station centrale de génétique animale, Centre national de recherches zootechniques (Central Station of Animal Genetics, National Centre of Zootechnical Research), Jouy-en-Josas, France.
Hála K.	Ústav experimentální biologie a genetiky, Československá akademie věd (Institute of Experimental Biology and Genetics, The Czechoslovak Academy of Sciences), Flemingovo nám. 2, Praha, Czechoslovakia.
Hall J. G.	Animal Breeding Research Organisation, Agricultural Research Council, 6 South Oswald Road, Edinburgh, Scotland.

Harboe M.	The University Institute for Experimental Medical Research, Ullevaal Hospital, Oslo, Norway.
Hašek M.	Ústav experimentální biologie a genetiky, Československá akademie věd (Institute of Experimental Biology and Genetics, The Czechoslovak Academy of Sciences), Flemingovo nám. 2, Praha, Czechoslovakia.
Hernandez C.	Universidad de la Habana (The University of Havana), Habana, Cuba.
Hesselholt M.	Afdeling for Fysiology og Endocrinologi Samt Blodtypeforskning, Den Kgl. Veterinaer og Landbohöjskole (The Royal Veterinary and Agricultural College, Department of Physiology, Endocrinology and Blood Group Research), Bülowsvej 13, Kobenhavn, Denmark.
Hess J.	Institut für künstliche Besamung (Institute of Artificial Insemination), Schönow bei Bernau, G. D. R.
Hložánek I.	Ústav experimentální biologie a genetiky, Československá akademie věd (Institute of Experimental Biology and Genetics, The Czechoslovak Academy of Sciences), Flemingovo nám. 2, Praha, Czechoslovakia.
Hojný J.	Laboratoř fysiologie a genetiky živočichů, Československá akademie věd (Laboratory of Physiology and Genetics of Animals, The Czechoslovak Academy of Sciences), Liběchov, Czechoslovakia.
Hort J.	Ústav experimentální biologie a genetiky, Československá akademie věd (Institute of Experimental Biology and Genetics, The Czechoslovak Academy of Sciences), Flemingovo nám 2, Praha, Czechoslovakia.
Horváth I.	Orszagos Mestérséges Termékenyítési Központ (State Centre of Artificial Insemination), Remény 42, Budapest, Hungary.
Hradecký J.	Laboratoř fysiologie a genetiky živočichů, Československá akademie věd (Laboratory of Physiology and Genetics of Animals, The Czechoslovak Academy of Sciences), Liběchov, Czechoslovakia.
Imlah P.	Blood Group Research Unit, Royal School of Veterinary Studies, Edinburgh, Scotland.
Ingram D. G.	Division of Immunology, Ontario Veterinary College, Guelph, Canada.
Iványi P.	Ústav experimentální biologie a genetiky, Československá akademie věd (Institute of Experimental Biology and Genetics, The Czechoslovak Academy of Sciences), Flemingovo nám. 2, Praha, Czechoslovakia.
Josipovič D.	Veterinarski zavod Slovenije (Veterinary Institute of Slovenia), Ljubljana, Yugoslavia.
Jovanovič V.	Poloprivredni Fakultet (Agricultural Faculty), Novi Sad, Yugoslavia.
Jover A.	Departamento de zootecnia, Facultad de Veterinaria (Zootechnical Department, Veterinary Faculty), Cordoba, Spain.

Kaczmarek A.	Wyzsza Szkola Rolnicza, Katedra Szczegolowej Hodowli Zwierzat (Agricultural College, Faculty of General Breeding of Animals), Poznaň, Poland.
Kaminski M.	Laboratoire d'Histophysiologie, Collêge de France (Laboratory of Histophysiology, University of France, Paris, France.
Kevek J.	Veterinární fakulta (Veterinary Faculty), Košice, Czechoslovakia.
Klein J.	Ústav experimentální biologie a genetiky, Československá akademie věd (Institute of Experimental Biology and Genetics, The Czechoslovak Academy of Sciences), Flemingovo nám. 2, Praha, Czechoslovakia.
Knížetová F.	Ústav experimentální biologie a genetiky, Československá akademie věd (Institute of Experimental Biology and Genetics, The Czechoslovak Academy of Sciences), Flemingovo nám. 2, Praha, Czechoslovakia.
Končar L.	Poloprivredni Fakultet (Agricultural Faculty), Novi Sad, Yugoslavia.
Kopečný V.	Laboratoř fysiologie a genetiky živočichů, Československá akademie věd (Laboratory of Physiology and Genetics of Animals, The Czechoslovak Academy of Sciences), Liběchov, Czechoslovakia.
Koubek K.	Vysoká škola zemědělská (Agricultural College), České Budějovice, Czechoslovakia.
Koutková M.	Ústav experimentální biologie a genetiky, Československá akademie věd (Institute of Experimental Biology and Genetics, The Czechoslovak Academy of Sciences), Flemingovo nám. 2, Praha, Czechoslovakia.
Kovácz G.	Department of Animal Nutrition, College of Veterinary Medicine, VII Rottenbiller 23, Budapest, Hungary.
Kraay G.	Stichting Bloedgroepen Onderzoek (The Foundation for Blood Group Research), Duivendaal 5, Wageningen, The Netherlands.
Křen V.	Ústav pro obecnou biologii, Lékařská fakulta Karlovy University (Institute of General Biology, Medical Faculty, Charles University), Albertov 4, Praha, Czechoslovakia.
Krummen H.	Institut für Tierzucht, Universität Bern (Institute of Animal Breeding), Neubrückstr. 10, Bern, Switzerland.
Lang B. G.	Blood Group Research Unit. Royal School of Veterinary Studies Veterinary Field Station, Roslin, Midlothian, Scotland.
Larsen B.	Afdeling for Fysiologi og Endocrinologi Samt Blodtypeforskning, Den Kgl. Veterinaer og Landbohöjskole (The Royal Veterinary and Agricultural College, Department of Physiology, Endocrinology and Blood Group Research), Bülowsvej 13, Kobenhavn, Denmark.
Lazar P.	Veterinarski zavod Slovenije (Veterinary Institute of Slovenia), Ljubljana, Yugoslavia.
Leiva G.	Universidad de la Habana (The University of Havana), Habana, Cuba.

14

Lie H.	Institut for Indremedisin. Norges Veterinaerhogskole (Institute of Internal Medicine, Norwegian Veterinary College), Ullevalsvn 72, Oslo, Norway.
Lindström G.	Keinosiemennysyhdistysten Liitto, Korkeavuorenkatu 6. A. 7., (Livestock Breeding Research Organization), Helsinki, Finland.
Lipecka C.	Wyžsza Szkola Rolnicza, Katedra Szczegolowej Hodowli Zwierzat (Agricultural College, Faculty of General Breeding of Animals), Lublin, Lesznyňskiego 9. Poland.
Madeyska-Lewandowska A.	Zaklad Hadowli Doswiadczalnej Zwierzat, Polska Akademie Nauk (Institute of Experimental Breeding of Animals, The Polish Academy of Sciences), Nowy Swiat 72, Warszawa, Poland.
Mácha J.	Katedra genetiky, Vysoká škola zemědělská (Department of Genetics, Agricultural College), Zemědělská 3, Brno, Czechoslovakia.
Martínek L.	Ústav experimentální biologie a genetiky, Československá akademie věd (Institute of Experimental Biology and Genetics, The Czechoslovak Academy of Sciences), Flemingovo nám. 2, Praha, Czechoslovakia.
Matoušek J.	Laboratoř fysiologie a genetiky živočichů, Československá akademie věd (Laboratory of Physiology and Genetics of Animals, The Czechoslovak Academy of Sciences). Liběchov, Czechoslovakia.
McDermid E. M.	Thornber Brothers Limited, Mytholmroyd, Halifax, Yorkshire, England.
Mervartová H.	Ústav experimentální biologie a genetiky, Československá akademie věd (Institute of Experimental Biology and Genetics, The Czechoslovak Academy of Sciences), Flemingovo nám. 2, Praha, Czechoslovakia.
Míček J.	Výzkumný ústav drůbeže (Poultry Research Institute), Ivanka pri Dunaji, Czechoslovakia.
Mikešová M.	Státní plemenářská správa (State Breeding Administration), Řepy u Prahy, Czechoslovakia.
Millot P.	Laboratoire des Groupes Sanguins des Bovidés (Laboratory of Blood Groups of Cattle), Jouy-en-Josas, France.
Moustgaard J.	Afdeling for Fysiologi og Endocrinologi Samt Blodtypeforskning, Den Kgl. Veterinaer og Landbohöjskole (Department of Physiology, Endocrinology and Blood Group Research, Royal Veterinary and Agricultural College), Bülowsvej 13, Kopenhavn V, Denmark.
Naik S. M.	Indian Cancer Research Center, Parel, Bombay 12, India.
Nikolajczuk M.	Wyžsza Szkola Rolnicza (Agricultural College), Wroclaw, Wornicka 29.
Novotný S.	Laboratoř fysiologie a genetiky živočichů, Československá akademie věd (Laboratory of Physiology and Genetics of Animals, The Czechoslovak Academy of Sciences), Liběchov, Czechoslovakia.
Okerman F.	Rijksstation voor Kleinveeteelt (State Station of Small Animal Breeding), 5 Dorp, Controde, Belgium.

Oosterlee C. C.	Stichting Bloedgroepen Onderzoek (The Foundation for Blood Group Research), Duivendaal 5, Wageningen, The Netherlands.
Oprescu S.	Institute of Biology of the Academy of the Rumanian People's Republic, Lt. Lemnea Street 16, Bucuresti, Rumania.
Papp M.	Department of Animal Nutrition, University of Veterinary Medicine, VII. Rottenbiller 23, Budapest, Hungary.
Pavlok A.	Laboratoř fysiologie a genetiky živočichů, Československá akademie věd (Laboratory of Physiology and Genetics of Animals, The Czechoslovak Academy of Sciences), Liběchov, Czechoslovakia.
Perramon A.	Station centrale de génétique animale, Centre national de recherches zootechniques (Central Station of Animal Genetics, National Centre of Zootechnical Research), Jouy-en-Josas, France.
Petrovský E.	Ústav experimentální biologie a genetiky, Československá akademie věd (Institute of Experimental Biology and Genetics, The Czechoslovak Academy of Sciences, Flemingovo nám. 2, Praha, Czechoslovakia.
Pilz J.	Institut für künstliche Besamung (Institute of Artificial Insemination), Schönow bei Bernau, G. D. R.
Podliachouk L.	Institut Pasteur (Pasteur Institute), 25 Rue du Docteur Roux, Paris, France.
Rapacz J.	Wyzsza Szkola Rolnicza (Agricultural College), Al. Mickiewicza 21, Kraków, Poland.
Rodero A.	Departamento de zootechnica, Facultad de Veterinaria (Zootechnical Department, Veterinary Faculty), Cordɔba, Spain.
Salerno A.	Instituto di zootecnia generale, Facolta di agraria dell' Universita di Napoli (Institute of General Zootechnics, Agricultural Faculty of the University of Naples), Porɔici (Napoli), Italy.
Sedláková E.	Laboratoř fysiologie a genetiky živočichů, Československá akademie věd (Laboratory of Physiology and Genetics of Animals, The Czechoslovak Academy of Sciences), Liběchov, Czechoslovakia.
Schmid D. O.	Institut für Blutgruppenforschung (Institute of Blood Group Research), Haydnstr. 11, München, G. F. R.
Schröffel J.	Laboratoř fysiologie a genetiky živočichů, Československá akademie věd (Laboratory of Physiology and Genetics of Animals, The Czechoslovak Academy of Sciences), Liběchov, Czechoslovakia.
Schultz J.	Medizinische Tierklinik, Karl-Marx Universität (Medical Animal Clinic, The University of Karl-Marx), Leipzig. G. D. R.
Schwerdtner H.	Zuchthygienedienst (Service of Breeding Hygiene), Potsdam, G. D. R.

Soos P.	Orszagos Mesterséges Termékenyítési Központ (State Centre of Artificial Insemination), Remény 42, Budapest, Hungary.
Sorokovoj P.	Vsesojuznyj naučno-issledovatelskij institut životnovodstva (All-Union Scientific and Research Institute of Animal Breeding), Dubrovica, Moskovskoj oblasti, SSSR.
Stojanovič Z.	Poljoprivredni Fakultet (Agricultural Faculty), Novi Sad, Yugoslavia.
Stormont C.	Department of Veterinary Microbiology, School of Veterinary Medicine, University of California, Davis, U. S. A.
Stukovsky J.	Orszagos Mesterséges Termekenyítési Központ (State Centre of Artificial Insemination), Remény 42, Budapest, Hungary.
Szeniawska D.	Zaklad Hodowli Doswiadczalnej Zwierzat, Polska Akademia Nauk (Institute of Experimental Animal Breeding, The Polish Academy of Sciences) Now Swiat 72, Warszawa, Poland.
Šereda L.	Státní plemenářská správa (State Breeding Administration), Řepy u Prahy, Czechoslovakia.
Šiler R.	Ústřední ústav živočišné výroby (Central Institute of Animal Production), Uhřiněves, Czechoslovakia.
Šmerha J.	Vysoká škola zemědělská (Agricultural College) Praha, Czechoslovakia.
Štark O.	Ústav pro obecnou biologii, Lékařská fakulta Karlovy University (Institute of General Biology, Medical Faculty, Charles University), Albertov 4, Praha, Czechoslovakia.
Tichonov V. N.	Institut citologii i genetiki Akademija nauk (Institute of Cytology and Genetics. The Academy of Sciences), Novosibirsk, SSSR.
Tucker E. W.	Institute of Animal Physiology, Babraham, Cambridge, England.
Varga M.	College of Agriculture, Gödöllö, Hungary.
Valenta M.	Laboratoř fysiologie a genetiky živočichů, Československá akademie věd (Laboratory of Physiology and Genetics of Animals, The Czechoslovak Academy of Sciences), Liběchov, Czechoslovakia.
Vasilev C.	Selskostopanski naučnoizsledovatelski institut (Agricultural Scientific and Research Institute), Ruse, Bulgaria.
Vojenčiak J.	Veterinární fakulta (Veterinary Faculty), Komenského 69, Košice, Czechoslovakia.
Widdowson R. W.	British Oil and Cake Mills, Pig Advisory Dept. Unilever Ltd., London, England.
Zavřel F.	Laboratoř fysiologie a genetiky živočichů, Československá akademie věd (Laboratory of Physiology and Genetics of Animals, The Czechoslovak Academy of Sciences), Liběchov, Czechoslovakia.
Želev A. I.	Institut po životnovodstvu (Institute of Animal Breeding), Kolarograd, Bulgaria.

Żurkowski M.   Polska Akademia Nauk. Zaklad Hodowli Doswiadczalnej
                        Zwierzat. (The Polish Academy of Sciences, Institute of
                        Experimental Animal Breeding), Nowy Swiat 72, War-
                        szawa, Poland.

OPENING ADDRESSES

J. MATOUŠEK

Laboratory of Physiology and Genetics of Animals, Czechoslovak Academy of Sciences, Liběchov

Ladies and gentlemen, dear friends,

Allow me to welcome you here today in the name of all workers of our laboratory and to extend sincere greetings to you all. We do hope and wish that you will enjoy your stay in Prague and that you will take home with you pleasant memories of Prague and of this conference.

In spite of many hours of effort devoted to the preparation of this conference, we know that there are certain faults in our arrangements and, accordingly, we earnestly request that you bear in mind that we are amateurs in this kind of work and forgive us for our errors.

And now allow me to ask Dr. Hašek to address you on behalf of the Czechoslovak Academy of Sciences.

M. HAŠEK

Czechoslovak Academy of Sciences, Prague

Ladies and Gentlemen,

In the name of the Czechoslovak Academy of Sciences, it is a very pleasant duty to welcome here all the distinguished colleagues from abroad who have shown so much enthusiasm in answering the invitation and coming to this symposium in Prague.

It is an honour to have among us so many specialists in the field of animal blood group research. I wish that this Conference should not only review a large body of new facts as follows from the programme, but also the discussions should be most fruitful and lead to an outline of further tasks in this, from many sides, so useful discipline of science. We are very pleased that the first conference after the establishment of the European Society for Animal Blood Group Research takes place in Prague.

The Czechoslovak Academy of Sciences and especially the laboratory directed by Dr. Matoušek feel greatly honoured by this decision. I hope that this Conference will contribute to further cooperative effort and exchange of information which are so important for the advancement and application of animal blood group research.

I conclude by asking Mr. President of the European Society for Animal Blood Group Research, Mr. Mikael Braend, whom I am particularly glad to welcome among us, to take over the chairmanship and open the Conference.

INTRODUCTION

M. BRAEND

President of the European Society for Animal Blood Group Research

Mr. Hašek, Mr. Matoušek, Ladies and Gentlemen,

I thank the two representatives from the Czechoslovak Academy of Sciences most heartily for their kind words of welcome. We are very grateful and much indebted to the Czechoslovak Academy of Sciences because they took the responsibility of arranging this conference.

This conference is called the 9th European Conference of Animal Blood Groups and the first of our Society. This may seem strange to those of you who do not know the historical background of our Society. I shall therefore briefly give you some information about this. About ten years ago a few of us after some basic training in America started to work on cattle blood groups in our respective countries. Due to the nature of these special characters of cattle red blood cells we found it necessary with close contacts and cooperation especially in the production of anti- and test sera. We also found it necessary to meet frequently and in 1954 we had our first meeting in Copenhagen. Although we were not more than four at that meeting it has now got the designation the First European Conference on Animal Blood Groups. In the years to follow our group increased very rapidly and so did our activities. We developed more and more extensive comparison and collaboration programmes. Two years ago we found us in a situation where we in many respects acted and behaved as a society even although being only an informal collaborating group of people. At our preceding conference in Ljubljana therefore we decided to establish a Society and called it "European Society for Animal Blood Group Research". We made a preliminary constitution and by-laws in Ljubljana. These we have improved to the best of our ability and to which we will return later at this conference.

This was a look backwards. It is also of importance to consider where we stand today and if possible how it will be in the future. Our field covers according to our Constitution genetically determined characters of animal tissues and fluids. If we look at the programme we can see that it is really a fact. At this conference we will have reports on all kinds of domestic animals which have been studied as to a large variety of genetic systems, but more and more will come. Advanced techniques will be taken into use. It is not possible to tell yet how far it will come and if these characters of animal tissues and fluids will be shown to be of greater importance than they are today is too early to predict. At any rate, all the secrets we are going to disclose and all the knowledge we are going to accumulate in this field will sooner or later in some respects or other be of importance as long as they represent fundamental research.

Today we are going to learn more about recent developments in our field and in the related ones and in the 2 days to come even more. We should therefore all be able to benefit to a great extent from this conference. Already now I therefore want to express our very best thanks to those people of the Czechoslovak Academy of Sciences who have arranged this conference in Prague. We know that they have had very much hard work to do. But I am certain that you all agree that they have done an excellent job. I also want to thank for the friendliness and hospitality offered to us upon our arrival in Prague. I do not know all those who have been working hard for long time in preparing this conference, but I am quite certain that we should thank most Mr. Matoušek. We are very grateful to you, Mr. Matoušek.

On behalf of our Committee and our Society, it is a great pleasure to wish you all welcome to this conference. Our Committee hopes that you all will feel at home in our Society. Do not be too formal. We want to have the same spirit as it has always been and which resulted in that we were called "a group of friends".

And now it is time to start with the scientific programme.

BLOOD GROUPS IN CATTLE

DEVELOPMENTS OF BLOOD GROUP STUDIES IN CATTLE

J. BOUW

Blood Group Laboratory, State Agricultural College, Wageningen

Introduction

In considering the developments of the studies on blood groups in cattle we can conclude that the number of references to the publications, presented on this subject before 1940, is increasing gradually. Without disregarding the value of these reports, we may state here that the fundamental work which has been developed in this field in the Department of Genetics at the University of Wisconsin is to be considered as the basis for blood grouping work in cattle, as this is performed now in a large number of laboratories all over the world.

After the pioneer work at Wisconsin which was reported by Ferguson (1941), Ferguson, Stormont and Irwin (1942), Stormont (1950, 1952) and Stormont, Owen and Irwin (1951), the research in this field has found its distribution through the U. S. A. and Canada in the period from 1945—1950.

A number of young scientists, mainly from the north-western parts of Europe, have after 1950 passed a schooling in the laboratories at Madison-Wisconsin and at Davis-California and thereafter have started their own work in the home-countries. During the last decade the number of persons joining this group has been increasing rapidly. At this moment, over 25 laboratories including more than 50 scientists distributed all over the world are collaborating in the blood grouping work of cattle in the European Society of Animal Blood Group Research (E. S. A. B. R.).

When considering this group of people we can observe a large variety in backgrounds of education, national and individual characters and scientific interests. The aims of the work performed in various laboratories can be found to vary from pure scientific purposes to a series of other objectives. Furthermore, we can conclude that the environments in which the work is done and the tasks with which the investigators are charged are differing greatly from one laboratory to another.

Another aspect of variation is the subject in which the laboratories and the workers in these laboratories are interested. The main interests of the workers are varying from immunological to serological and genetic aspects, while those who are less interested in the "old subjects" can focus their attention to biometrical and animal breeding aspects of the blood grouping work in cattle populations. For those who wish to enter other details in the field of genetically determined characters in body tissues and fluids, the use of methods for electrophoresis is offering interesting possibilities.

A consideration of the reports which have recently been presented and the variations in persons and institutes raises the question whether it will be possible to follow in the future all newly established findings in detail. An even more important question calling for attention here is, whether it will be possible and also desirable to maintain a common working basis with uniform methods and nomenclature.

In this report a number of developments connected with the principles of the work on blood groups in cattle will be reviewed. In connection with these developments it will be considered if and to what extent the collaboration of laboratories and workers in the field are desirable.

Since Stormont (1962) presented a review on the current status of blood group systems in cattle and since the developments in the studies of protein polymorphism in cattle will be reported by other authors at this conference, these subjects will not be discussed here.

The applications of the data of our research are by now penetrating more and more into the fields of animal breeding and population genetics. Since the scope of this report does not permit to discuss these relatively new subjects in detail, the developments in this field will be reported elsewhere.

Production of reagents

After Landsteiner's discovery of blood group in man in 1900, various investigations have been made, among others by Ottenberg and Friedmann (1911), Little (1929), Schermer and Otte (1953) and Tolle (1953) to detect blood groups also in cattle by using naturally occurring isoagglutinins as reagents. In these years when the common knowledge of the blood groups as well as the equipment were relatively scarce, the observations made by these and various other authors have no doubt been of great interest, and have certainly required immense and laborious exertions.

After the introduction of isoimmune-haemolysins for the detection of cattle blood groups by Ferguson (1941) and Ferguson, Stormont and Irwin (1942), the observation was soon made that with this method a large number of strongly reactive and highly specific reagents could be produced in cattle blood serum.

By re-reading now Ferguson's first report of 1941 we can find that his first series of immunizations generally consisted of four intravenous injections of one liter citrated blood at weekly intervals. Since these amounts of blood and possibly also of sodium citrate caused rather severe symptoms of anaphylactic shock, the injected amounts of blood have been gradually decreased.

Stormont (1950) reported a decrease in the amount of blood to be injected to 250 ml. and in some cases to 25 ml. Neimann Sørensen (1958) reports about weekly injections of 100cc citrated blood and 50cc in Jersey cattle. Bouw (1958) and Braend (1959) mentioned injections of 50cc only.

From the discussions at the first meetings of the workers on cattle blood groups in Europe we remember well that this decrease in the amount of injected blood had not yet dissipated the fear for the symptoms of anaphylactic shock and their possible consequences.

To prevent at least production of antibodies against blood plasma constituents and possibly also the implications from the injections of large amounts of sodium citrate, various laboratories are now using the method described by Ferguson et al. (1942) for immunization of rabbits by injections of only washed erythrocytes.

From the literature and the discussions to this subject we can further conclude that variations in the techniques as described by Ferguson were related not only to the amounts and qualities of the injected material, but also to the time intervals between the injections. Apart from necessary changes in the intervals owing to the fixed working schedules in the laboratories and in the herds, investigations have been done to see whether changes in these time intervals can lead to higher concentrations of antibodies and possibly also to the production of reagents for unknown factors.

In this respect, mention can be made of the experiments of Bouquet (1963) which were confirmed in our laboratory: the animals possessing relatively high concentrations of antibodies in their blood serum were re-injected. In various cases it was observed that re-injections in these periods can very well lead to a rapid decrease in the concentrations of antibodies in the blood serum. In our opinion, this fall in concentration can be explained by assuming that neutralization of antibodies and antigens has taken place in the animal's body after the last injection.

Most workers in the field have by now experienced that re-immunizations of the animals 6 months or more after the first immunization period usually resulted in antisera with higher concentrations of antibodies.

In view of these experiences, it seems advisable to use the first set of injections only for sensitization of animals as long as such a procedure is practicable. The risks from the mentioned neutralization effects can be escaped when the time interval between the first and second and eventually a third re-immunizing injection does not exceed 24 hours each.

For judging the results of immunizations, serum samples are usually taken maximally 7 days after the last injection. In this respect, attention may be drawn to the observation made by Ehrlich and Morgenroth (1900) already 60 years ago that the blood serum of a goat, which had been immunized with the goat blood, reacted with the donor red blood cells 13 days after the last injection.

In the production of specific antibodies for the detection of cattle blood groups, attention has been focused mainly on isoimmune haemolysins during the past decades. Moreover, various attempts to obtain specific reagents from other sources have been made in the past 20 years.

Of these sources can be mentioned in the first place the sera of cattle which have not been immunized previously with cattle blood. Of the antibodies occurring in the normal cattle sera anti-J has attracted much attention of a large number of scientists. Interesting reports on serologic relationships of this anti-J with naturally occurring antibodies in body fluids of man and various species of animals have been reported, among others by Ycas (1949), Neimann Sørensen et al. (1954), Tucker (1962) and Stone (1962).

Mitscherlich, Tolle and Urbaschek (1956) and Rendel, Tolle and Neimann Sørensen (1957) observed that various blood factors detected with immune haemolysins also reacted specifically by agglutination with a few out of several thousands of normal cattle sera. In our laboratory four of such sera have been used for a short period of time. However, three of them were found to loose their specificity within a year's time.

The production of specific reagents for cattle blood in other species of animals has been started already by Ferguson et al. (1942), who injected intravenously washed cattle erythrocytes into rabbits. Since the results of the first trials were rather successful, most workers in the field have used this procedure for some time. Twenty years of experiments have disclosed that specificity and storing qualities of heteroimmune haemolysins for cattle blood produced in rabbits do not usually reach the levels of those of isoimmune sera. We must conclude now that apart from a small number of special reagents, most laboratories are using rabbit sera only as long as certain reagents cannot be produced in cattle.

Although better results were expected from immunization of animals more closely related to cattle, as far as specificity is concerned, immunization of goats and sheep with cattle erythrocytes as reported by Neimann Sørensen (1958) has so far not resulted in many applications.

Interesting data about the production of heteroimmune-haemolysins for cattle blood can be found in a table by Stormont (1962) giving the sources of the bovine blood-typing reagents used in Davis, California.

The usefulness for the blood groups of natural substances like seed extracts (lectins) reacting specifically with blood groups, has been reported by Ren-

konen (1948) in man, in fishes by Sprague (1961), and in chickens by Scheinberg and Reckel (1961) and Borel (1962). Reports about the usefulness of
such substances for the detection of cattle blood groups have not yet been
presented as far as it is known to the author.

Since Stormont (1962) has presented a survey of the reagents which are
now in use in various laboratories, there is no need to go into this subject in
detail.

As to the possibilities for the production of reagents for the detection and
differentiation of so far unknown blood group factors, it can be concluded
here that after the establishment of the first 45 odd reagents, produced in the
first period of work at Wisconsin, various authors have questioned whether
the limits of further differentiations were reached or not.

If we now consider the regular increase in the number of reagents during
the past 15 years (in this period at least 25 new internationally accepted or
acceptable reagents were added), we must conclude that predictions in this
respect are rather risky unless more fundamental knowledge of the variations
in the structure of the red blood cells is available.

Variations in the immunization techniques and the use of cattle belonging
to breeds, which have not yet been for this purpose, can, of course, be very
profitable for the production of new specificities of antibodies. We may
conclude here, however, that most of the recently introduced so-called "new
factor reagents" have been produced with the "classical" methods and breeds.

The consideration of these data is, to the author's opinion, demonstrating
clearly that there is no reason to fear that the limits of the possibilities for
further developments in this field are reached.

Specificities of reagents

The so-called "reference tests", as they have been performed now in six
successive years, revealed a number of interesting aspects for the study of
specificities of reagents detecting clearly distinguishable antigenic factors.

Until 1962, the reagents presented in table 1 have been selected as references.

In 1963 an anti-B reagent from Ljubljana, Yugoslavia, anti-G from
Finland and anti-K from France could be appointed as references, while in
1964 two or three more reagents can possibly be selected.

For the selection of reagents which could be accepted as references, the
reagents of various laboratories with closely parallelling specificities have
been compared in tests with blood samples from animals of as many different
breeds of cattle as possible.

Table 1

Reagent	Producing laboratory	Year of acceptance	Reagent	Producing laboratory	Year of acceptance
A_2	Norway	1959	I_1	Denmark	1960
K'	Denmark	1959	T_1	Denmark	1960
I'	Czechoslovakia	1959	O'	Sweden	1960
Y'	Sweden	1959	C_3	Sweden	1960
C_1	Sweden	1959	X_2	Netherlands	1960
R	South Africa	1959	S_1	Yugoslavia	1960
W	Netherlands	1959	Q	Netherlands	1961
F_1	Yugoslavia	1959	Y_2	Norway	1961
V_2	Norway	1959	B'	Czechoslovakia	1961
L	Denmark	1959	Z	Poland	1961
M	Denmark	1959	H	Germany	1962
S_2	Sweden	1959	D'	Netherlands	1962
U'	Germany	1959	L'	Finland	1962
Z'	Czechoslovakia	1959	U_1	Germany	1962

After the last selection, we may say that a reference is available of more than half the total number of reagents, which are generally in use now.

Table 1 demonstrates that the number of selected references is decreasing yearly. The experiences with tests also showed that the difficulties with the selections are gradually increasing. This increase in difficulties can be partly explained by the fact that the number of breeds, whose blood samples are used, is gradually increasing. In the last reference test, for example, we observed a perfect parallellism between the reaction patterns of various reagents for factor P. The specificities of these P reagents were, however, found to differ markedly in tests with the blood samples originating from cattle belonging to the "Africaner" breed.

In this respect, mention may also be made of the fact, that the selection of a reference reagent for factor J' had been possible already several years ago when no blood samples from Germany and France carrying the pheno-group $OE_2'J'$ had been introduced in these tests. The investigations performed by Erhard and Grosclaude (1964) have now revealed that the difference in specificities of the various J' reagents is most certainly due to the existence of two subtypes of J', J_1' and J_2'.

To our encouragement, we have been able to conclude from the last reference test that the selection of a reference reagent for a factor like G', which has come into the picture only a few years ago, is very well possible. In this comparison, a perfect parallellism between 4 reagents for factor G' could be observed in a test in which 247 blood samples, originating from 17 different laboratories, were used.

A consideration of the difficulties with the selection of reference reagents demonstrates that the main problems are connected with the specificities of the reagents for those factors which are known to be of more than one type. Trials for the selection of reference reagents for the subtypes of the factors O and E', (three linear subtypes had been introduced) have so far been unsuccessful.

From this point of view, it is very well explainable that especially the later established laboratories, in which even other breeds of cattle are in use, encounter great difficulties in their attempts to produce reagents reacting specifically with various subtypes of such factors.

A consideration of the problems involved in the selection of reference reagents and the recent establishment of various types of serological relationships of blood group factors demonstrates, however, that also the "older" laboratories will most certainly have to face a number of new problems in this subject.

From the experiments with the reference tests we must conclude that most laboratories are usually able to produce pure reagents, also for the subtypes in so far that the sera react only with the blood possessing the blood group factor as such and possibly with the substances which are serologically closely related to these factors.

In considering the results of the most recent reference tests we must conclude that for the so-called linear subtypes a considerable variety of specificities is observable when the reagents for such types are compared in tests in which blood samples of a wide variety of cattle breeds are used.

The available data show that the description of subtypes of various blood group factors in cattle by Stormont (1950) has been extremely valuable for a correct understanding of the blood grouping work. For the laboratories working with breeds of cattle of different origins, however, it will be extremely difficult to produce reagents for these subtypes which parallel in specificity those originally produced in the U. S. A.

The data from the recent reference tests also show that various laboratories have recently produced reagents with specificities of non-linear types for a number of factors, in which until now only linear subtypes have been described. The most striking examples for these phenomena are the reagents for the factors O and E', 3 linear subtypes of which have been generally accepted.

The further development of the investigations will most certainly disclose that similar findings of linear and non-linear subtypes will be made in the near future for a number of blood group factors, only one type of which is known until now.

In this respect, mention can be made of some interesting reports by Stormont et al. (1961) and by Grosclaude et al. (1963) about the specificities of sera reacting with various substances controlled by the S locus. The investigations of these authors and their co-workers have disclosed clearly that antisera detecting substances which are controlled by the S-locus can contain one specificity, which cannot be narrowed by any kind of absorption, while the same specificity is reacting with substances which have so far known as different blood group factors of the S-system. Stormont (1950) produced a reagent which (following his nomenclature)*) reacted with the phenogroups SH′ and U_2 thereby establishing a close serological relationship between factors S and U. Sera with the same specificities have now also been produced in France and in the Netherlands.

We will express our hope here that Grosclaude (1964) will disclose at this meeting some more of his observations on the diversities of specificities of antisera detecting various substances controlled by the S-locus.

At this moment, we can conclude that the investigations on serological relationships of blood group factors as discussed here indicate that Landsteiner's (1900) statement that antibodies are not absolute in their specificities does also apply to the reagents detecting blood group factors in cattle.

A consideration of these and various more or less similar findings in the S system raises the question of whether such sera with broader specificities have not been produced for the B blood group system to which much more attention has been paid until now. To answer this question, we can conclude here that most workers in the field of cattle blood groups have mainly focussed their attention to those antisera, out of which reagents could be produced, which were comparable in specificity with the original American standards.

Most workers in the field are well aware of the fact that they have stored in their freezers a number of sera containing specificities which are not generally in use. Although a precise examination of these sera will usually not be helpful for a direct application of the work, such analysis can possibly offer a number of interesting aspects on serologic relationships of blood group factors of the B-system. In such investigations, serologic relationships of factors like, for example, O and A′, can be revealed while relationships between such factors can be explained very well by assuming that such factors are in fact subtypes of one more complex factor.

A consideration of the various findings concerning the specificities of antibodies for blood group factors in cattle shows that an interesting field of research is still to be explored by those who are interested in the scientific background of the blood grouping work.

---

*) Unfortunately, we have not yet reached an agreement about the nomenclature of the factors in this system. According to the European nomenclature this reagent is detecting the phenogroups S_1 and U′.

Structures and genetic control of blood groups

In the discussion of the specificities of reagents mention has been made of the fact that one reagent (anti-SU) can react specifically with more than one serologic factor (S_1 and U'). Such observations are raising the question whether also the more classical reagents which we are very much inclined to consider as specific for only one factor, are, in fact, detecting more than one serological factor. By taking into account Landsteiner's statement abour specificities of antibodies and also the recent findings of the blood groups of the S system in cattle, we must conclude that such possibilities have to be considered seriously.

From this point of view, we have to face the question whether, for example, a reagent like anti-Y_2, when reacting with the blood group GY_2E_1', is detecting absolutely the same serologic substance as when reacting with the group BOY_2D'.

Although we must admit that the answers to such questions cannot yet be completely satisfactory, a more profound understanding can be reached of the problems related to the selection of reference reagents.

Apart from the question of identity of serologic factors in different blood group structures, the question of identity of the blood groups as such is attracting more and more attention in recent years.

Comparisons of the blood groups in different breeds of cattle are repeatedly raising the question whether the blood groups established in different breeds of cattle are indeed identical when the red blood cells react with the same reagents.

Considering the fact that a blood group like GY_2E_1' has a rather high frequency in most double purpose breeds of cattle in the U. S. A. and in Western Europe, for example, may easily lead to certain conclusions in the field of population genetics and animal breeding.

For these reasons, it may be valuable to consider whether or not all substances on the red blood cells reacting specifically with the reagents for the factors G, Y_2 and E_1' are indeed identical.

In such investigations, the so-called "new factor reagents" for blood group factors controlled by the B-locus can help to demonstrate that differences in the substances designated GY_2E_1' exist not only between different breeds of cattle, but even within certain breeds.

An example of such differences can be described here for the well-known B-group BOY_2D'. Neimann Sørensen (1958) reported a clear distinction between the groups BOY_2D' and BOY_1D', which could be made by means of reliable reagents for the subtypes Y_2 and Y_1. Bouquet (1964) observed a marked difference between the groups BO_3Y_2D' and BO_1Y_2D' in Belgian cattle breeds as a result of the use of a series of reagents detecting different subtypes of factor O. In our laboratory, a new factor reagent called anti-H_4, makes a clear distinction between the groups BOY_2D' and $BOY_2D'H_4$ in the same breeds of cattle.

Such examples show clearly that we have to be careful not to conclude too easily that the genetically determined characters of erythrocytes are indeed identical in different animals when they are characterized by the reactions with the same specificities of antibodies.

Although the serological data can reveal a number of interesting aspects of the structures of the blood groups, the backgrounds of genetic control of these substances can also contribute to the development of our knowledge.

A consideration of the opinions about the genetic control of the blood groups shows that most workers in the field have until recently accepted more or less the hypothesis that the complex blood groups, as they are known, for instance, in the systems B and C, are controlled by single genes. According to this theory each blood group in a system is corresponding to only one gene, while each of the various groups in this system is considered to be controlled by one allele out of a multiple allelic series at the corresponding locus of the chromosomes. Recent observations reported among others by Stormont et al. (1964), Bouw et al. (1964) and Nasrat (1964) suggest strongly, however, that various irregularities in the transmission of these complex blood groups are contradictory to this assumption. In the reported data various cases can be found in which transfers of blood group factors from one blood group complex to the other have taken place. According to the present state of our knowledge, such transfers of factors from one complex to another can be explained satisfactorily by assuming that the transferred factors are in fact controlled by separate genes. A crossing over between the genes controlling such factors and the gene(s) controlling the remaining part(s) of the blood group complex will in such cases result in the transfer of one or more factors from one group to the other.

In the author's opinion, a conclusion that certain blood group complexes are controlled by a series of closely linked genes does not necessarily mean that all factors, which are serologically distinguishable, have a separate genetic control. From the serological point of view, the question can be put forward, for instance, whether serologically related substances like linear and non-linear subtypes of the same factor are not controlled by the same gene. In that case the blood groups are in fact controlled by a limited number of genes each of which is controlling a smaller or larger part of the serologically determined substance.

A critical study of the reported data and viewpoints on the inheritance of the blood groups will lead to the conclusion that a large number of problems has still to be solved in this field.

It may be stated here that Stormont's (1955) report on "linked genes, pseudo-alleles and blood groups" has made interesting contribution to the understanding of this subject. However, this author also showed in his report that data and viewpoints from other fields of genetic research can be extremely useful for the improvement of our knowledge of the blood groups.

The problems of the inheritance of the complex blood groups can finally be brought closer to its solution, when all workers in the field will start to focus their attention to this subject and report all their well confirmed cases of irregular transmissions.

International collaboration

A serious consideration of the described differences in methods, materials and objectives in the various laboratories raises the question whether an international collaboration in blood group studies is desirable and possible.

As far as the desirability of such collaboration is concerned, we can conclude that the needs for contacts are varying strongly from one laboratory to another — depending mainly upon the state of development. It can be stated here, however, that almost all laboratories working in the complex field of cattle blood groups have started with the assistance of one or more other laboratories. It would be waste of time and money for the institutes intending to start investigations in this field if the existing possibilities for guidance and material help were not used.

Although the established laboratories will possibly be able to maintain the level once reached and develop their own methods and opinions, an isolation in this field of work will most certainly involve a severe risk for sterility.

In the introduction it has been stated already that the results of the blood grouping work in animals attract the attention of more and more workers in related fields of research. But the complex nature of our work is a serious barrier for those who are interested. If no uniformity is pursued in our methods and nomenclature, these barriers will increase and can easily lead to confusion. From this point of view, we must realize that many of the developments of the work can benefit from the interest of others and that, also for these reasons, the promotion of our common objectives is of particular importance.

As far as the possibilities for collaboration are concerned, we may say that the most important basis for the existing collaboration was that a relatively large number of workers in the field were able to pass a basic schooling in one of the first established laboratories in the U. S. A. and most of them maintained a close cooperation after this training period. These facts have strongly influenced the existing uniformity in methods and in nomenclature until now.

The increase in the number of persons working in the field and the recent developments initiated, in part, by the later established laboratories lead to various differentiations.

In the light of these developments, we can be grateful that the Food and Agricultural Organisation of the United Nations (F. A. O.) has instituted a panel of blood group scientists charged with the study of some aspects of collaboration.

At its first meeting, the F. A. O. panel (1963), has made up the recommendation that the countries intending to establish new laboratories in the field of animal blood group research should consider carefully their reasons for doing so and secure adequate facilities for basic scientific training followed by at least one year's experience in an established laboratory.

The panel appointed one sub-committee to consider a guidance publication primarily for the countries interested in initiating this kind of research, and one sub-committee which was requested to prepare suggestions for a uniform system of nomenclature both for cellular antigens and biochemical polymorphism.

Furthermore, the panel expressed its wishes that a comprehensive list of co-designations of the blood groups of the B system in cattle should be prepared and all reference sera for cattle blood grouping work should be stored in the laboratory of the secretary of the E. S. A. B. R.

The proposed objectives of the European Society for Animal Blood Group Research (E. S. A. B. R.) are presented in the Constitution and By-laws of this Society. The objectives are mainly related to the continuation of a close cooperation of the workers in the field.

As a result of the increase in the number of participating persons and laboratories, the technical and administrative affairs of this Society have to be handled by a committee composed of representatives of the various fields of animal blood group research. The merits of this Society will be determined mainly by the interest in cooperation of its members.

In the author's opinion, a complete freedom of each member to propose his own ideas and methods within the Society is indisputable. However, considerations of the described variations in persons and work and possible benefits for the members themselves, are drawing attention to the fact that uniformity in methods and nomenclature are indispensable for a sound development of the blood grouping work in cattle.

Summary

A survey is presented about the development of the investigations on the fundamental aspects of the blood grouping work in cattle.

It was shown that — although a constructive and practical basis has been laid — a large number of research projects in this field are still in full progress.

The investigations in the immunological and serological fields showed that a number of new complications with regard to the production of specific reagents appeared after the initiation of studies of so far not tested breeds of cattle.

Recent observations on transmisision of blood group complexes from parents to offspring have suggested that interesting new aspects can be revealed in the field of the inheritance of the blood groups by closer studies of the available data.

A number of variations in the interests, materials, methods and objectives of the various laboratories have been described. With the great increase in the number of persons entering the field of animal blood group research severe risks for desintegration have been introduced.

For the benefit of both the workers in related fields, interested in the developments and results of the blood grouping work and the investigators in the field itself, an appeal has been made to pursue as much uniformity as possible in methods and nomenclature.

References

Bouquet, Y. (1963). Personal communication.
Bouquet, Y. (1964). Bloedgroepen Onderzoek op Belgische Rundveepopulaties (in press).
Borel, J. F. (1962). Report 8th Animal Blood Group Conference, Ljubljana — Yugoslavia.
Bouw, J. (1958). Blood Group Studies in Dutch Cattle Breeds. Veenman — Wageningen.
Bouw, J., G. E. Nasrat and C. Buys (1964). Genetica, 35, 47.
Braend, M. (1959). Blood Group of Cattle in Norway. Scandinavisk Bladvorlag — Oslo.
Ehrlich, P. und J. Morgenroth (1900). Über Hämolysine. Berl. Klin. W. schr., 37, 453.
F. A. O. (1963). First meeting of the F. A. O. panel of blood group scientists. Food and Agricultural Organization of the United Nations. Rome — Italy.
Ferguson, L. C. (1941). J. Immunol., 40, 213.
Ferguson, L. C., C. Stormont and M. R. Irwin (1942). J. Immunol., 44, 147.
Grosclaude, F. (1964). Rep. 9th Animal Group Conf., Prague — Czechoslovakia.
Grosclaude, F. et P. Millot (1963). Ann. Biol. Anim. Bioch. Biophys., 2, 119.
Landsteiner, K. (1900). Zentralbl. f. Bakt. u. Parasit. k., 27, 357.
Little, R. B. (1929). J. Immunol., 17, 377.
Mitscherlich, E., A. Tolle und B. Urbaschek (1956). Über die mit normalen Rinderseren nachweisbaren Blutgruppenfaktoren des Rindes.
Nasrat, G. E. (1964). Rep. 9th Animal Blood Group Conf., Prague — Czechoslovakia.
Neimann Sørensen, A. (1958). Blood Groups of Cattle. A/S Carl Fr. Mortensen København.
Neimann Sørensen, A. J. Rendel and W. H. Stone (1954). J. Immunol., 73, 407.
Ottenberg, R. and Friedmann, S. S. (1911), J. Exp. Med., 13, 531.
Rendel, J., A. Tolle und Neimann Sørensen (1957). Z. f. Tierzücht. u. Zücht. biol., 70, 21.
Renkonen, K. O. (1948). Ann. Med. Exp. biol. Fenn., 26, 66.

Scheinberg, S. L. and R. P. Reckel (1961). Poultry Sci., 40, 689.

Schermer, S. und E. Otte (1953). Z. f. Immun. u. Exp. Ther.; 110, 296.

Sprague, L. M. (1961). Genetics, 46, 901.

Stone, W. H. (1962). Ann. N. Y. Acad. Sci., 97, 269.

Stormont, C. (1950). Genetics, 35, 76.

Stormont, C. (1952). Genetics, 35, 76.

Stormont, C. (1955). The Amer. Naturalist, 89, 105.

Stormont, C. (1962). Ann. N. Y. Acad. Sci., 97, 251.

Stormont, C. R. D. Owen and M. R. Irwin (1951). Genetics, 36, 134.

Stormont, C., W. J. Miller and Y. Suzuki (1961). Genetics, 46, 541.

Stormont, C. and B. Morris (1964). Immunogenetics Letter, 3, 130.

Tolle, A. (1953). Titererhöhung und Konservierung der Isohämantikörper zu Blut-gruppenuntersuchungen beim Rind. Weende — Göttingen.

Tucker, E. M. (1962). Rep. 8th Animal Blood Group Conf., Ljubljana — Yugoslavia.

Ycas, M. (1949). J. Immunol., 73, 407.

BLOOD GROUP STUDIES ON B-GROUPS IN POLISH RED CATTLE

J. RAPACZ, L. DOLA and J. JAKÓBIEC

Department of Cattle Breeding, College of Agriculture, Kraków

The number of blood groups discovered in cattle during the past 25 years leads to continually increasing use of these well-known genetic markers for both scientific and practical purposes. The investigations by Ferguson (1941) revealed considerable differences in blood antigenic factors between the various breeds. Evidence for extensive series of alleles in two systems, the B and C, has been given by Stormont, Owen and Irwin (1951).

The great number of blood antigenic factors inherited in various combinations, especially in the B system, and their simple mode of inheritance provide unusual advantage for the study of genetic differences, structure and origin of breeds. This has been emphasized in a number of investigations performed by Owen et al. (1947), Stormont et al. (1951), Neimann-Sørensen (1958), Rendel (1958), Bouw (1960), Neimann-Sørensen and Spryszak, Gasparski et al. (1960) and Braend et al. (1962).

It is the purpose of this report to present some additional data on the genetic structure of blood groups in the Polish Red Cattle in the regions of Kraków and Rzeszów.

The Polish Red Cattle are included into the brachycerous type and form a variety which presumably since praehistoric times has been spread over a large part of Europe. From olden times Poland has not been rich in cattle. During the past century, however, cattle became of great importance and their population increased considerably. It is generally assumed (Jakóbiec 1958, Szczekin-Krotow 1957) that cattle were of native origin at that time.

The first breeding practice with the Polish Red Cattle was started by the association of breeders in Kraków. It was founded in 1894 with an aim to improve productivity of this cattle by means of selection within the breed and in some herds by crosses with imported breeds. Several breeds were introduced into Poland: Black- and White Lowland, Brown Swiss, Simental Danish Red, Angeln and Red Friesian Cattle. The last breed has been introduced into the region of Kraków, however, the larger breeding herds have been formed by animals of native origin from small farms. In some parts of the country the imported cattle have rapidly been eliminated by environmental conditions (climate, poor soil and feeding deficiency). Very hard time for cattle breeding in Poland during World War I and II brought to ruin the pedigree herds and most of animals with higher production.

Varied mixture of material and particularly the environment created certain types of cattle. Southern Poland has "the hillside variety" of the Polish Red Cattle spread on hills, mountains and some in the lowlands with rather poor soil where cattle nearest to the native origin frequently occur. Until about the middle of this century, they were triple-purpose type. At the present time, double-purpose type is most numerous in this region. Over 95% of that variety of cattle which belong to small farms are estimated as a "pure" breed. About one per cent of 700 000 of animals (it makes about 45% of all the cattle in these two regions) are registered in breeding books.

During the course of four years' investigations, 1 289 blood samples were tested for genetic studies of blood groups; it is about 17% of registered breeding cattle. The following reagents of our own preparation were used during the first three years for all blood samples and in the last year 13 additional reagents were used: first group; A_2, Z', B, G_2, I_1, O_1, P, Q, T_1, Y_2, B', D', E_2', E_3', G', $Kr2$, I', J', C_1, C_2, R, W, X_1, F, V, J_2, L, S_1, U_1, U', Z / M; second group; A_1, G_1, I_2, O_3, O_x, A', $Kr3$, L', X_2, J_1, S_2, $Kr5$, $Kr4$.

The number of cattle tested consisted of 135 bulls from six insemination stations; another 108 breeding bulls from places with lower numbers of breeding cattle; 1,046 dams and offspring. Cases with questionable parentage and the crosses with Danish Red Cattle have not been included into the number of tested animals. Over 33% of sampled animals (410) could not be recognized with respect to B-groups; for 329 individuals both alleles were established and for 530 cattle only one allele could be classified.

This numerous group of individuals, which have not been classified according to the B-alleles, formed cows and offspring, whose parents, both or one of them, have not been tested or their genotype was unknown. Numerous families with few members were very difficult to genotype. In respect to the bull side, from three to eight offspring for each bull have been tested with the exception of five bulls where over ten offspring were tested. In order to give a representative picture of the distribution of the blood groups in Polish Red Cattle, we tried to cover all the regions where they occur. The intensity of breeding cattle varied considerably in different populations from 0 to 65%. The greatest difficulty in genetic analysis of tested cattle were single, unrelated families; they provided many rarely occurring alleles which have not been found by Neimann-Sørensen and Spryszak (1959) and Gasparski et al. (1960). It appears that the gene frequency cannot be estimated correctly in tested population because of 33% of animals with unknown genotype and over 40% with only one established allele. The further difficulty was a great variation of alleles; e.g. on ten farms in the very closed area with three cows per farm, on an average, 21 different B-alleles were recognized; the reverse phenomenon was observed on a pedigree state farm where over 130 individuals were tested and only 11 B-alleles were noted.

In all tested cattle we were able to establish 132 (table 1) different B-alleles. The alleles presented in table 2 were found to be most frequent.

Since some of the B-groups presented in table 1 may be incomplete concerning the composition of antigenic factors, some B-alleles which are not presented here have to be confirmed and further investigations are under

Table 1

B — alleles in the Polish Red Cattle

1. B	48. $G\ O_1\ Y_2\ O'Y'$	93. $Q_1\ I'$
2. B G	49. $G\ O_1\ Y_2\ Y'$	94. $Q_1\ O'$
3. $B\ G\ I_1\ O_1\ A'$	50. $G\ O_1\ T_1\ Y_2\ E'Kr2\ Y'$	95. $Q_1\ E'Kr2\ O'$
4. $B\ G\ O_1$	51. $G\ O_1\ T_1\ E'K'Kr2$	
5. $B\ G\ Q_1\ Y_2\ D'I'Y'$	52. $G\ O_1\ E'$	96. T_1
6. $B\ G\ Q_1\ E'G'Kr2$	53. $G\ O_1\ E'Kr2\ O'$	97. $T_1\ B'$
7. $B\ G\ Q_1\ E'G'Kr2O'$	54. $G\ O_1\ E'O'$	98. $T_1\ B'O'$
8. $B\ G\ Q_1\ E'O'$	55. $G\ Q_1$	99. $T_1\ E_3'\ I'$
9. $B\ G\ Y_2\ A'$	56. $G\ T_1B'$	100. $T_1\ I'O'$
10. $B\ G\ Y_2\ D'$	57. $G\ Y_2\ E'$	101. $T_1\ O'$
11. $B\ G\ Y_2\ E'G'Kr2\ O'$	58. $G\ B'$	
12. $B\ G\ Y_2\ E'G'Kr2\ O'\ Y'$	59. $G\ E_2'$	102. Y_2
13. $B\ G\ A'D'G'\ Kr2\ O'$	60. $G\ E'G'Kr2$	103. $Y_2\ B'D'E'O'Y'$
14. $B\ G\ E'$	61. $G\ E_2'Kr2$	104. Y_2D'
15. $B\ G\ O_x\ E'A'G'Kr2$	62. $G\ E_2'O'$	105. $Y_2\ D'I'$
16. $B\ G\ E'I'O'$	63. $G\ E_2'\ Kr2\ O'$	106. $Y_2\ D'I'G'$
17. $B\ G\ E'O'$		107. $Y_2\ E_2'\ G'Kr2\ Y'$
18. $B\ G\ O'$	64. I_1	108. $Y_2\ E'Kr2\ Y'$
19. $B\ I_1$	65. $I_1\ O_1$	109. $Y_2\ E_2'\ J'$
20. $B\ I_1\ O_1\ Q_1$	66. $I_1\ Q_1$	110. $Y_2\ E_2'\ O'$
21. $B\ I_1\ Q_1$	67. I_1E_3'	111. $Y_2\ G'Kr2\ Y'$
22. $B\ I_1\ Q_1\ O'$	68. $I_1\ E_3'\ G'Kr2$	112. $Y_2\ O'$
23. $B\ I_1\ Y_2$	69. $I_1\ K'$	113. $Y_2\ O'Y'$
24. $B\ I_1\ Q_1\ I'J'O'$	70. I_2	114. $Y_2\ Y'$
25. $B\ O_1$	71. $I_2\ D'E'G'Kr2$	115. $A'O'$
26. $B\ O_1\ P\ E'$		
27. $B\ O_1\ Q_1$	72. O_1	116. $B'E'Kr2\ O'$
28. $B\ O_1\ Y_2$	73. $O_1\ Q_1$	
29. $B\ O_1\ Y_2\ A'E'G'Kr2$	74. $O_1T_1E_3'$	117. D'
30. $B\ O_1\ Y_2\ D'$	75. $O_1\ Y_2\ A'$	118. $D'E'G'Kr2$
31. $B\ O_1\ D'$	76. $O_1\ Y_2\ D'$	
32. $B\ O_1\ E'G'Kr2$	77. $O_1\ Y_2\ Y'$	119. E_2'
33. $B\ Q_1$	78. $O_1\ E_3'$	120. $E_3'\ G'Kr2$
34. $B\ Q_1\ G'Kr2$	79. $O_1\ I'$	121. $E'Kr2$
35. $B\ T_1$	80. $O_1\ E'O'$	122. $E'I'Kr2$
36. $B\ T_1\ B'E'Kr2\ O'$	81. $O_1\ J'K'$	123. $E'O'$
37. $B\ Y_2$	82. $O_x\ B'E'O'$	124. $E'G'Kr2\ O'$
38. $B\ Y_2\ E'G'O'$		125. $E'Kr2\ O'$
39. $B\ Y_2\ E'G'Kr2\ Y'$	83. P	
40. $B\ Y_2\ E'Y'$	84. $P\ Q_1$	126. $G'Kr2$
41. $B\ E'$	85. $P\ Q_1\ E'I'$	127. $Kr2$
42. $B\ G'Kr2\ O'$	86. $P\ Y_2$	
	87. $P\ E_3'$	128. I'
43. G	88. $P\ E_3'\ I'Kr2$	
44. $G\ I_1$	89. $P\ I'$	129. O'
45. $G\ O_1$		
46. $G\ O_1\ Q_1$	90. Q_1	130. b
47. $G\ O_1\ Y_2$	91. $Q_1\ A'E'Kr2$	131. $B\ Q\ I'Kr3$
	92. $Q_1\ G'O'$	132. $B\ Kr3$

Table 2

Most frequent B-groups in the Polish Red Cattle

B-alleles	No. of animals	B-alleles	No. of animals
b	199	I_2	10
Y_2	110	O'	10
G_2E_2' Kr2 O'	102	G I_1	10
I_1 E_3' G'Kr2	100	Y_2 G'Kr2 Y'	9
O_x B'E'O'	99	B O_1 Y_2 D'	8
Y_2 Y'	56	B G	7
O_1	30	G O_1	7
I'	26	G E' G'Kr2	6
G	24	O_1 Q_1	6
B O_1	18	E'G'Kr2	6
B	12	B Y_2	4
G O_1 Y_2 Y'	12		

way, we have decided not to interpret our results, with respect to the B gene frequencies, breed structure, origin, similarities and differences from other breeds, until the investigations will be finished; consequently the B-groups are given in straight numbers of occurrence.

Numerous data of our studies showed that two new antigenic factors Kr2 and Kr3 determined in the presented B-groups belonged to the B system and were members of the 35 and 2 B-alleles, respectively. The Kr2 showed a strong relation with the G' and for a long time (from 1958—61) we could not separate the anti-G' from anti-Kr2. At the present time we know an allele ($B^{Y_2,D'I'G'}$) where the G' occurs without the Kr2. This allele was often found in our Friesian Cattle, but only once in Polish Red Cattle.

According to the frequency of other blood groups, the most striking example was noted for the M blood factor which was completely absent from Polish Red Cattle but was observed in crosses with Danish Red Cattle in 93% of cases.

References

Bouw, J. (1960). Z. Tierz. Zücht. Biol. 74, 248—266.

Braend, M., Rendel, J., Gahne, B., Adalsteinsson, S. (1962) Hereditas 48.

Dola, L. Allele w układzie B u bydła rasy nizinnej czarne-białej z rejonu krakowskiego (in press).

Ferguson, L. (1941). J. Immunology 40, 213—242.

Gasparski, J., Rapacz, J. and Rendel. J. (1960). Roczniki Nauk Rolniczych 76 B-3: 565—568.

Jakobiec, J. (1958). Miedzynarodowe Czasopisme Rolnicze 1, 106—116.

Neimann-Sørensen, A. (1958). Blood groups of cattle. Copenhagen, p. 177.

Neimann-Sørensen, A. and Spryszak, A. (1959). Anim. Prod. 1, 179—188, 1959.

Owen, R. D., Stormont, C. and Irwin, M. R. (1947). Genetics 32, 64—74.

Rendel, J. (1958), Acta Agric. Scand. 8, 191—215

Stormont, C., Owen, R. D. and Irwin, M. R. (1951). Genetics 36, 134—161.

Szczekin-Krotow, Wł. (1957). Przeglad Hodowlany 9, 24—41.

BLOOD GROUP STUDIES ON PINZGAU-CATTLE

L. ERHARD and D. O. SCHMID

Institute for Animal Blood Group and Resistance Research — Livestock Breeding Research
Organization, Munich

In addition to our immunogenetic studies on highland cattle, German spotted breed, German Brown-Swiss, German Yellow cattle, Murnau-Werdenfels cattle (Schmid 1962, Schmid 1963, Schmid and Erhard 1963) and Podolic steppe cattle, which is an ancestor of the highland breeds (Schmid and Mancic 1964), we have now studied the blood groups of Pinzgau cattle.

Bos primigenius of the brachycerous type is the ancestor of the Pinzgau cattle (Scheuch 1925), which belongs to the great group of spotted breeds and lives in the Alps since more than 2,000 years. The Pinzgau cattle is an important and adaptive highland breed with 1.560,000 animals in total. Mainly Pinzgau in the region near Salzburg in Austria and also the Chiemgau area of Bavaria near the Austrian frontier and the southern part of Tyrol are the breeding centres of the Pinzgau cattle. There are 450,000 animals, which is nearly one third of the whole lifestock. Roumania (Transylvania, Bukovina, West Carpathia), Slovakia, Italy (Southern Tyrol), Yugoslavia (Slovenia, Bosnia and Croatia) and since 1902 South-West Africa are also the breeding districts.

The basic colour of the Pinzgau cattle is chest-nut brown, the colour varies from light to dark brown, along back, croup and belly and also on the forelegs and shanks you can see white spots. The parts of the body without hairs are mostly flesh-coloured. The muzzle of dark brown red animals is dark grey or nearly black. The palate and the tongue are also light grey spotted.

As the spotted breed, the Pinzgau cattle is also a combined beef and milk type. A good constitution and a high vitality characterize this breed, which has a remarkable adaptability, hardiness and resistance for overcoming all climate, even in the tropics, soil and keeping handicaps and safeguarding the economic production. The Pinzgau cattle is highly fertile, early mature, with a remarkable capacity for meat production and best meat quality. In 1963 the average production of cows recorded throughout a year in the Chiemgau area was 3,671 kg of milk at 4·01 % resulting in 147 kg of butterfat. The Pinzgau cattle live mostly in small herds. Their claws are very solid, of a quality, which makes possible to overcome all the difficulties in climbing and marching for a long time in the mountains. This ability is highly appreciated in the East, South-East and South-West Africa.

Two large Pinzgau herds of the Bavarian Breeding Centre in the castle farm Herrenchiemsee and in the jail farm Bernau with altogether 390 cattle and 40 cattle from the Pinzgau herd-book organization Salzburg were available for our blood-group studies. The animals were mostly offspring of 13 bulls.

By means of the immune haemolytic test and 47 monovalent, internationally approved blood-group test sera anti-A, H, Z', B, G, K, I_1, I_2, O_1, O_3, P, Q, T_1, T_2, Y_1, Y_2, B', D', E_2', E_3', G', I', J', K', O', Y', C_1, C_2, R, W, X_2, L', E, F, V, J, L, M, S, H', U_1, U_2, Mü-1 (U_x), U', Z, R' and S' we determined the blood type of each animal and in addition the haemoglobin type and the serum transferrin type by starch gel electrophoresis. Many of the blood typed animals are highly related so that we have not computed the gene frequencies of the different blood groups and give only qualitative results. On the other hand, the high degree of relationship of the animals permits to study closed families through many generations and gives the opportunity of studying the genetics of B-phenogroups. The gene frequency study in individual blood-group factors to characterize one cattle breed is insignificant, but the phenogroup analysis often gives important insights into the blood-group structure of individual cattle breeds.

With the exception of the B-system, we could not find any typical deviations from the other blood-group systems in Pinzgau cattle as compared with other breeds. The same finding was made before in other highland breeds.

The B-system of cattle alone contains more than 30 different blood-group factors. By means of many thousand progeny and family investigations in the USA and Europe almost 300 characteristic, non-segregating factor combinations were determined within the B-system. These factors are always transmitted and are considered as "phenogroups". The B-system of cattle is, therefore, the greatest series of defined multiple alleles, controlling the formation of cellular antigens of one species.

Owen, Stormont and Irwin (1947) were the first to find the variations in the antigenic structure of red cells in cattle of two different breeds, the Holstein-Friesian and Guernsey breed. In subsequent years, a number of scientists tried to solve the problem of characterizing the different breeds by investigations of blood groups. The conclusions, resulting often from mathematical analysis, to which too much significance is often given, are not always correct.

The differences in the antigenic structure of the red blood cells can be proved by the results obtained with different highland breeds. However, we think that some blood groups, the phenogroups of which are specifically connected with a certain breed, are not generally specific for a certain breed. During the investigations on paternity cases, the phenogroups are often proved in a cattle breed, in which they could not be recognized for many years, and can be found even in quite another cattle breed. The frequencies of different genetic units may show as well pronounced differences with some populations of one cattle breed and different breeds.

The decisive importance of the phenogroups is therefore not the serological characterization of different breeds. There may be indications concerning the development and the relationship between various breeds, but they cannot be considered without reservation. Analysis of phenogroups is much more a question of immunogenetic typing when determining the zygosity of multiples, the infertility of the female offspring of multiple birth and above all paternity cases where the phenogroups are highly valuable as a direct and indirect proof for an exact determination of complicated cases (Schmid 1962).

During our genetic investigations we studied 277 offspring with their mothers of 13 bulls and observed 45 phenogroups of the blood-group system B. Earlier, we were already able to prove 38 of these phenogroups in cattle of other highland breeds. The table shows a comparison between the non-segregating inherited phenogroups of the Pinzgau breed and other breeds as German spotted breed, German Brown Swiss, German Yellow cattle and Murnau-Werdenfels cattle. The following phenogroups of the Pinzgau cattle were identified for the first time:

$$BI_1Y_1B' \qquad\qquad PE_2'$$
$$GPE_3'I' \qquad\qquad B'E_2'$$
$$PQE_3'L'O'$$

It must be pointed out that the Pinzgau breed has two phenogroups, which were described only in the Lowland cattle until now. These phenogroups were even considered to be characteristic for these lowland cattle. Thus, we obtained further evidence against the theory of characteristic phenogroups for every breed. These groups are $BGKO_3Y_1B'E_3'G'K'O'Y'$ (B_{28}) which were up to now known only with the Jersey and Guernsey breed and the group $BGKY_2A'$ (B_{171}) of the Holstein-Friesian cattle.

The question is open whether the Pinzgau cattle has a special position among the highland breeds because of the blood-group serological proof of unknown B-phenogroups not recognized up to now with highland and other breeds, or whether it is justified to speak of completely the same phenogroups, when identifying genetic relationships between different highland breeds, characteristic for every single animal of the same or other particular cattle breed. I must say that the B-phenogroups are used to prove the genetic relationship between the different highland breeds.

It was striking that in the phenogroups $BGKY_1E_3'G'O'$, $GPE_3'I'$ and GI' the blood-group factor G was reacting with our monovalent anti-G reagent very weakly in the lytic test.

45

B-System Code No.	Phenogroups	Pinzgau cattle	Spotted breed	Brown Swiss	Yellow cattle	Murnau-Werden-fels cattle
	b	+	+	+	+	+
164	B	+	+	+	+	+
	BIY_1B'	+	−	−	+	−
28	$BGKO_3Y_1B'E_3'G'K'O'Y'$	+	−	−	−	−
	$BGKY_1E_3'G'O'$	+	−	+	−	+
171	$BGKY_2A'$	+	−	−	−	−
	$BGKB'O'$	+	−	+	−	−
	$BGKE_2'O'$	+	−	+	+	+
	$BGKO'$	+	+	−	+	+
3	BO_1	+	+	+	−	+
8	BO_1Q	+	−	−	+	+
112	BO_1E_3'	+	+	+	+	−
	BE_3'	+	−	+	+	−
159	G	+	+	+	+	+
140	GO_1	+	+	+	+	+
	GO_1Y_1	+	+	−	−	−
141	GO_1E_3'	+	+	+	−	−
	$GPE_3'I'$	+	+	+	−	−
194	GE_2'	+	+	+	+	+
196	GI'	+	+	−	+	+
	$I_1O_3E_3'G'I'K'$	+	−	−	−	+
202	I_2	+	+	+	+	+
45	O_1	+	+	+	+	+
54	$O_1T_1E_3'K'$	+	+	+	+	+
144	O_1E_2'	+	+	+	+	+
212	O_3	+	+	+	+	−
	$PQE_3'I'$	+	+	+	+	−
	$PQE_3'O'$	+	−	−	−	−
	PE_2'	+	−	−	−	−
	PE_3'	+	+	−	−	−
71	$QD'E_2'$	+	+	−	−	−
146	QE_3'	+	+	+	+	−
	$QE_3'J'$	+	−	−	+	−
82	Y_2	+	+	+	+	+
245	Y_2I'	+	+	−	−	−
161	Y_2Y'	+	+	+	+	+
	$B'E_2'$	+	−	−	−	−
137	D'	+	+	+	+	−
246	$D'E_3'$	+	+	+	+	−
85	E_2'	+	+	+	−	+
86	$E_2'I'$	+	+	+	+	−
	$E_2'J'$	+	+	+	−	−
	$E_3'G'$	+	+	−	+	−
89	I'	+	+	+	+	+
	$I'O'$	+	+	+	+	−
	J'	+	+	−	−	−
96	O'	+	+	+	+	+

In some animals, this factor, which must have been present according to the parentage, could be found only by absorption. In the above mentioned phenogroup $BGKY_1E_3'G'O'$, the factor K was always present very strongly in contrast to the other described BGK phenogroups. In agreement with earlier observations, also in the actual family study in Pinzgau cattle, we could determine phenogroups in the A, C and SU blood-group system only in some cases, because the high frequencies of the single-group factors in these systems do not permit an exact phenogroup analysis even in the case of big family examinations.

As in the spotted breed, Brown Swiss, Yellow cattle, Murnau-Werdenfels cattle, Vorderwälder and Hinterwälder cattle, our tested highland breeds (Schmid 1963), haemoglobin and transferrin polymorphism were also established in Pinzgau cattle. The haemoglobin genes Hb^A, Hb^B, and the transferrin genes Tf^A, Tf^D, Tf^E are also present in Pinzgau cattle.

Summary

In investigations of the blood-group structure in Pinzgau cattle, 45 blood-group-factor combinations or phenogroups were established in the B-system as non-segregating genetic units. Five of these phenogroups could not be observed in cattle of highland breeds until now.

As in other highland breeds, polymorphism exists in haemoglobin and serum-transferrins.

References

Erhard, L., Schmid. D. O., Blutgruppenuntersuchungen beim Pinzgauer Rind. Züchtungskunde (in press).

Holz, G. (1957). Monographie des Pinzgauer Rindes. Inaug. Dissertation Hochschule für Bodenkultur Wien.

Owen, R. D., Stormont, C., Irwin, M. R. (1947). Genetics 32, 64.

Santner, Th. (1955). Über das Pinzgauer Rind in Österreich. Arbeitsgemeinschaft der Pinzgauer Rinderzuchtverbände Salzburg.

Scheuch, R. (1925). Untersuchungen über die Abstammung und Rassenzugehörigkeit der Pinzgauer Rinder. Arbeiten der Lehrkanzel für Tierzucht an der Hochschule fur Bodenkultur Wien, Wien.

Schmid, D. O. (1962). Mh. Tierheilkunde 14, 158.

Schmid, D. O. (1963). XI. Intern. Congress of Genetics Den-Haag-Scheveningen.

Schmid, D. O. (1963). Zeitschrift für die Tierzüchtung und Züchtungsbiologie 79, 286.

Schmid, D. O., Erhard, L. (1963). Züchtungskunde 35, 300.

Schmid, D. O., Mančić D., Zeitschrift für Tierzüchtung und Züchtungsbiologie (in press).

Sieblitz, K. (1962). Die Rinderrassen in Süddeutschland. Handbuch der Tierzüchtung III. Band Parey-Verlag Hamburg-Berlin.

STUDIES ON BLOOD GROUPS IN CATTLE, HORSES AND PIGS

M. HESSELHOLT, B. LARSEN, P. B. NIELSEN and B. PALLUDAN

Department of Physiology, Endocrinology and Blood Groups, The Royal Veterinary and Agricultural College, Copenhagen

A. Cattle

1. Test for linkage of the genes controlling transferrins, haemoglobins and blood groups

In cattle at least 10 genetic systems of blood groups are known (Stormont, 1962). The more recent discoveries of various inherited biochemical characters such as transferrins, haemoglobins, lactoglobulins, casein, etc. have provided additional markers on the cattle chromosomes. Studies on linkage between loci for some of those traits were reported. Morton et al. (1956) studied the possibility of linkage between various blood group systems and were able to exclude close linkage of the loci investigated. Datta and Stone (1963a) excluded linkage of the order of 10 % crossing over between the transferrin locus and the loci controlling the A, B, C and FV blood group systems.

In the following, the results from test for linkage of the genes controlling blood groups and the loci for transferrin and haemoglobin types will be reported.

The material studied consists of sire families typed in connection with a progeny testing program. The transferrin and haemoglobin types were determined as described by Moustgaard and Møller (1962) and Brummerstedt-Hansen et al (1963). For blood typing our standard battery of reagents was used and tests were made for factors at the A, B, C, FV, L, M, SU, and Z systems.

Linkage tests were performed by the sequential probability ratio test described by Morton (1955, 1957). Only families of sires heterozygous at both loci under test were used. Double and single back crosses were the only informative matings. The lod scores (Z_1) from each family were accumulated ($\sum Z_1$) and tested for significance using the criteria recommended by Morton (1955).

The results obtained are given in table 1. From the table it appears that the recombination values of the genes controlling the transferrin types and

Table 1

Results from tests for linkage of the genes controlling transferrins, haemoglobins and blood groups. The ΣZ_1 values correspond to the recombination values given

Loci under test	No. of sire familes	No. of progeny	The recombination value	For $\Sigma Z_1 = $ *)
			%	
Transferrins versus A-syst.	27	133	30	−2·3563
Transferrins versus B-syst.	83	588	30	−5·9272
Transferrins versus C-syst.	63	415	30	−7·3159
Transferrins versus FV-syst.	16	104	20	−4·6342
Transferrins versus L-syst.	24	99	20	−3·4314
Transferrins versus M-syst.	31	158	20	−3·7916
Transferrins versus SU-syst.	59	315	20	−9·4683
Transferrins versus Z-syst.	24	111	10	−6·5591
Transferrins versus haemoglobins	6	39	5	−4·7874
Haemoglobins versus A-syst.	4	32	—	− ·1663†)
Haemoglobins versus B-syst.	12	104	20	−4·7885
Haemlglobins versus C-syst.	10	69	20	−3·3021
Haemoglobins versus FV-syst.	4	35	10	−4·2824
Haemoglobins versus L-syst.	5	21	—	−1·7216†)
Haemoglobins versus M-syst.	4	38	20	−2·8553
Haemoglobins versus SU-syst.	8	45	20	−2·2076
Haemoglobins versus Z-syst.	7	37	10	−3·4134

*) When $\Sigma Z_1 \leq \log B = -2$, the true recombination value is greater than recombination value under test.

†) ΣZ_1 for 5% of recombination.

the blood groups are greater than 20 %, except for the Z system, where only 10 % is excluded. For the transferrins versus the A, B and C systems, the recombination value is probably greater than 30%. Between the genes for haemoglobin and the B, C, M and SU blood group systems less than 20 % of crossing over are excluded. Also for the remaining comparisons a close linkage could be excluded, except for the haemoglobins versus the A and L blood group systems, where the material was insufficient for any conclusion even at a crossing over value of 5 %.

2. Investigations on cattle from the Middle East and India

During the past two years tests for blood-, transferrin- and haemoglobin types were performed on cattle from India, Cyprus, Egypt and Syria. To get some information about the distribution of the different blood group factors, the transferrins and haemoglobins among the cattle in those countries, the material so far collected has been examined, and a comparison was made to the red Danish dairy cattle (RDM).

For the National Dairy Research Institute, Karmal, India, tests were made on 99 Thaparkar and 85 Sahival cows. From Cyprus 147 animals were typed and from a similar breed in Egypt 87 animals are blood typed. Finally 119 animals of the Damascus breed in Syria have been tested. A slightly different number of animals were tested for transferrin and haemoglobin types.

The two breeds from India originate from Zebu and the Damascus, Cyprus and Egyptian cattle are of Zebu and Bos taurus brachycerous origin (Mason, 1957). The RDM to which a comparison is made, is originating from Bos taurus brachycerous.

Due to lack of family material and because of the small number of animals tested from each breed, which may not be fully representative, attempts have not been made to calculate the frequencies of the alleles supposed to be present in the A, FV and SU systems. For the more simple systems, and in the systems where the genotype can be observed directly, gene frequencies are computed.

In table 2 the relative frequencies of the phenotypes observed in the A system are given. The D factor is not included, because all breeds are not

Table 2

The percentage of animals showing the various phenotypes in the A system. The D factor is not included

Phenotype	Cyprus cattle	Egyptian cattle	Damascus cattle	India		RDM
				Thaparkar	Sahival	
A	5	7		1	2	35
AH	64	54	64	30	72	10
AZ'	5	1	5	37	2	
AHZ'	12	5	17	19	11	
H	7	9	9	7	8	16
—	7	24	5	5	5	39

tested against exactly the same D reagent. In the RDM four phenotypes and four alleles are known in the A system disregarding the D factor. These are A^A, A^{AH}, A^H and A^-. In addition to these, the other breeds on table 2 appear to have the allele $A^{AZ'}$, known from Jersey. The frequency of $A^{AZ'}$, seems to be highest in the Thaparkar breed, somewhat lower for the Cyprus, Damascus and Sahival cattle, and lowest among the Egyptian cattle. The allele A^{AH}, which has a very low frequency in the RDM, seems to be rather frequent among the other breeds in table 2.

Table 3 shows the relative frequencies of the phenotype observed in the FV system. In the cattle so far studied with V_1 and V_2 reagents, the V_1 has

Table 3

The phenotypes observed in the FV system (percentage)

Phenotype	Cyprus cattle	Egyptian cattle	Damascus cattle	India		RDM
				Thaparkar	Sahival	
F_1F_2	33	28	18	97	76	85
F_2	1	2	1	1		
V_1V_2	6	9	12			
V_2		1				
$F_1F_2V_1V_2$	31	32	31			15
$F_1F_2V_1$	4	5	1			
$F_1F_2V_2$	13	9	14	2	24	
$F_2V_1V_2$	8	10	15			
F_2V_2	3	3	8			

appeared as a subtype of V_2. As may be seen from table 3, this seems not to be the case in the Cyprus, Egyptian and Damascus cattle since animals giving reaction with V_1 but not with V_2 are found (phenotype $F_1F_2V_1$). Those animals were retested with the same results using V_1 and V_2 reagents prepared from different antisera. Also apropriate absorption experiments were carried out, and it was shown that blood from animals of phenotype $F_1F_2V_1$ was able to remove V_1 antibodies, but not the V_2. The two factors may, therefore, occur as different factors in the Cyprus, Egyptian and Damascus cattle. These observations on V_1 and V_2 are similar to those reported by Stormont et al. (1961) for S_1 (or S) and S_2 (or H') in studies of American buffalo. They found that the buffalo had S_1 but not S_2, which has not been observed in domestic cattle.

Among the RDM only two alleles are known in the FV system. These are F^{F1F2} and F^{V1V2}. From the material available it has not been possible to rule out the alleles present in the FV system of the other breeds in table 3. However,

Table 4

The phenotypes observed in the SU system (percentage)

Phenotypes	Cyprus cattle	Egyptian cattle	Damascus cattle	India *)		RDM
				Thaparkar	Sahival	
S_1S_2	26	23	22	77	82	34
S_2	19	25	12			26
$S_1S_2U_1$	10	14	21	9	18	
S_2U_1	29	28	41			
S_1S_2U'	7	2	2	13		8
S_2U'	6	3				5
S_2U_1U'	1	5	2			
U'	3					11
—						16

*) Differentiation between S_1 and S_2 is not made because of contaminated S_1 reagent.

the phenotypes observed in the Cyprus, Egyptian and Damascus cattle suggest that alleles as F^{F1F2}, F^{F1}, F^{F2V2}, F^{V1V2} and maybe F^{V2} may occur. In addition, a mechanism responsible for the appearance of V_1 without V_2 has to be postulated.

The phenotypes observed in the SU system are given in table 4. For the breeds from India differentiation between S_1 and S_2 is not made since non-specific reactions with S_1 were observed in those breeds. The U_1 factor, which is practically absent from the RDM, is seen to appear frequently among the other breeds of table 4, whereas the U′ is less frequent. Animals without any factors in the SU system are not observed among the Middle East and Indian cattle, but occur frequently in the RDM. In table 5 the frequencies of the genes

Table 5

The frequencies of the genes in the J, L, M, Z and R′S′ systems

	Cyprus cattle	Egyptian cattle	Damascus cattle	India		RDM
				Thaparkar	Sahival	
J	·18	·30	·22	·23	·45	·15
L	·40	·53	·50	1·00	·54	·07
M	·02	·03	·02	·06	·02	·18
Z	·65	·72	·68	·90	·89	·19
R′	·31	·28	·26	not tested		·32
S′	·69	·72	·74	·90	·76	·68

in the J, L, M, Z and R′S′ systems are given. The table shows that the L and Z factors appear with high frequencies, whereas the M factor has a very low frequency compared to the RDM. The two breeds from India were not tested for R′. Among the Cyprus and Egyptian cattle the R′ and S′ behaved as expected according to the two allele hypothesis. Among the Damascus cattle, however, a significant deviation was found in the R′S′ system, suggesting that either genetic equilibrium is not present or that the two allele hypothesis is not true in that breed. The haemoglobin and transferrin systems do, however, show genetic equilibrium in the Damascus cattle.

In the B and C systems any conclusions as to the existing alleles could not be made. It was noted, however, that the X_1 factor of the C system appeared with very high frequencies. The R factor showed a low frequency as in most other breeds so far studied.

By starch gel electrophoresis of the above mentioned material the same transferrin and haemoglobin phenotypes, which occur in Danish and other European cattle breeds, were observed. In some cases the transferrin phenotype AD exhibited an atypical type of pattern. In the "atypical" AD phenotype the C band was separated into two distinct bands. Similar phenotypic variants have been described as occurring in some samples from Norwegian cattle (Gahne, 1961). Whether this new type is an inherited variant has not yet been investigated.

In table 6 the gene frequencies are given and compared with those found in the Red Danish Cattle. It can be seen from the table that the frequency

Table 6

Transferrin- and haemoglobin gene frequencies observed in cattle from Cyprus, Syria, Egypt. India and Denmark (RDM)

Country	No. of animals	Gene frequencies				
		Tf^A	Tf^D	Tf^E	H^A	H^B
Denmark (RDM)	1132	0·47	0·40	0·13	1·00	0·00
Cyprus	133	0·27	0·52	0·21	0·88	0·12
Syria	116	0·36	0·32	0·32	0·89	0·11
Egypt	114	0·36	0·38	0·26	0·86	0·14
India (Thaparkar)	98	0·41	0·25	0·34	0·89	0·11
India (Sahival)	85	0·02	0·31	0·66	0·60	0·40

of the Tf^A gene is relatively low in the cattle from the Middle East and India, while the Tf^E frequency is relatively high. This distribution is very pronounced in the Indian Sahival breed. In this breed a very high frequency of haemoglobin B was observed when compared with European cattle breeds.

3. Test for cattle twins of the same sex

As a tool for detection of monozygotic twins for experimental purposes, 169 twin pairs of the same sex were examined for blood and transferrin types during the last two years. When possible also the parents of the twins were typed, and their blood types were used for the detection of displaced mosaicism by absorption experiments (Larsen, 1960). With a few exceptions, the twins were of the red Danish dairy cattle (RDM). Before tests for blood and transferrin types some selections for monozygosity were made by means of morphologic characters. The majority of the twin pairs were typed during the first few months after birth.

Table 7 shows the results of intra-pair comparison of blood and transferrin types of the 169 twin pairs. Only 21 pairs (12 %) showed blood and transferrin types indicating monozygosity. About the same frequency of monozygotic pairs might be expected to be found in a random sample of twins of the same sex (Johansson & Venge, 1951; Rendel, 1958). The effect of the morphological selection seems, therefore, to have been small.

Table 7

Intra-pair comparison of blood and transferrin types of 169 cattle twins of the same sex

Blood types	Transferrins		Total
	Same	Different	
Same with mosaicism	79	55	134
Different for the J factor, mosaicism	5	3	8
Different, no mosaicism	2	2	4
Same, mosaicism not detected	21	2	23
Total	107	62	169

Disregarding the 21 pairs which might be monozygous, 58% of the dizygotic pairs are of the same transferrin type. Using the gene frequencies for the transferrin types for the RDM breed given in table 6, the expected proportion of offspring in a family of two having the same transferrin type can be calculated to 51%. The discrepancy between the observed 86 pairs and the expected 75 pairs with the same transferrin type might be due to transferrin chimerism as discussed by Datta and Stone (1963b). However, going through 60 dizygotic pairs from table 7, where both parents were typed for transferrins, a close agreement was found between the observed 32 pairs and expected 30 pairs having the same transferrin type. The material examined gives no evidence as to the existence of mosaicism in the transferrin types.

B. Horses

Blood groups of the Icelandic horse

Since the beginning of the present century studies on equine blood groups have been under way in several countries. Podliachouk has isolated 13 specific agglutinins corresponding to the blood group factors A—M (Podliachouk, 1957). Specific isoimmune agglutinins and isoimmune hemolysins have been described by Lehnert and Franks (Lehnert, 1939; Franks, 1962). Recently, a report was presented by Stormont et al. concerning the results from blood typing of horses with sixteen specifically different equine blood typing reagents. It appeared that genes of six loci were involved in the genetic control of the sixteen blood group factors (Stormont et al., 1963).

Preliminary Danish investigations of normal horse sera showed that the "naturally" occurring isoagglutinins seemed to be but slightly active and gave non-specific agglutination reactions. Furthermore, as their number seemed to be limited, it was decided to produce and isolate, by means of iso- and hetero-immunizations, specific antibodies to be used in blood typing of horses. Agglutinating immune sera were produced from 25 isoimmunizations and 11 heteroimmunizations (of which 2 were in cattle and 9 in rabbits). From these sera 7 blood typing reagents were isolated. All the specific immune antibodies were found to be complete agglutinins applicable to the direct agglutination test in saline medium (Hesselholt, 1961).

In 1961, at the Pasteur Institute, comparative studies were made of French and Danish reagents, and a nomenclature was established (Podliachouk and Hesselholt, 1962). This nomenclature will be used in the following report.

The object of the following investigation was to elucidate the mode of genetic transmission of the blood type factors which could be identified by means of the above blood-typing reagents.

Table 8

Reagents used for blood-typing of horses

Reagent	Nature of antibody	Source of antibody
A	agglutinin hemolysin	isoimmune isoimmune
C	agglutinin hemolysin	heteroimmune (cattle) isoimmune
D	agglutinin	isoimmune or swine normal serum (Podliachouk & Hesselholt, 1962)
*E	agglutinin	horse normal serum
G	agglutinin	isoimmune
H	agglutinin	isoimmune
*J	agglutinin	horse normal serum
K	agglutinin	isoimmune
Da 1	agglutinin	isoimmune
Da 2	agglutinin	isoimmune

* The E and J reagents were kindly provided by Dr. L. Podliachouk.

The work was done in Iceland with financial support of the Danish State Research Foundation (Statens almindelige Videnskabsfond).

Material and methods. Blood samples from a family material comprising 10 stallions, 92 mares with 112 of their progeny and 278 randomly selected horses, all of the Icelandic horse breed, were blood typed.

The direct agglutination test in saline medium was performed as earlier described (Podliachouk and Hesselholt, 1962).

Table 9

Inheritance of the single blood type factors. A study of 112 matings

Blood group factor	Mating type	No. of matings	No. of progeny +	−
A	+ × +	21	15	6
	+ × −	41	20	21
	− × −	40		40
C	+ × +	45	44	1
	+ × −	56	39	17
	− × −	11		11
D	+ × +			
	+ × −	17	9	8
	− × −	95		95
E	+ × +	7	7	
	+ × −	35	20	15
	− × −	70		70
G	+ × +			
	+ × −	7	2	5
	− × −	105		105
J	+ × +	1		1
	+ × −	17	7	10
	− × −	94		94
K	+ × +	6	5	1
	+ × −	42	27	15
	− × −	72		72
Da 1	+ × +	46	38	8
	+ × −	52	20	32
	− × −	14		14
Da 2	+ × +	2	1	1
	+ × −	34	18	16
	− × −	76		76

The blood-typing reagents used in this study are presented in table 8. Da 1 and Da 2 are preliminary designations which will be used until comparison tests are performed.

Results. Regarding the results from the family material first, it was important to know whether the ten blood group factors were transmitted as dominant or recessive characters. A priori, it may be supposed that all ten blood group factors were transmitted as dominant characters, since the great majority of the blood group factors known in man, cattle and pigs are inherited as dominant characters. Since the question of unifactorial, dominant inheritance is a very important one, not only in relation to various practical aspects, the material was classified in table 9 in order to elucidate this further.

It can be seen in the table that matings of the type positive × positive and the type positive × negative may result in both positive and negative offspring, while matings of the type negative × negative gave only negative offspring.

Because of the low frequency of some blood factors, not all the mating types or progeny types were actually found. For example, the H antigen was not observed in the Icelandic horse. Despite this, the results confirm that each of the other blood group factors is transmitted as a dominant, unifactorial character.

Furthermore, studies were performed in order to elucidate the genetic relationship and the inheritance laws of the various blood group factors. The problem of allelism or non-allelism is best solved by segregation studies within relevant families. In table 10 the results from double backgross matings are presented. The data in this table seem to show non-allelism for the genes

Table 10

Distribution of progeny after 14 double back-cross matings. Only factor designations relevant to the test for allelism are indicated

Mating type	+	+	+ +	− −
A/−, E/− × −/−, −/−	3	2	6	3
A/−, Da 1/− × −/−, −/−	1	3	1	3
A/−, Da 2/− × −/−, −/−			14	12
C/−, D/− × −/−, −/−	2	2		1
C/−, E/− × −/−, −/−	1		1	1
C/−, G/− × −/−, −/−	1	1	1	1
C/−, J/− × −/−, −/−	1	1	1	1
C/−, K/− × −/−, −/−	2	2	1	5
C/−, Da 1/− × −/−, −/−	2		1	1
D/−, Da 1/− × −/−, −/−	1	1	1	3
E/−, J/− × −/−, −/−	1	2	1	1
E/−, Da 1/− × −/−, −/−	2	1	2	2
J/−, K/− × −/−, −/−	1	4	2	1
J/−, Da 1/− × −/−, −/−	1	2	1	1

involved in the matings, except for A and Da 2. A and Da 2 are inherited together. All Da 2 positive animals have been found to be A positive as well. Furthermore, absorptions have shown that A and Da 2 represent a serological sub-group system. Due to extremely high or low frequency of some antigens, not all combinations are presented in table 10.

In order to study further any possible serological or genetic association between the above blood factors, the results from blood-typing of 278 horses selected at random were treated in two-by-two contingency tables. The results are given in table 11. Where the values were too small for the application of the traditional X^2-test, Fisher-Yates exact treatment of two-by-two tables was applied (Andresen, 1963). No apparent association was found between any of the eight blood group factors treated in table 11.

Table 11

X^2- and P^*-values from 2×2 tables after blood typing of 278 randomly selected horses

	A	C	D	E	G	J	K
C	0·26						
D	0·02	0·52					
E	0·81	0·15	0·79				
G	$P^* = 0\cdot51$	1·78	$P^* = 0\cdot31$	$P^* = 0\cdot24$			
J	3·02	0·01	$P^* = 0\cdot25$	0·36	$P^* = 0\cdot37$		
K	0·59	0·05	0·95	0·03	$P^* = 0\cdot24$	0·01	
Da 1	0·02	0·01	0·09	2·82	2·36	1·12	2·74

The above results confirm that the occurrence of the blood group facort A, C, D, E, G, J, K, Da 1 and Da 2 is transmitted as a dominant, unifactorial character. Furthermore, data are presented which tend to show that 8 loci are involved in the genetic control of the blood factors mentioned. Thus 8 blood group systems have been established. The A and Da 2 antigen form a sub-group system.

C. Pigs

The M system. By isoimmunization of pigs of Danish Landrace, Danish white and black race and Pietrain race three antigen factors have been established, Ma, Mb and Mc (Nielsen 1961, 1964).

On the basis of genetic and statistical investigations in the Danish Landrace of the M system, it has been demonstrated that the system is comprised by at least four allele genes M^a, M^b, M^c and M^-, and that the system is in genetic equilibrium (tables 12 and 13). Within the Pietrain race and the Danish white and black race, two further alleles have been demonstrated, M^{ab} and M^{bc}.

Coupling studies have proved that the M-system is not sex-linked hereditarily and besides, the M system is apparently transmitted independently of the blood group systems A, E, F, G, H, I, J, K, L, and N.

Table 12

The distribution of Ma, Mb and Mc to the offspring from matings double heterozygous × double recessive

Mating type	No.	Offspring			
M(a+b+) × M(a−b−)	8	26 M(a+b−)	17 M(a−b+)	0 M(a+b+)	0 M(a−b−)
M(a+c+) × M(a−c−)	8	22 M(a+c−)	16 M(a−c+)	0 M(a+c+)	0 M(a−c−)
M(b+c+) × M(b−c−)	13	31 M(b+c−)	26 M(b−c+)	0 M(b+c+)	0 M(b−c−)

Table 13

The observed and expected distribution of the M phenotypes in the Danish Landrace

Pheno-type	Ma	Mb	Mc	Mab	Mac	Mbc	M−	Total
Observed	34	242	170	14	12	68	208	748
Expected	33·7	240·9	170·5	15·2	11·3	68	208·5	748
Chi-square for 1 d. f.							$X^2 = 0.15$	

The A system. The initial investigations of the occurrence of A substance in serum and on erythrocytes and anti-A have been made on a material comprising 35 boars, 47 sows and their 180 offspring.

The result of this research work is stated in table 14. As it will be seen, the material can be divided in 3 classes: 1) those having A substance on the red cells and in the serum, 2) those having A substance in the serum only, 3) and those having anti-A in the serum.

Table 14

The distribution of A substance and anti-A in pigs of Danish Landrace

No. of animals	Erythrocytes reacting with anti-A	Serum	
		A substance (inhib. tit.)	anti-A
46	+	+ (1 : 4 − 1 : 128)	.
30	.	+ (1 : 1 − 1 : 8)	.
186	.	.	+

It is further seen that the concentration of A substance must exceed a certain threshold value (tit. 1/4—1/8) before the red cells become A positive. An examination of the families investigated seems to indicate that the concentration of A substance in serum is controlled by multiple alleles, but that another genetic mechanism is involved too.

References

Andresen, E. (1963). A Study of Blood Groups of the Pig, Munksgaard, 34.

Brummerstedt-Hansen, E., J. Moustgaard & I. Møller (1963). Annual Yearbook, Royal Vet & Agric. College, 13.

Datta, S. P. & W. H. Stone (1963a). Nature 199, 1209.

Datta, S. P. & W. H. Stone (1963b). Proc. Soc. Exp. Biol. 113, 756.

Franks, D. (1962). Ann. N. Y. Acad. Science, vol. 97, 235.

Gahne, B. (1961). Anim. Prod. 3, 135.

Hesselholt, M. (1961). Annual Report, Inst. Sterility Research 96.

Johansson, I. & O. Venge (1951). Z. f. Tierzücht. u. Zücht. biol. 59, 389.

Larsen, B. (1960). Annual Report. Inst. Sterility Research 79.

Lehnert, E. (1939). Ein Beitrag zur Kenntnis der Blutgruppen des Pferdes. Diss. Almquist & Wiksell, Uppsala.

Mason, I. (1957). A World Distionary of Breeds, Types and Varieties of Livestock. Comm. Agric. Bur., Edinburgh.

Morton, N. E. (1955). Am. J. Human Genetics 7, 277.

Morton, N. E. (1957). Am. J. Human Genetics 9, 55.

Morton, N. E. W. H. Stone & M. R. Irwin. (1956): Genetics 41, 655.

Moustgaard, J. & I. Møller (1962). Annual Report, Inst. Sterility Research 175.

Nielsen, P. B. (1961). Acta Vet. Scand. 2, 246.

Nielsen, P. B. Annual Report, Inst. Sterility Research 1964 (in press).

Podliachouk, L. (1957). Les Antigenes des Groupes Sanguins des Equides et leur Transmission Hereditaire. Thes.

Podliachouk, L. & M. Hesselholt (1962). Immunogenetics Letter 2, 69.

Podliachouk, L. & M. Hesselholt (1962). Ann. Inst. Pasteur 102, 742.

Rendel, J. (1958). Acta Agric. Scand. 8, 162.

Sprague, L. M. (1958). Genetics 43, 906.

Stormont, C. (1962). Ann. New York Acad. Sci. 97, 257.

Stormont, C., W. J. Miller & Y. Suzuki (1961). Evolution XV, 196.

Stormont, C., Y. Suzuki & E. A. Rhode (1963). Proc. XI Int. Congr. of Genet., The Hague, The Netherlands, 192.

RESEARCH WORK ON THE J SYSTEM OF INSEMINATION BULLS IN THE GERMAN DEMOCRATIC REPUBLIC

R. EBERTUS

Institute for Artificial Insemination, Schönow near Bernau

Up to the present time our Laboratory was mainly occupied with the production of test sera. This work did not yet permit to carry out additional comprehensive research work. However, there was an opportunity to carry out research work concerning the J system of insemination bulls in spring of 1964 without an important expense of labour. To provide the basis for further investigations, our object was to clear the situation with regard to the J substance in the blood plasma (J^S) and the presence of anti-J in the cattle breeds existing in our country. The animals investigated originate exclusively from the pedigree breeding and can disclose information on their conditions in this respect.

The J system has been explored in its principle in a large number of investigations. Stormont (1949) and Stone and Irwin (1954) were able to demonstrate that some animals, in which J is missing on the erythrocytes, have the J substance dissolved in their blood plasma. The concentration of J dissolved in the plasma shows significant variations which are probably under genetic control according to Stone and Irwin (1954) and Jamieson (1960). No erythrocyte mosaicism exists for the factor J^{cs} (Stone, Stormont and Irwin quoted according to Osterhoff and Rendel, 1954).

Individuals having no J substance may possess naturally occurring J antibodies. Elliot and Ferguson (1956) found anti-J in 23·5 % of cases during an investigation of 634 sera. As with all normal antibodies, the titre of anti-J is subject to seasonal fluctuation and is higher in summer than in winter (Stormont, 1952, Stone, 1953, Braend, 1959).

Material and Methods

Serum of 1,008 bulls was investigated. The animals belonged primarily to the Friesian breed (DS) and to the brindled highland cattle (DF). A few bulls belonged to the German red cattle (DR), the Franconians (DG), to the Jersey breed and to cattle produced by cross-breeding, specially, brindled highland cattle and Jersey cattle. The blood specimens were taken by the Veterinary Examination Boards and Veterinary Sanitary Boards and dispatched to us without any addition. The blood was taken within the routine sanitary bulls service which is carried out in our country two times a year.

Immediately after the arrival of specimens, the serum was separated from the blood clot and subsequently inactivated in a water-bath for half an hour at 56°C. Prior to preparation of the tests, the sera were kept at −20°C.

The sera were tested by the inhibition test for the determination of J substance in the plasma and the haemolysis test for the detection of normal antibodies.

The inhibition test was made only qualitatively, using undiluted serum with erythrocytes which were strongly J-positive. The anti-J-serum used had a titre of 1 : 6 and belonged to one batch.

The haemolysis test was also prepared with undiluted serum. Each serum was tested by means of erythrocytes from 20 known individuals. These erythrocytes had to be exchanged several times so that several investigation groups existed. We thus had the possibility of observing the behaviour of 30 J-positive erythrocytes. The various groups under study were set up in such a way that approximately half of them were J-positive.

All sera, in which clearly visible reactions took place — although in traces only — were considered to be positive. This interpretation appeared to us to be justified, particularly, because all sera, except for three, in which J substance was found, showed absolutely negative reactions.

Results

603 = 59·83 % out of 1,008 sera specimens examined possessed the J substance and 405 = 40·17 % were negative. 295 = 72·84 % of sera without J substance showed anti-J, whereas 110 = 27·16 % had neither J substance nor antibodies. 4 sera also haemolyzed J-negative erythrocytes completely or in part. Three of them had J substance dissolved in the plasma, whereas one serum was without J substance. These sera were not considered, since they still required thorough investigations.

Table 1 shows the distribution of J^s according to breeds. Detailed calculations were carried out by us only with insemination bulls of the DS and DF breeds which are most represented in our country. The other breeds were not included on account of the small number involved.

The remarkable difference of 25·65 % in the presence of J^s between the two breeds is statistically highly significant (P < 0·001).

J^{cs}, whose frequency is lower than J^s, was found in a small number of bulls. Although we did no quantitative studies, we got some information that the erythrocytes of individuals, whose sera inhibited anti-J incompletely and has accordingly a lower concentration of J substance, reacted negatively with anti-J serum in all cases.

Table 1

J^s distribution

Breed	Number	J^s-positive	J^s-negative	Frequency J^s	Gene frequencies	Standard deviation
DS	806	509	297	0·6315	$J^s = 0.3930$ $j = 0.6070$	±0·0140
DF	144	54	90	0·3750	$J^s = 0.2094$ $j = 0.7906$	±0·0255
DG	12	9	3	0·7500	—	—
DR	13	9	4	0·6923	—	—
Jersey	11	6	5	0·5454	—	—
DF/Jersey	7	5	2	0·7142	—	—
Others	15	11	4	0·7333	—	—
	1·008	603	405	0·5983		

Table 2

Anti J

Breed	Number	Anti-J	Without antibodies	Frequency anti-J	Frequency without antibodies
DS	297	210	87	0·7071	0·2929
DF	90	71	19	0·7889	0·2111
DG	3	3	0	—	—
DR	4	3	2	—	—
Jersey	5	4	1	—	—
DF/Jersey	2	2	0	—	—
Others	4	3	1	—	—
	405	295	110		

Table 2 shows the occurrence of anti-J in the J^s-negative insemination bulls.

As with the analysis of J^s, only the two most represented breeds were taken into consideration for calculation of frequencies.

The difference between both breeds with respect to the formation of antibodies was not proved statistically.

Braend (1959) found in J-negative cows an anti-J frequency of 26·76 % in summer and 11·11 % in winter. The result of his investigation with regard to bulls is remarkable since it shows a frequency of 46·43 % in summer and 2·94 % in winter (in one out of 34 individuals examined). During his investigation Braend used serum dilutions of 1 : 2, 1 : 4 and 1 : 8 and undiluted sera in a few cases only. Apart from differences due to the breed, the different methods used in the investigations may also be responsible for the great difference.

Not all out of the 295 sera carrying antibodies haemolyzed all J-positive erythrocytes. Fig. 1 shows the character of the sera.

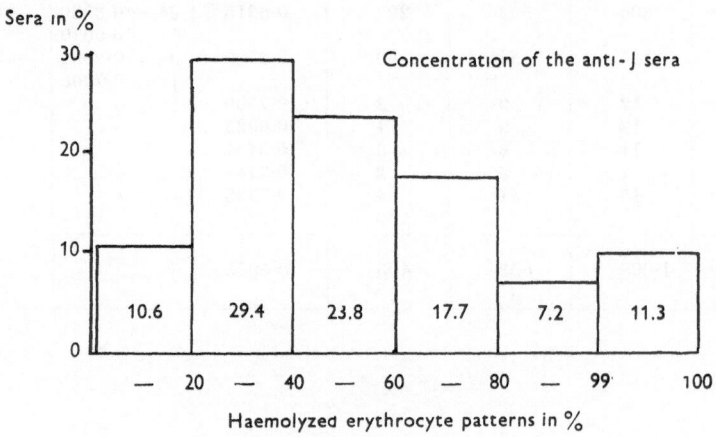

Fig. 1.

Only 11·3 % of these sera haemolyzed all J-positive erythrocytes.

The erythrocytes from 30 individuals carrying antigen J showed a very different behaviour. One specimen only was haemolyzed by all antisera. 11 erythrocytes were haemolyzed by 70 to 90 % of antisera, another 11 by 30 to 60 %, whereas 7 reacted seldom, namely, up to 29 % of antisera.

This result demonstrates that the antigen J on the erythrocytes varies remarkably with regard to its strength. Further investigations in this field will be necessary to obtain additional information in this respect.

Summary

1,008 insemination bulls were examined for J substance in the blood plasma and for anti-J production in February 1964. The two cattle breeds to be found most frequently in the G. D. R. — the Friesian (DS) and brindled highland cattle (DF), were compared. The presence of J^s in the DS breed is significantly higher than in the DF breed.

No differences between both breeds exist with regard to the production of anti-J. The content of antibodies was predominantly low. Only 11·3 % of all sera were strong.

Relations appear to exist between the strength of the J substance in the plasma and the presence of J on the erythrocytes.

The strength of J^{cs} shows a comparatively high variation.

References

Braend, M. (1959). Blood groups of cattle in Norway. Skandinavisk Bladforlag.
Elliot, A. and Ferguson, L. C. (1956). J. Immunol. 76, 78—82.
Jamieson, A. (1958). Animal Breeding Research Organisation Report. Edinburgh.
Neimann-Sörensen, A., Rendel, J. and Stone, W. H. (1954): J. Immunol. 73, 407—414.
Osterhoff, D. und Rendel, J. (1954). Zschr. Tierz. u. Zücht. biol. 63, 1—20.
Stone, W. H. (1956). J. Immunol. 77, 369—375.
Stone, W. H. and Irwin, M. R. (1954). J. Immunol. 73, 397—406.
Stormont, C. (1949). Proc. Nat. Acad. Sci. 35, 232—237.

THE INHERITANCE OF BLOOD GROUPS IN THE BLOOD GROUP SYSTEM C IN CATTLE

G. E. NASRAT*)

Blood Group Laboratory, State Agricultural University, Wageningen

In the Netherlands, a study has been carried out recently on the phenogroups of the C-system and their frequencies in Dutch cattle breeds. Nearly 50 phenogroups were established, of which those with a frequency of more than 1 % have been reported by Nasrat et al. (1964). The first report on the C-system was presented by Stormont et al. (1951) on the distributions of 22 C-system phenogroups in four breeds of cattle in U. S. A. In 1959, information was given by Stormont on the frequencies of 24 C-system phenogroups in three breeds of cattle in U. S. A. Later, in 1962, 35 phenogroups in the C-system have been reported by Stormont.

The transmission of the C-groups has been found to follow usually the rules of inheritance of antigenic complexes. Some incidental irregularities in transmission have been found in our material. Two cases with clear genetic evidence will be presented and discussed. This may lead to an understanding of the nature of the C-locus.

Material and Methods

The blood typing reagents used in this study are designated: anti-C_1, -C_2, -E, -R, -W, -X_1, -X_2, -L' and (-H_6). The methods of reagent production and of testing used, followed those described by Ferguson (1941) and Bouw (1958). The latter reagent (H_6), which is not known internationally, has been produced in the Netherlands two years ago out of iso-immune sera produced in Belgium, and was found valuable for the establishment of various C-groups.

The factors controlled by this locus were found to be inherited in various combinations with each other. The study of the blood groups of large families has resulted in the establishment of at least 28 blood groups of the C-system with a frequency of more than 1 % in Dutch cattle. Frequently occurring

---

*) On leave from the Faculty of Agriculture, University of Cairo, Egypt, U. A. R.

blood groups of the C-system in Dutch cattle are: C_1E, C_1WX_2, EW, C_1, X_2, E, L', RW and (H_6). In this report the following irregularities will be presented:

1st case:

In the study of a family of cows, six generations of animals have been tested. After a normal transmission in two generations, the cow 526638 was found to have transmitted the factors controlled by the C-locus in an irregular way. The data on the exceptional origin of the C-group of the bull 5151,080 are presented in figure I.

Figure I.

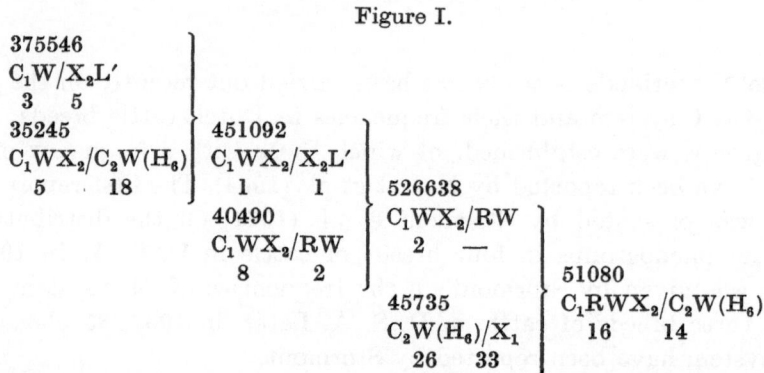

Note: Numbers beneath the complexes present the number of offspring having this complex.

Of the bulls 35 245, 40 490 and 45 735 and of the cows 35 5546, 451 092 and 526 638, large numbers of related animals and offspring partly with their dams have been tested. On these bases, the C-groups of both the bulls and the cows could be established without difficulties.

For verification, 30 offspring of the bull 51 080 were tested, 16 of them were found to have inherited the new complex C_1RWX_2 and 14 the other group $C_2W(H_6)$.

Repeated tests and all possible absorption were performed in order to exclude errors.

2nd case:

It was found in this case that the bull 42 789 has transmitted the factors controlled by the C-locus in an irregular way as shown in figure II.

Figure II.

412795
C_1WX_2/C_1
  1   2

42789
$C_2W(H_6)/RW$
  10   10

56372
$C_1RW(H_6)$
$(C_1/C_2RW(H_6))$

Of the bull 42,789, 20 other offspring with their dams have been tested, on which basis, the C-group genotype of this bull could be established as $C_2W(H_6)/RW$.

Of the cow 412,795, several related animals including 3 offspring were tested. From that, the C-groups of this cow could be established as C_1/C_1WX_2.

The son 56,372 has inherited from his sire as well R as (H_6). Unfortunately, no offspring of this bull could be tested. From the available data it can be concluded that his most likely genotype is $C_1/C_2RW(H_6)$.

Discussion

There has been much reiteration on the inheritance of the human Rhesus blood-group system by both British and American scientists. On the one hand, Fisher in 1944 (cited by Race 1944) conceived the idea that the genes controlling the human Rhesus factors are members of three closely linked pairs of genes. This concept is known as the Fisher-Race hypothesis.

On the other hand, Wiener (1943) advanced a different hypothesis to explain the genetic mechanism controlling the Rh-system. He assumes that the Rh-system is based on a series of multiple alleles of a single gene.

Blood groups in animals have demonstrated that also these various, more or less, complicated blood group systems can be detected.

In pigs, 12 blood group systems have been established (Andresen and Baker 1964). By his work on blood group systems in pigs, Andersen (1963) has followed the Fisher-Race hypothesis for the E-system. Three apparent irregularities in the transmission in one of these systems (E-system) have been observed (Rasmusen 1963) which could be explained as a result of crossing-over.

In sheep, at least seven blood group systems have been established, some of which, especially the B-system, seem to be very complex giving at least 52 phenogroups detected by Rasmusen (1962). He explained such complexes on the basis of multiple allelism.

In chicken, Briles et al. (1950) reported the existence of two independent autosomal loci (A and B) determining agglutinogens on the red blood cells. There were nine multiple alleles at the A-locus and five at the B-locus. Now the minimum number of blood group loci currently is 11 (Briles 1962). Most, if not all, loci have multiple alleles, of which the B-locus appears to show extreme multiple allelism.

Until 1962, 11 independent blood group systems in cattle have been established. The factors controlled by each locus were found to be transmitted in various combinations with each other. Two obvious explanations for the transmission of these complexes can be presented.

1. One is based on the assumption that the antigenic factors are each due to individual genes very closely linked, so that the various antigenic factors appear to be inherited in certain constant combinations. For example, the antigenic eomplex C_1WX_2 may be inherited as a unit because of very close linkage between the genes producing each of the antigenic factors C_1, W and X_2. This point of view has been approached by Irwin (1949) and Bouw et al. (1964) for certain blood group factors of the B-system in cattle. The number of factors identified in any system specifies the minimum number of genes involved in the control of that system. Similar cases of closely linked genes have been reported in both plants and animals.

2. The data may be explained by hypothesizing that each antigenic complex is produced by one of a series of multiple alleles. For example, a single gene produces an antigenic substance or substances capable of reacting with each of the serological reagents C_1, W and X_2. This view has been suggested by Stormont (1955) for the explanation of the inheritance of the groups in the the various blood group systems in cattle. In this case, the number of groups identified in any system specifies the minimum number of alleles involved in the control of that system.

Although the inheritance data presented in this report do not rule out any one of the above two explanations, I am inclined to emphasize the first explanation based on the assumption that the antigenic factors are each due to individual genes very closely linked.

The detection of various factors within a given phenogroup at the same age of embryologic development could be interpreted on the basis of this phenogroup being controlled by one gene and behaving as a physiologic unit. Shaw and Stone (1962) found that, at a given age, the various factors comprising a phenogroup of the B-locus were not developed simultaneously. They interpreted their data by favouring the linked gene hypothesis, and concluded that, in so far as age of development is concerned, the B-locus of cattle does not behave as a physiologic unit.

Datta and Stone (1963) have noted differences in the degree of reaction of the same blood factor as it occurs in different phenogroups or in the same phenogroup in different cattle. They suggested that such variations may result from intra-allelic interaction, e. g. position effect of closely linked genes.

This is not surprising since anomalies in transmission of blood groups have been observed in the V-system by various investigators (Stormont 1955; Rendel 1958; Datta et al. 1959; Stormont and Suzuki 1962; Moustgaard and Neimann-Sørensen 1962; Lie and Braend 1963; Stormont et al. 1964 and Bouw et al. 1964). These data suggest to my opinion that the B-phenogroups are controlled by a complex locus consisting of a cluster of closely linked genes.

This hypothesis has been favoured by Bouw et al. (1964) for certain blood group factors of the B-system in cattle.

As for the C-system, two clear cases, which could be explained as due to crossing-over, have been reported in this study. In the first case, the bull 51,080 has inherited from his dam 526 638 both C-groups C_1WX_2 and RW giving rise to new complexes C_1RWX_2 and the C-group $C_2W(H_6)$ from his sire. In the second case, the bull 56 372 has inherited from his sire 42 789 as well R as the group $C_2W(H_6)$ and from his dam 312 795 the group C_1.

Stone (personal communication) found in his material that not all of the factors of the C-locus are fully developed at a certain age, and that there are variations in the reactivity of the same factors which he assumes to be due to falling in different phenogroups. He suggested that, due to inefficiency of phenogrouping of the C-locus, it does not show the same degree of variations as in the B-locus in the development of the factors of a phenogroup.

These observations may offer an explanation that the C-locus does not behave as a physiologic unit, and would therefore favour the hypothesis based on the assumption that the C-phenogroups are controlled by a complex locus consisting of a cluster of closely linked genes.

Summary

Data on 2 cases of abnormal transmission of blood groups in the blood group system C in cattle have been presented.

On the basis of these cases, various possible explanations for the background of the transmission of the blood groups in this system were discussed. The closely linked genes hypothesis was found to be the most working one.

Acknowledgement

The author wishes to express his sincere gratitude to Dr. J. Bouw, director of the Stichting Bloedgroepen Onderzoek, Wageningen, Holland, for his keen interest and criticism during the course of this study.

References

Andresen, E. (1963). Munksgaard, Copenhagen.
Andresen, E. and L. N. Baker (1964). Genetics 49 : 379—386.
Bouw, J. (1958). H. Veenman & Zonen, Wageningen.

Bouw, J., G. E. Nasrat and C. Buys (1964). Genetica 35 : 47—58.

Briles, W. E. (1962). Ann. N. Y. Acad. Sci. 97 : 173—183.

Briles, W. E., W. H. McGibbon and M. R. Irwin (1950). Genetics 35 : 633—652.

Datta, S. P. and W. H. Stone (1963). J. Immunology 90 : 857—864.

Datta, S. P., W. H. Stone, W. J. Tyler and M. R. Irwin (1959). Genetics 44 : 504.

Ferguson, L. C. (1941). J. Immunology 40 : 213—242.

Irwin, M. R. (1949). Quart. Rev. Biol. 24 : 109—123.

Lie, H. and M. Braend (1963). Immunogenetics Letter 2(1) : 23—26.

Moustgaard, J. and A. Neimann-Sørensen (1962). Immunogenetics Letter 2 (7) : 62—64.

Nasrat, G. E., G. J. Kraay and J. Bouw (1964). Immunogenetics Letter (in press).

Race, R. R. (1944). Nature 153 : 771—772.

Rasmusen, B. A. (1962). Ann. N. Y. Acad. Sci. 97 : 306—319.

Rasmusen, B. A. (1963). Immunogenetics Letter 3(2) : 31—33.

Rendel, J. (1958). Acta Agr. Scand. 8(3) : 191—215.

Shaw, D. H. and W. H. Stone (1962). Proc. Soc. Exp. biol. & Med. 111 : 104—111.

Stormont, C. (1955). Amer. Nat. 89 : 105—116.

Stormont, C. (1959). Proc. Int. Congr. Genetics 1 : 206—224.

Stormont, C. (1962). Ann. N. Y. Acd. Sci. 97 : 251—268.

Stormont, C., B. Morris and B. W. Gregory (1964). Immunogenetics Letter 3 (3) : 130—133.

Stormont, C., R. D. Owen and M. R. Irwin (1951). Genetics 36 : 134—161.

Stormont, C. and Y. Suzuki (1962). Immunogenetics Letter 2 (7) : 80—81.

Wiener, A. S. (1943). Proc. Soc. Exp. Biol. N. Y. 54 : 316—319.

BOVINE ISOHAEMOLYSINS SEEMING TO HAVE SEVERAL SPECIFICITIES

P. MILLOT

Laboratory, Pasteur Institute, Paris, Laboratory of Blood Groups of Cattle, Jouy-en-Josas

When describing new alleles at the S locus, we have pointed out (1) that some antibodies of the S system are strictly specific for a given factor (anti-U_1, anti-S, anti-U', anti-U", etc.), while others have a double and even a triple specificity, as they react simultaneously with several factors of this system (anti-U_1U'/= anti-U_2), anti-SU'(= anti-S_2), anti-SU", anti-SU'U_1.

The immune sera we have obtained, or their absorbed fractions, frequently contain a mixture of these antibodies. When we absorb, for example, a fraction containing the two antibodies — anti-U_1U'(anti-U_2) and anti-U_1, with U_1 erythrocytes on the one hand, and with U' erythrocytes on the other, the former absorbs the two antibodies, while the latter absorbs exclusively the anti-U_2 with a double specificity (U_1U'), but not the anti-U_1. The antigens U_1 and U', therefore, seem to behave as subgroups with respect to this fraction, therefore, they were called U_1 and U_2 by Stormont (1961).

This example is exactly applicable to other fractions encountered in our studies; they contain 2 antibodies: anti-S + anti-SU', or anti-S + anti-SU", U' and U" seem to be, each in turn, the subgroup 2 of the same S antigen.

These are not serological subgroups similar to O or E' subgroups of the B system, for instance, because total absorption of antibodies anti-O or anti-E' can often be made by means of the O_3 or O_3' red cells, whose reactivity is weak, provided that a sufficient number of red cells is used. Therefore, in the case of true subgroups, we have a cross-reaction of sub-group antigens with the same antibody. Within the S system, on the contrary, there is no cross-reaction of subgroup antigens with the same antibody and the asorption by means of the subgroup 2 erythrocytes of the antibody strictly specific for the subgroup 1 does not occur, whatever may be the number of erythrocytes used; we have not subgroups but a mixture of 2 antibodies, one of them being specific for one factor and the other for several factors at the same time; the latter cannot be separated from the first, because of the lack of erythrocytes suitable for absorption.

The subject of the present communication is a somewhat similar case of false subgroups in the B system, which seems to us comparable to the subgroups in the S system.

The immunization of a cow of the Bazadaise breed against the red cells of its calf could not give any known haemolysin. It produced an unknown haemolysin revealing a new antigen which is named "antigen 81" and a fraction which was simultaneously anti-81 and anti-G'.

On the other hand, a Montbeliard cow immunized against the red cells G' of the Friesian breed gives an anti-G' and the preceding fraction: anti-81 + + G'. We have called temporarily the anti-81: anti-G_2' specific or anti-G" and anti-G_2' the fraction with double specificity anti-G'G", encountered with anti-G" or with anti-G' in the 2 cows mentioned above.

We have observed that the anti-G_2' is absorbable without discrimination by G' and by G", while the two strictly specific anti-G' or anti-G" antibodies do not permit any cross-absorption. The group of antibodies: anti-G', anti-G_2' or G", anti-G_2', (G'G") seems, at first, comparable to the group of antibodies: anti-U_1, anti-U_2 specific or U', anti-U_2 (U_1U') of the S system or anti-S, anti-U", anti-SU" of the same system.

After 6 months further injections were made to the Bazadaise cow 81, always with the same red cells. We quickly obtained the anti-G" (of better titre) and anti-G'G" of lower specificity as it reacted this time with several red cell suspensions which were neither G' nor G". Those red cells had already absorbed the first produced antibody but they were not haemolysable. We named them temporarily G_3'. Of the 75 animals of different breeds, examined by means of anti-G', anti-G", anti-G' + G" (anti-G_3') we find: 11 G', 9 G", 3 G'G" and 8 G_3', those last being negative for the 2 factors G' and G".

From the genetic viewpoint, the G' and G" factors, studied with strictly specific antibodies, are distinctive, but can, in some breeds, be found together in the same phenogroup, and consequently, are controlled by the same allele. Two breeds were studied up to now: the Salers breed from the French province Auvergne and the French Brown Swiss. The alleles of these 2 breeds, which include G' and G", are the following:

Breed	Total No. of animals	Alleles with G' (G_1')	No.	Alleles with G" (G_2')	No.
Salers	102	BGKE"O_4'E_2'G'I'O'	5	IE$_1$"E_2'G" Y_2D'E_3'G"	15 7
Brown Swiss	80	BGKO_4'B'E_3'G' B$O_3$$Y_2$A'A"$E_2$'G'	1 2	IE$_2$"E_3'G'G" IE$_2$"Y_2G"Y' (1) E_1'G"	15 2 4

(1) this allele without G" also exists (encountered twice).

It can be seen that the G'' allele, most frequently found in the Salers breed: $IE_1''E_2'G''$ is nearly the same in the Brown Swiss breed, but in the latter contains simultaneously G' and G'': $IE_2''E_3'G'G''$. It should be pointed out that we encountered the first phenogroup in the Norman breed, the second in the Montbeliard breed, but our results are still fragmentary.

The genetic study of the supplementary G_3' specificity is in progress. We have already noted that it frequently forms a phenogroup with E' which can be compared with the $E_1'G''$ phenogroup of the Brown Swiss breed. It seems, therefore, that we have here a secondary specificity closer to the G'' than to the G', but these results must still be confirmed.

Discussion

The antibody anti-G_2' or $G'G''$ seems first to be similar to double specificity antibodies of the S system (anti-U_1U', anti-SU', anti-SU''). It is, however, different from these antibodies because:

1. The last correspond to antigens which are never encountered together in one phenogroup.

2. They have no supplementary specificity with regard to the 2 primary factors (the anti-U_2 does not haemolyze red cells other than the U_1 or U').

On the contrary, the antibody observed in the B group detects first the 2 specificities G' and G'' (sometimes found together in the same phenogroup) which could be called G_1' and G_2' and then a secondary specificity: the G_3'.

In fact, it probably detects the specificity G''' different from G' or G'', but linked usually to the one or the other of these antigens, or to both of them, probably within the same phenogroup. This specificity can exist separately, especially linked to the E' factor.

The hypothesis, according to which a special antigenic receptor G''' exists, is most likely, considering what we know about the B system. It is, however, interesting to note that after the first immunization, the specificity revealed by the described antibody is shared by the antigens G' and G'' and that only absorption or a stronger immunization reveals the secondary specificity without G' and G''. When examining the immunological reactions, we wondererd whether this enlarged specificity, which includes more frequently G' and G'', could not correspond to a primitive antigenic structure serving, in a way, as a substratum for a secondary differentiation.

The old hypothesis of partial mutations ending as a progressive differentiation of some of primitive B genes should be revised according to this observation. It should perhaps explain the comparable case of some series of the B phenogroups which have the same antigenic fundamental structure:

BGK ... BOY ... O_1TE_3' ... etc., the phenogroups, in each series, being more strongly differentiated by means of one or several factors:

$$O_3TE_3' - O_1TE_3'I' - O_1TE_3'K' - O_1TE_3'I'K'.$$

The differentiation of the gene at the S locus, in spite of the existence of new phenogroups described by Grosclaude and by us, seems, on the contrary, less marked, as the specificities common to 2 or 3 factors are not found separately. It is possible that these shared specificities correspond, within the S system, to distinct unknown antigens linked to those that we already know (this appears likely, because we have found an anti-U_2' which detects a subgroup of U'). Now the antigen U_2' does not react with anti-U_2 (U_1U'). We wonder whether in this case the genetic process would not tend to suppress the less differentiated antigenic complexes in the S system or, conversely, the multi-differentiated ones, while they would be retained in the B system.

Summary

A new antigen G'' of the B system is described owing to the following antibodies obtained by means of isoimmunization:
1. the strictly specific antibody anti-G'',
2. an antibody which seems to be shared by G' and G''.

After stronger immunization, the latter shows a secondary specificity, which is linked to G' and G'' and can be detected only by absorption after the first immunization. Comparison is made between this antibody of the B system common to 2 specificities and similar antibodies of the S system.

References

An. biol. anim. bioch. biophys. (1963) 3 (2), 119—124.
Genetics (1961) 46, (5), 541—551.

STUDIES ON THE S BLOOD-GROUP SYSTEM IN FRENCH CATTLE BREEDS

F. GROSCLAUDE

Central Station of Animal Genetics, National Centre of Zootechnical Research, Jouy-en-Josas

1. Introduction

For several years a great deal of work has been devoted to the B system of bovine blood groups. The interest shown in this system is due to its remarkable variability and the easy interpretation of genotypes which facilitate theoretical and practical genetic studies. However, in addition, other systems offer a real interest for theoretical considerations and also in the field of practical application. The latter applies to the case of the S system.

The primary work by Stormont (1950) and Stone and Miller (1953, 1961) was followed by a basic analysis by Stormont et al. (1961) of the S system using four antibodies: anti-S_1, anti-U_1, anti-U', and anti-H'. They described five phenogroups and therefore five alleles: $S^{S_1H'}$, $S^{U_1H'}$, $S^{U'}$, $S^{H'}$ and S^s, the latter determining the absence of reaction with the known reagents. Recently, we have described (Grosclaude et Millot, 1963; Grosclaude, 1963) three new antibodies related to this system — anti-U", anti-U_2^p and anti-S" which permitted, in addition to the first four antibodies, a distinction of ten alleles of the S system in French cattle breeds. Since then, new observations have been made in our laboratory. The aim of the present report is to give the current status of our knowledge of that system.

2. Results

a) General description of the system

Table 1 gives the list of the reagents now used in our laboratory, the list of the phenogroups found up to the present time in French cattle breeds and the reactions given by these phenogroups with the above mentioned reagents.

With regard to the reagents, we have introduced (Grosclaude et Millot, 1963) a distinction between "primary" and "secondary" antibodies. We term a "secondary" antibody an antibody which reacts with blood samples

Table 1

Relationships existing between the 12 antibodies and 15 phenogroups of the S system of cattle blood groups; plus and minus signs indicate positive and negative reactions, respectively. N. B., anti-(U_1U_1') and anti-(S_1U_1') are called anti-U_2 and anti-S_2, respectively by Stormont et al. (1961). It has not yet been determined whether the phenogroups corresponding to the S 13 and S 14 alleles possessed the factor H′

Phenogroups	Alleles	Primary antibodies								Secondary antibodies			
		S_1	S''	U_1	U_2	U_1'	U_2'	U''	H'	(S_1U_1')	$(S_1U_1'U_1)$	(U_1U_1')	(S_1U'')
—	s	—	—	—	—	—	—	—	—	—	—	—	—
H'	S 1	—	—	—	—	—	—	—	+	—	—	—	—
S_1H'	S 2	+	—	—	—	—	—	—	+	+	+	—	+
U_1'	S 3	—	—	—	—	+	+	—	—	+	+	—	—
U_1H'	S 4	—	—	+	+	—	—	—	+	—	+	+	—
$U_1U''H'$	S 5	—	—	+	+	—	—	+	+	—	+	—	+
$U''H'$	S 6	—	—	—	—	—	—	+	+	—	—	—	+
U_2H'	S 7	—	—	—	+	—	—	—	+	—	—	—	—
$S_1S''H'$	S 8	+	+	—	—	—	—	—	+	+	+	—	+
$S''H'$	S 9	—	+	—	—	—	—	—	+	—	—	—	—
U''	S 10	—	—	—	—	—	—	+	—	—	—	—	—
U_2'	S 11	—	—	—	—	—	+	—	—	—	—	—	—
$U_2'U''$	S 12	—	—	—	—	—	+	+	—	—	—	—	—
$U_1'U''[H']$	S 13	—	—	—	—	+	+	+	?	+	+	+	+
$U_2U''[H']$	S 14	—	—	—	+	—	—	+	?	—	—	—	+

possessing at least one of the factors of a given set, with all the bloods of this type, and with these bloods only. Thus, for instance, anti-(U_1U_1'), called anti-U_2 by Stormont et al., reacts with bloods possessing either U_1 or U_1', or these two factors together, and with these bloods only.

If we possess the corresponding primary reagents, a secondary reagent is of no use for the determination of the antigenic formulae. To reduce the symbols of notation, we decided to recall the type of nomenclature used primarily by Stormont (1950). In this nomenclature a secondary antibody is designated by joining the symbols of the corresponding primary factors. It must be emphasized that this nomenclature does not mean that a secondary antibody possesses several specificities; on the contrary, it is more logical to admit that it possesses one specificity corresponding to a common property linked to the primary factors, to which it is related. A set of primary factors and a corresponding secondary factor form what is usually called "a non-linear sub-group system".

Since our previous publications, a new primary antibody, anti-U_2' has been introduced. This reagent is as yet an experimental one. Its titre is low (1/8), and it is contaminated with anti-H (of the A system) which is known to be impossible to eliminate completely. However, some information is known about this antibody and the corresponding factor. Until now, the factor has been observed in two phenogroups, where the H′ factor is absent. Matoušek et al. (1962) mention the production of an anti-U_2' reagent, but his reagent has not yet been compared with ours.

With regard to the nomenclature, we have simplified in U_2 our previous notation U_2^β. Confusion should be avoided with anti-(U_1U_1') called anti-U_2 by many laboratories. On the other hand, it must be emphasized that the factor U_2, when not U_1, does not react with the secondary antibodies related to U_1. Similarly, the factor U_2', when not U_1', does not react with the secondary antibodies related to U_1'.

In the case of the $U_1'U''$ and U_2U'' phenogroups, for the present, we have not been able to determine whether they possess the factor H'.

b) Allelic frequencies in five French breeds

Table 2 gives the allelic frequencies estimated in 5 French breeds, except for those of $S^{H'}$ and S^s alleles, which we preferred not to estimate, since our anti-H' reagent has not been always satisfactory. The samples studied are

Table 2

Estimations of allelic frequencies of the S system in five French cattle breeds, except for alleles $S^{H'}$ and S^s

Phenogroups	Flamande breed	Normande breed	Charolaise breed	Blonde d'Aquitaine breed	Salers breed
S_1H'		0·01	0·03	0·02	
U_1'	0·10	0·12	0·05	0·04	0·02
U_1H'	0·01	0·02		0·04	0·09
$U_1U''H'$			0·35	0·21	0·13
$U''H'$			<0·01	0·06	0·14
U_2H'					0·03
$S_1S''H'$	0·03	0·20	0·04	0·05	0·02
$S''H'$	0·03	0·02	<0·01		
U''	0·01				
U_2'	0·04	0·01	<0·01		
$U_2'U''$			<0·01		
$U_1'U''$ [H']			<0·01	<0·01	
U_2U'' [H']			<0·01		<0·01
Number of cows	215	211	195	122	89

relatively limited, because the production of our latest reagent is recent, so that we could not test a larger number of samples with the completed set of reagents.

Important differences appear between the five breeds; the allele $S^{U_1 U''H'}$ is frequent in the three breeds of the southern half of France, but absent from the two breeds of the northern half.

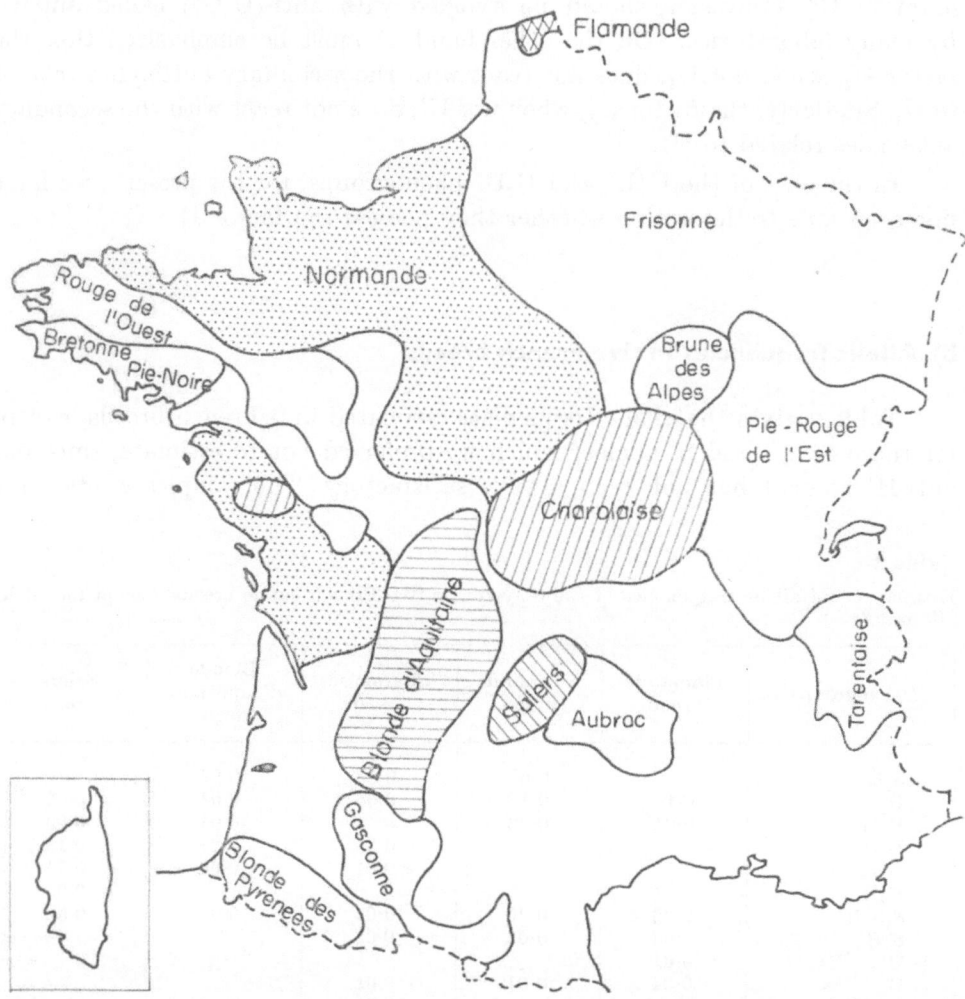

Fig. 1. Areas of dominance of the main French cattle breeds (simplified from Quittet, 1963) including the 5 breeds studied in our work.

The absence of this allele from the Normande, Flamande, and even Hollandaise breeds has been checked with a much higher number of analyses than that reported in table 2, since anti-U″ is the oldest of our new reagents. For instance, more than 2,000 analyses have been made in the Normande breed with this reagent. Allele $S^{U''H'}$ is frequently found in Salers breed, while $S^{U'}_2$ may be found not infrequently in the Flamande breed. In the five breeds, the $S^{S}{}_2{}^{S''H'}$ allele is found more frequently than the $S^{S_1H'}$ allele.

Some alleles have been found only in rare cases. For instance, $S^{U'}{}_2{}^{U''}$ has only been observed in a Charolais bull and some of its progeny.

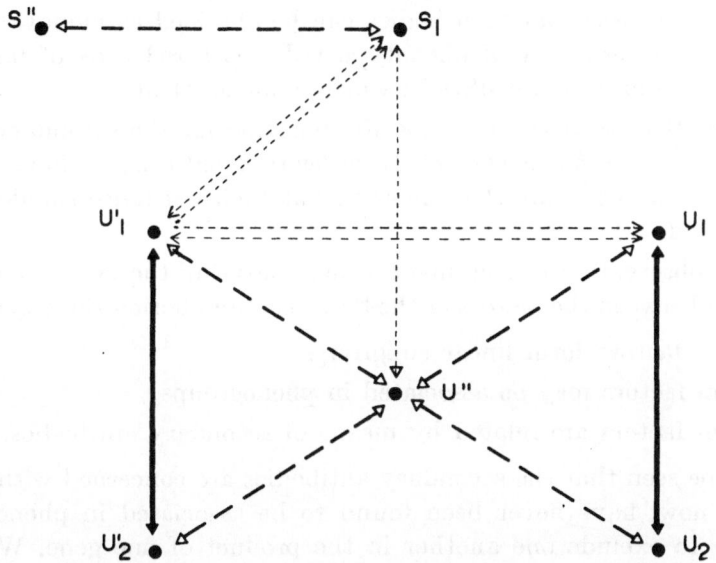

Fig. 2. Relationships existing between the antigenic factors of the S system, H′ excepted:

    ◄——►   factors forming linear sub-group systems
    ◄— —▷   factors which may be associated in phenogroups
    ←— —→   factors related by secondary antibodies.

3. Discussion

Our data, although calculated from small samples, and limited to five of our breeds, show clearly the importance of a closer analysis of the S system of cattle blood groups. Breed differences exist, which have to be explained; moreover, the introduction of new reagents is very useful for practical applications.

From the theoretical viewpoint, the S system shows a particular feature, the existence of secondary antibodies, which have not yet been observed in the other systems. What is the meaning of these antibodies?

First, it can be seen, from previous discussion by Stormont (1955), that the production of such an antibody as anti-$(U_1U'_1)$ is possible only because there is at least one more allele than those determining the factors U_1 and U'_1; this means that this antibody detects a property common to the phenogroups possessing U_1 and U_1', but always missing in the phenogroups without U_1 or U_1'. The four secondary antibodies described until now thus express a series of relationships existing between the phenogroups of the system. The adoption of a specific designation for the reactions given by the secondary antibodies should thus be logical. This conception is supported by the fact that, under certain conditions, secondary antibodies can be obtained as monovalent. For instance, immunizing a cow of phenotype U_2H' with red cells of phenotype $U_1U''H'$, we produced an anti-(S_1U''), which is monovalent.

However, the utilization of a specific notation for the secondary factors would give a complex nomenclature. In order to avoid this, we have adopted a nomenclature which omits these factors, but their existence should always be kept in mind.

Another observation can be made when analyzing the types of relationships existing between the factors of the S system, numbering three types:

1. certain factors form linear subgroups,
2. certain factors may be associated in phenogroups,
3. certain factors are related by means of secondary antibodies.

Now it may be seen that the secondary antibodies are concerned with factors which, until now, have never been found to be associated in phenogroups, thus seeming to exclude one another in the product of one gene. We could be tempted to suppose that factors related by means of secondary antibodies are determined by the same site in the gene, while factors associated in phenogroups are determined by different sites. Looking at the secondary antibodies it can be shown that S_1, U_1, U_1' and U'' should then be determined by the same genic site but, since phenogroup $U_1U''H'$ exists, the hypothesis cannot be accepted. It is possible that it would be better to look for an explanation at the level of the molecular configuration of antigenic properties of red cells.

A parallel may be drawn between the antibodies of the S system of cattle, on the one hand, and the antibodies of the La system of pig blood groups, on the other. The set anti-S_1, (S_1U_1'), $(S_1U_1'U_1)$ and U_1', seems to be similar to the set anti-La_1, La_2, La_3', and La', described by Brucks (1964). This author points out the analogy, but she does it between U_1, (U_1U_1') and U_1' on the one hand, and La_1, La_2 and La' on the other. Whatever it may be, it will be interesting to examine thoroughly the possibility of homology between the two loci.

Finally, because the factor S_1 has never been found without the factor H' most of the European laboratories have changed the nomenclature from H' to S_2. But another factor, U_1 has also never been observed without H', thus the factor H' seems to be something other than a simple sub-group of S_1. Until additional observations are made, it seems preferable to keep the notation H'.

Summary

Eight monovalent reagents are in use at present in the French laboratory of Jouy-en-Josas for the S system of cattle blood groups: anti-S_1, U_1, U_2, U_1', U_2', U'', S'' and H'.

Until now, these reagents have permitted the distinction of fifteen phenogroups determined by the alleles of this system. The frequencies of these alleles are estimated in five breeds: Flamande, Normande, Charolaise, Blonde d'Aquitaine, and Salers from limited but randomly drawn samples. Important differences exist between these breeds, the most characteristic difference being concerned with allele $S^{U_1 U'' H'}$, absent from the breeds of the northern half of France (Flamande and Normande) but frequently found in the breeds of the southern half (Charolaise, Blonde d'Aquitaine and Salers).

The existence of "secondary" antibodies: anti-$(U_1 U_1')$, anti-$(S_1 U_1')$, anti-$(S_1 U_1' U_1)$ and anti-$(S_1 U'')$ is also peculiar to the S system. The significance of these antibodies is discussed.

References

Brucks, R. (1964). Z. Tierzüch. Züchtungsbiol. 80, 66—80.

Grosclaude F. (1963). Ann. Biol. Anim. Bioch. Biophys. 3, 433—435.

Grosclaude F., Millot P., (1963). Ann. Biol. Anim. Bioch. Biophys. 3, 119—124.

Matoušek, J., Čuta, J., Schröffel J. (1963). Živočišná výroba 36, 531—534.

Quittet, E. (1963). Races bovines francaises. La Maison Rustique, Paris, 2ème éd. 80 pp.

Stone, W. H., Miller, W. J. (1963). Genetics 38, 693, (abst.).

Stone, W. H., Miller, W. J. (1961). J. Immunol. 86, 165—169.

Stormont, C. (1950). Genetics 35, 76—94.

Stormont, C. (1955). Am. Naturalist 89, 105—116.

Stormont, C., Miller, W. J., Suzuki A. (1961). Genetics 46, 541—551.

ASPECTS OF RELATIONSHIPS BETWEEN GENETICALLY DETERMINED CHARACTERS IN CATTLE

G. J. KRAAY

Blood Group Laboratory, State Agricultural College, Wageningen

Until now, the study of blood groups in cattle has resulted in the detection of at least 11 loci on the chromosomes controlling the blood groups (Stormont 1962). This number will certainly increase in the coming years. Furthermore, some other loci are known, which control electrophoretically detectable variations in substances in the blood, such as haemoglobins (Bangham 1957), transferrins (Ashton 1957, Hickman and Smithies 1957), postalbumins (Gahne 1963a) and phosphatase (Gahne 1963b). It is not likely that the limits of the possibilities are reached, since more substances are known showing a variation the genetic control of which is not yet fully understood, such as haptoglobins (Kristjansson 1962) and slow alpha 2-globulins (Ashton 1958, Gahne 1962).

Many of the mentioned loci are independent of each other, suggesting that they are located on different chromosomes. Although tests for independence are not carried out for all these loci, the assumption that about half the number of chromosomes of cattle is marked by these loci seems not unlikely.

As a result of this, these loci can serve as tools for the elucidation of the inheritance of characters, which are controlled by loci on the same chromosomes. It is possible that the blood group genes can be used as markers of genes controlling characters, which are not easily detectable or are detectable only in adult animals. In this case, the blood groups will be of value for the cattle breeders as an aid in the selection.

In this paper, a brief discussion will be presented of the possibilities and the limitations of the studies on the genetic relationships between blood groups and other characters of cattle.

Two characters can be related genetically to each other by pleiotropy or linkage.

Pleiotropy is the phenomenon that 2 (or more) characters are affected by the same gene. In this case, the relation between the characters is absolute. An individual has both characters or none of them. When an animal has that gene in homozygous condition, all the offspring will inherit both characters; when the individual is heterozygous for that gene, half the offspring will receive both characters and the other half none of them.

Linkage is the phenomenon that 2 characters are controlled by the genes of 2 loci, located close to each other on the same chromosome. The genes tend to be inherited in blocks rather than to assort independently.

Linkages are broken, when homologous chromatids break in the region between the 2 loci and exchange corresponding parts in meiosis. This process of crossing over is presented schematically in figure 1.

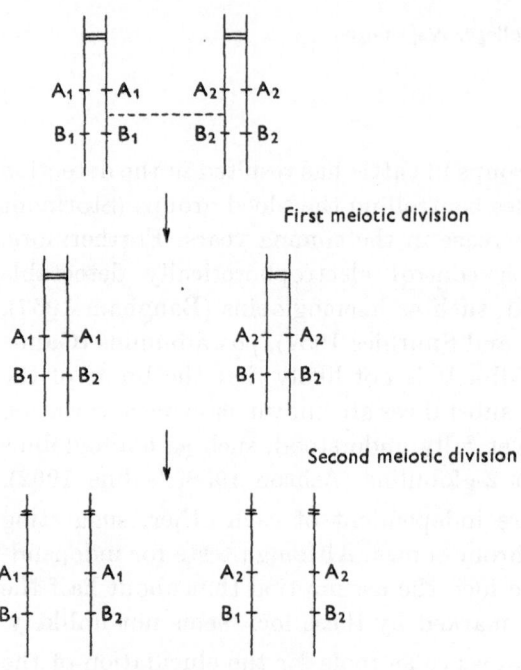

First meiotic division

Second meiotic division

Fig. 1. Schematic presentation of linkage and crossing over in meiosis.

In this figure the combinations of the gene A_1 with B_1 and of A_2 with B_2 are shown. Through crossing over these combinations are broken and gametes will be found with the new combinations of the genes: A_1B_2 and A_2B_1. The frequency of these recombinations is dependent on the distance between the 2 loci: the closer the distance, the lower the frequency.

From the point of view of population genetics, linkage has some aspects which are important for the studies on the relations between characters in cattle populations.

1. Crossing over gives rise to recombinations of the genes only in individuals which are heterozygous on both loci.

2. Consequently, linkage can only be detected and measured in the offspring of doubly heterozygous individuals. When dominance is involved, only those mating types are conclusive, which result in segregation of both characters. More details in this respect are given by Mather (1951). A clear demonstration of linkage, involving 2 blood group loci of the pig, was recently reported by Andresen and Baker (1964).

3. There are 2 types of double heterozygotes: A_1B_1/A_2B_2 and A_1B_2/A_2B_1. It can be shown that in a random mating population after a number of generations a linkage equilibrium will exist. In that situation the 2 types of double heterozygotes are equally frequent and the characters caused by the linked genes are distributed independently (Li, 1955). The number of generations, required to reach the equilibrium state, is inversely proportional to the frequency of recombination.

4. Consequently, when an association between two traits is found in a population, this is not an evidence for linkage. When two traits are correlated by linkage, sibships in which the traits are positively correlated and in which the traits are negatively correlated will be found in equal numbers.

As cattle breeders are mostly interested in characters like milk production, composition of the milk, fertility, body conformation etc., the questio ben whether a genetic relation between these characters and blood groups can of found, is of practical importance.

Neimann Sørensen and Robertson (1961) concluded that blood grouping has only a very limited value in predicting the breeding value of an animal as compared with the information obtainable by more conventional methods. Their reasons for this are that these characters are controlled by the genes of a large number of)oci. Consequently, each gene can have only a rather small influence. Further, these characters are greatly influenced by the environment, so that the variation due to genetic factors is only a part of the total phenotypic variance. Keeping this in mind, strong relations between blood groups and characters of economic value caused by linkage or pleiotropy can hardly be expected, especially for characters with a low heritability.

When an association between blood groups and other traits is found in a population, this can be caused by:
1. pleiotropy,
2. common environmental influences,
3. physiological relations,
4. subdivision of the population into heterogeneous groups and even sampling errors could be mentioned in this respect.

Pleiotropy is discussed before and needs no more comments.

Environmental factors, influencing both characters in the same direction, can be left out of the picture, since blood groups are not influenced by the environment (except the blood group J).

Physiological interactions between blood groups and other characters in cattle can be used as an explanation for possible associations. Since nothing is known about the physiological function of the blood groups in cattle, it is still a hypothesis that must be tested. Furthermore, the question can be raised in how far physiological relationships and pleiotropy are different phenomena.

Subdivision of a cattle population into heterogeneous groups is a rather common case. A population can be divided into geographical subgroups, with only a low rate of exchange of breeding animals between them. Between the subgroups differences in frequencies of the blood groups can arise as a result of the separation. The fact that such differences sometimes can be rather large is demonstrated in the Dutch cattle breeds by Kraay and Bouw (1964) and Nasrat et al. (1964).

When different purposes are pursued at the same time in the different groups by selection, an association between blood groups and other characters will be induced in the population as a whole.

An other demonstration of such coincidental relations between blood groups and other characters can be given with the following hypothetical example. A population of 800 cows, sired by a number of bulls, is available for a study of the relationship between blood groups and the fat content of the milk. The mothers of the cows formed a homogeneous group and had a mean fat content of the milk of 4 %. The fathers had a breeding value equal to the population mean, except one. This bull had a much higher breeding value; his daughters have a mean fat content of the milk of 4·20 %. This bull also had a certain blood group, called B in heterozygous condition; the group B is not found in the other bulls.

The cow population is then divided into two groups: animals having the blood group B and animals lacking this group. For each of the two subgroups the mean fat content of the milk is calculated. The calculations are performed for various frequencies of the blood group B in the mothers of the cows, showing that the "effect" of the blood group is higher by lower frequencies. The number of offspring of the bull also influences the results of the calculations: the higher the number of offspring, the larger the "effect". The results are presented in table 1, where the frequency of blood group B is assumed to be 0, 5, 10 or 20 % and the number of offspring of the bull 5 or 10 % of the total material.

Table 1

Relation between a blood group "B" and the fat content of the milk, induced by one bull with genotype B/—

Number of daughters of the bull	Frequency of B	0		5		10		20	
		No.	fat%	No.	fat%	No.	fat%	No.	fat%
40	with B	20	4·20	59	$4·07^1$	98	$4·04^5$	176	$4·02^7$
	without B	780	$4·00^5$	741	$4·00^5$	702	$4·00^5$	624	$4·00^5$
	difference		$\overline{0·19^5}$		$\overline{0·06^6}$		$\overline{0·04}$		$\overline{0·02^2}$
80	with B	40	4·20	78	$4·10^8$	116	$4·07^6$	192	4·05
	without B	760	4·01	722	4·01	684	4·01	608	4·01
	difference		$\overline{0·19}$		$\overline{0·09^8}$		$\overline{0·06^6}$		$\overline{0·04}$

These two examples show that the structure of a cattle population and the breeding practices can induce relationships between blood groups and other characters. These disturbances can be avoided by performing the calculations within the subgroups; the most reliable results will be obtained within daughter groups of bulls, provided that the half-sib groups are large enough to reduce sampling errors to an acceptable level.

Summary and Conclusions

The value of blood groups as markers of chromosomes and tools for selection is discussed. Genetic relations between blood groups and other characters are caused by linkage or pleiotropy.

Pleiotropy causes a stable relation between two or more traits and is therefore rather easy to detect.

Linkage is only detectable in special types of families. Half the number of these families will show a positive relation and the other half a negative relation between the traits.

In a population a relation between two genetically determined characters can be caused accidentally by breeding and selection. Some examples of this are given for cattle populations.

It is concluded that studies of the correlations in cattle populations lead to the best results, when the calculations are performed within halb-sib groups.

References

Andresen, E. and Baker, L. N. (1964). Genetics 49, 379.
Ashton, G. C. (1957). Nature 180, 370.
Ashton, G. C. (1958). Nature 182, 193.
Bangham, A. D. (1957). Nature 179, 467.
Gahne, B. (1962). Report 8th Animal Blood Group Conference, Ljubljana, Yugoslavia.
Gahne, B. (1963a). Hereditas 50, 126.
Gahne, B. (1963b). Nature 199, 305.
Hickman, C. G. and Smithies, O. (1957). Proc. Genet. Soc. Canad. 2, 39.
Kraay, G. J. and Bouw, J. (1964). Immunogenetics Letter 3, 119.
Kristjansson, F. K. (1962). Report 8th Animal Blood Group Conference, Ljubljana, Yugoslavia.
Li, C. C. (1955). "Population Genetics". The University of Chicago Press.
Mather, K. (1951). "The Measurement of linkage in heredity", 2d ed., London, Methuen and Co.
Nasrat, G. E., Kraay, G. J. and Bouw, J. (1964). Immunogenetics Letter 3, 159.
Neimann-Sørensen, A. and Robertson, A. (1961). Acta Agric. Scand. 11, 163.
Stormont, C. (1962). Ann. N. Y. Acad. Sci. 97, 251.

INVESTIGATIONS ON THE BLOOD GROUPS OF WISENTS (BISON BONASUS) AND HYBRIDS IN COMPARISON WITH THE BLOOD GROUPS OF CATTLE

J. M. GASPARSKI

Polish Academy of Sciences, Institute of Experimental Breeding of Animals, Warsaw

The scope of research work of our Institute includes research studies on various blood traits in domestic cattle and domestic fowl, as well as in some other animal species, such as wisents and hybrids. The primary aim of carrying out the research studies on wisents, initiated by Professor Czaja, was to delineate their blood group traits and to determine the degree of genetic variation within that strongly inbred, relict population in Poland. In spite of a several years' work, the obtained results are rather scanty, because, up to now, only the individuals withdrawn from breeding have been used for investigations. Nevertheless, the methods and laboratory aids and appliances adapted to the research work as well as the wisent experimental material permitted to develop research studies, rather theoretical in nature, as a means of determining the existing relationship and differences between the wisent and domestic cattle blood properties, and detecting possible evolutionary changes in the blood traits.

The present results of investigations on the blood group traits of wisents and hybrids as compared with the blood group traits of domestic cattle permit to assume that the red cell antigens A, W, X_1, V_2, J, and Z are identical in individuals of the two investigated animal species and the red cell antigens in wisent blood are similar to antigens B, G, O, A', E', C, X, F, and L found in domestic cattle. By using the fractionated anti-wisent immune sera and cattle red cells, several antigenic factors were detected which seem to be "new" or anyway not identified yet.

In the presentation of our research work, I am going to concentrate but on that part which is connected with the specialized investigations on the domestic cattle blood group traits.

Haemolysis tests and absorptions carried out by using the domestic cattle blood-typing reagents and wisent red cells showed that in some cases, certain reagents of the same kind reacted with the same red cells of wisents, while some others did not. Nevertheless, all of them showed the same reaction in numerous tests performed with red cell samples of cattle of various breeds and strains. A characteristic example of that kind is given by the results obtained in testing wisent red cells with seven test sera identifying the antigenic factor F and with ten sera identifying the factor V in cattle blood. The differences in the reaction of anti-V sera with wisent red cells depend upon the kind of antibodies contained in the serum. In some cases, the antibodies correspond to one factor only, either to V_1 or V_2, in others to both of them at the same time. Since wisent red cells contain antigenic factors V_2, there are differences in their reaction with corresponding sera. The exceptionally rare occurrence of phenogroup V_2 in domestic cattle red cells did not permit to notice these differences, when testing the ten test sera mentioned above and domestic cattle red cells. As regards the cattle blood-typing F reagents, and taking into account the inadequate results obtained up to now, it is difficult to explain the cause of the differences in their reaction with wisent red cells. Nevertheless, it can be assumed that the wisent red cells contain antigens which are similar to red cell antigens of cattle and possibly to F_2 factors, too. It is worth while mentioning that in the haemolysis test two of the test sera did not react with the tested wisent red cells in any dilutions within the range from 1 : 1 to 1 : 6, but when absorbed with the wisent red cells, they did react with F homozygous red cells of cattle, in each particular case, yet, without producing any dosage reaction. Thus, the obtained results may suggest that the specific reactions with homozygous red cells may be not only quantitative but also qualitative in nature as a result of a specific antigen and corresponding specific antibodies. Further research in this field will most probably permit to determine more exactly the specificity of the observed reactions and define their proper characteristics.

The recently initiated studies on comparison of test sera showed already in preliminary tests that it will be possible to notice differences between the test sera B, O_3, E', L, and Z which are similar to those of the F or V serum.

Another problem to be solved relates to a phenomenon frequently encountered in our investigations, namely, the fact that many cattle blood-typing reagents of various kind do not react with wisent red cells in the haemolysis test; yet when subjected to absorption with those red cells, they lost completely their previous ability to react with corresponding red cells of domestic cattle. The results of the same kind of reaction were also noticed in haemolysis tests and absorption, when using anti-wisent immune sera with cattle red cells. The results of recently initiated preliminary tests with various kinds of normal sera, as complement, and trypsinized red cells as well as anti-globulin sera seem to give some hope for the possibility of finding an explanation and disclose the characteristics of the above reactions.

The interesting feature of the results obtained in the research on the wisent red cell antigenic factors of the FV system is the corroboration that $F_{Wi(wisent)} \approx F_{2(cattle)}$, $V_{Wi(wisent)} = V_{2(cattle)}$, "—" (no-factor) phenogroups occur in wisents. The occurrence of phenogroups was noted in the investigation on the red cells of hybrid individuals (cattle $\times$ wisents). These investigations made it possible to determine the diploid combination of the three post-ulated phenogroups in two wisents as F_{Wi}/V_2. There is another feature con-nected to some degree with the results of the above investigations. In the course of absorptions of various anti-wisent heteroimmune sera with cattle red cells, we were able to obtain from one of them the serum which reacted only with cattle red cells containing simultaneously both F and V antigens. This was observed in a group of animals from eight different breeds and strains in the total number of 468 test samples, including 274-FF, 147-FV, and 47-VV. The present state of our investigations makes it difficult to es-tablish what kind of antigenic specificity is identified by the obtained FV_{Wi} serum; nevertheless we can assume, as a working hypothesis at least that this serum identifies the specific heterozygous antigenic substances. These substances may be expected to occur as a result of the interaction of either suitable antigenic factors or phenogroups, or genes determining them.

One of the methods applied to our research on the relationship and dif-ferences between wisent and domestic cattle antigenic blood properties has been the examination of the effects produced by immunizing various animal species (cattle, sheep, goats, and rabbits) with wisent red cells. By using cattle red cells for absorption of those anti-wisent immune sera, we were able to obtain test sera identifying A, W_1, V_2, J, Z, A'-like, X-like, and specific FV antigens in cattle red cells as well as six other antigens, which are not yet identified. Thus from the total number of 44 sera elaborated up to now, derived from immunization with red cells from six individual wisents we were able to obtain sera for fifteen different cattle antigenic factors. The most frequently obtained sera were those containing antibodies against antigens A, next come those with antibodies against antigens X_1, V_2, and Z, and one or two against the other antigens mentioned above.

Furthermore, the examination of the immunization effects of cattle with wisent red cells attracts attention to some other interesting results.

Twenty-three individuals of domestic cattle were immunized with red cells from seven different wisent, at different seasons of the year. There was one donor for 2—6 individuals. After but one series of 3—4 intravenous or intravenous and intramuscular injections, 20—50 cc of 50% washed red cell suspension each, a positive result of immunization was obtained in all instances. The mean titration score of non-absorbed sera in the reaction with cattle red cells, were found within the range of 14—29, while scores for particular serum in relation to particular red cells were within the range of 10—19 for the weakest serum and 23—42 for the strongest. As regards the red cells of the wisent-donors or any other wisents, titration scores were relatively higher and found within the range of 22—54. The results obtained now in the investigation of anti-wisent cattle immune sera in haemolysis tests and absorptions with wisent red cells do not supply sufficiently adequate data for the determination whether these sera contain any heteroantibodies or whether there are other antibodies against antigenic factors specific for wisents. In spite of this, two facts are worth mentioning: a) Non-absorbed non-cattle (sheep, goat, rabbit) anti-wisent immune sera tested with numerous cattle red cell samples have produced a haemolytic reaction in all instances. Thus, they may be assumed to contain probably the same heteroantibodies against wisent and domestic cattle red cells. This suggestion may be based on the fact, that one non-absorbed serum out of 23 cattle anti-wisent immune sera reacted with each red cell sample of several wisent, except for two. b) By the absorption of cattle anti-wisent immune sera with wisent red cells, a serum was obtained which did not react with red cells used in absorption and with the numerous red cell samples of tested cattle blood, it did react with red cells of one of the tested wisents. The stated reaction suggests the necessity of further continuation of tests aimed to detect the antigenic factors specific for wisents.

By the immunization of cattle with wisent red cells in one individual out of twenty-three mentioned above, a certain serological specificity has been noticed. A cow, which after crossbreeding with a wisent gave birth to a hybrid-calf, had been kept with a dairy herd for over a year before her pregnancy and immunized with wisent red cells. And as a result of immunization, antibodies against A and X_1 antigenic factors were found in the serum of the average titration score. In the nineteenth month after the immunization and in the fifth month after giving birth to a hybrid-calf, the cow was to be reimmunized, but this time with red cells of her hybrid male offspring. However, before the intended reimmunization, haemolysins against red cells of the male offspring and against red cells of wisents and hybrids, and some red cells of cattle were found in the serum of this cow. The average titration score was 28 and antibodies corresponded to antigenic factors X_1. The cow was not reimmunized and her serum was systematically tested each month for a period of two years in order to determine its titre and the quality of antibodies contained. The presence of the same antibodies against X_1 factors was always noted. The mean titration score of the monthly tested serum samples varied between 12 and 28. The highest score amounting to 28 was found in the first months of the second effective crossbreeding of that cow with a wisent. From the moment when presumably the five-month-old foetus was aborted, the serum titration score began to decrease gradually during a 16-month-period falling to the lowest score of 12. In that period of 16 months the cow was pregnant again and gave birth to a calf which was an offspring of the crossing between that cow and bull of black and white lowland breed. The lowest score of antibodies was found in the serum withdrawn a few days after the effective crossing of the cow with the same wisent, that was previously used twice for the crossbreeding. In the second month of pregnancy, the titration score increased up to 24. Tests of the serum are still carried on. It is difficult to prejudge, at present, what is the reason of the occurrence of the observed specificity; it seems, however, that in this case the cow has become a regular producer of in a sense natural antibody identifying a certain cattle red cell antigenic factor.

To sum up my report, I wish to stress once more that I am far from drawing any formal conclusions. I do realize that my investigations have to be carried on and further developed by making use of supplementary devices. In view of what I said above I would like to take the opportunity and make a request to you, ladies and gentlemen, for your kind help and assistance in promoting the advancement of research in this field by creating the possibility for carrying out the experiments and tests of red cells of wisents and hybrids as well as anti-wisent immune sera in your laboratories, too.

Acknowledgement

I wish to thank most sincerely and to express my deep gratitude to M. R. Irwin, W. H. Stone and J. Rendel for their much valuable advice and help. I wish, also, to thank all my colleagues from Poland for their assistance and above all to express my appreciation to I. Wiatroszak[1]) and M. Krasińska[2]) for their extensive and generous contribution to the advancement and widening of the scope of my research work.

References

Czaja, M. and J. Gasparski (1960). Nature 185 (4707) : 185—186.
Gasparski, J. and S. Dubiski (1962). Ann. N. Y. Acad. Sci. 97 : 285—295.
Gasparski, J. and A. Gerner-Nowakowa (1963). Immunogenetics Letter 3/1 : 65—67.
Gasparski, J., M. Krasińska and E. Szynkiewicz (1963). Immunogenetics Letter 3/1 : 68—70.
Gasparski, J. and S. Dubiski. (1963). Acta Theriologica 7/16 : 317—320.
Gasparski, J., M. Krasińska and E. Szynkiewicz (1963). Acta Theriologica 7/15 : 311—315.
Gasparski, J. (1964). Biuletyn 4, Z. H. D. Z. P. A. N.

[1]) Department of General Animal Breeding, College of Agriculture, Poznań.
[2]) Mammals Research Institute, Polish Academy of Sciences, Białowieźa.

FLUCTUATIONS OF THE LEVEL OF CONGLUTININ IN BOVINE SERA

D. G. INGRAM and D. A. BARNUM

Ontario Veterinary College, Guelph, Ontario

Introduction

Conglutinin is a heat-stable globulin which occurs most abundantly in normal bovine serum. It is recognized by its ability to aggregate or conglutinate red blood cells which have been coated with antibody and have adsorbed or fixed complement.

The activity of conglutinin was first demonstrated early in this century and principles involved in the conglutination reaction were defined by Bordet and Streng (1909). These authors recognized the absolute necessity for complement in the conglutination reaction and thus clearly distinguished conglutination from agglutination since the agglutination reaction does not require complement.

Von Jettmar (1923) reported that during infections of cattle there was a drop in the level of conglutinating activity in the serum and that the level returned to normal after the recovery of the animal. Coombs (1941) tested serum samples from normal cattle and found conglutinin titres varying from 80 to 1,280.

The study reported herein was undertaken to find the fluctuations in conglutinin levels which occur in the serum of individual cows over a prolonged period and to discover, what factors contribute to these fluctuations.

Materials and Methods

Serum samples were obtained at monthly intervals from the cows of the Ontario Veterinary College Mastitis research herd. These animals were mostly of the Holstein breed, although a few Jersey and Guernsey animals were also included. Most of the cows were between 2 and 3 years of age when placed on the experiment, but several older animals were also used in these tests.

The cows were bled by jugular puncture and the serum was harvested within 6 hours. Serum samples were frozen immediately and stored at −20°C until tested. Serum was collected over a two year period and all serum samples collected from an individual cow over a one year period were tested at one time against a single standard preparation of reagents.

All sera were titrated for Forssman antibody by the method described by Coombs, Coombs and Ingram (1961) and for conglutinin by method IIb described in the same publication.

Results

During a two year period from December 1961 to November 1963, a total of 630 serum samples were collected from 36 cows. The distribution of the natural Forssman antibody titres in these sera is shown in Table 1. Most of the samples had titres between 80 and 320, but the overall variation was from 5 to 2,560.

Table 1

Levels of conglutinin and Forssman antibody in 630 serum samples from 36 normal cows

Serum Activity	Titre											
	0	5	10	20	40	80	160	320	640	1280	2560	5120
Conglutinin	6	7	7	15	31	58	110	173	135	64	22	2
Antibody	0	2	10	22	62	122	183	145	71	12	1	0

When the sera were titrated for conglutinin, the titres were distributed as shown in table 1. Most sera had conglutinin titres between 160 and 640 but the range was from 0 to 5,120. These data show that the distributions of the activities of both these serum substances form single normal curves. If the distributions of these activities are plotted by months, although the occasional sample caused slight irregularity, the distribution of titres were relatively smooth and normal for each month. No split- or double-peaks of activity of conglutinin or antibody were shown for any month during the test period.

During the course of these experiments, cows were followed through 43 calving periods. fig. 1 shows the mean titres of conglutinin in these animals beginning 3 months before calving and continuing for 5 months after calving. In two cases, no decrease in conglutinin activity was detected at calving time, however the pooled data show that the mean decrease in all animals was a little more than 4 two-fold dilutions.

Fig. 2 shows the mean monthly titres of conglutinin throughout the two-year period. It is apparent from these data that there is a seasonal variation in the level of conglutinin in cattle. The lowest mean levels occur in February of both years and the highest levels occur in August to October.

Similar data for the Forssman antibody activity are also shown in fig. 2 but in this case, there did not seem to be any consistent seasonal variation.

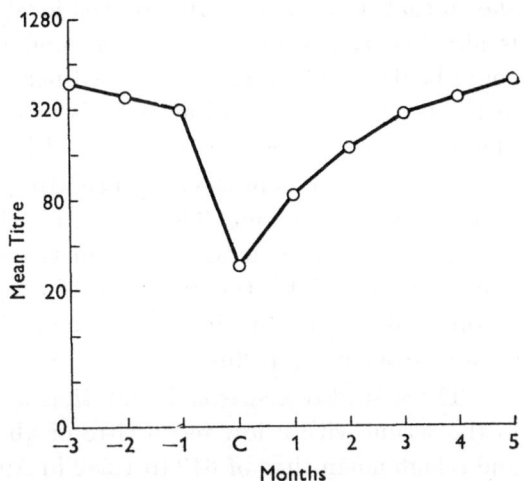

Fig. 1. Level of conglutinating activity in cattle serum near time of parturition.
Legend: Mean titres of 45 cows.

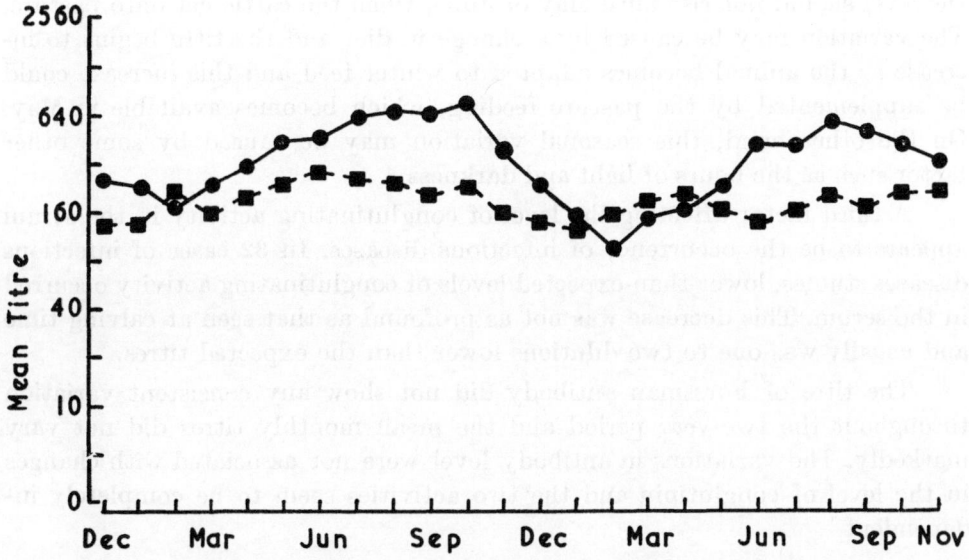

Fig. 2. Mean monthly levels of conglutinin and Forssman antibody in 36 normal cows.
Legend: ● — ● Mean titre of conglutinating activity ■ — ■ Mean titre of antibody activity.

Discussion

Conglutinin is a protein which occurs in normal bovine serum. Immuno-electrophoretic studies of this protein by Lackman and Richards (1964) have shown that it is a beta globulin. The level of serological activity of conglutinin is affected by a number of factors and in some individual cows no activity could be detected in serum obtained near the time of calving. This fluctuation in the level of serological activity is probably a result of the variation in the amount of this protein in the serum of individuals at different times.

A marked drop in the conglutinating activity of the serum occurs in most animals at calving time. The reason for this drop in activity is not known. It may be due to a general decline in the animal's protein reserves. However, it may be caused by the excretion of conglutinin into the colostrum. Normal bovine milk does not show any conglutinating activity, whereas colostrum does contain conglutinin.

There is also a seasonal variation in the level of conglutinating activity in the serum with a low mean titre of about 80 to 160 occurring in February and a high mean titre of 640 to 1,280 in August to October. In the herd studied the births of calves were quite evenly spread throughout the year and, there-fore, the low mean titres, which occurred in February, were not caused by a large number of calvings at this period. The cause of the low activity in February is not known. If it were due to a lower plane of nutrition in the winter, the level should not rise until May or June, when the cattle get onto pasture. The variation may be caused by a change in diet and the titre begins to in-crease as the animal becomes adapted to winter feed and this increase could be supplemented by the pasture feeding, which becomes available in May. On the other hand, this seasonal variation may be caused by some other factor such as the hours of light and darkness.

A third factor affecting the level of conglutinating activity in the serum appears to be the occurrence of infectious diseases. In 32 cases of infectious diseases studies, lower-than-expected levels of conglutinating activity occurred in the serum. This decrease was not as profound as that seen at calving time and usually was one to two dilutions lower than the expected titres.

The titre of Forssman antibody did not show any consistent variation throughout the two-year period and the mean monthly titres did not vary markedly. The variations in antibody level were not associated with changes in the level of conglutinin and the two activities seem to be completely in-dependent.

Summary

The level of conglutinating activity in the serum of cattle is influenced by at least three factors:

(1) The conglutinating activity of the serum drops markedly in most cows at calving time.

(2) A seasonal fluctuation in the level of conglutinating activity occurs with a high level in August to October and a low level in February.

(3) A decrease in the conglutinating activity of the serum of cattle occurs during or following infectious diseases, especially chronic infections.

References

Bordet, J. and Streng, O. (1909). Zbl. Bakt. (Orig.), 49, 260—276.

Coombs, R. R. A. (1947). The Conglutination and Sensitization Reactions. Dissertation to the University of Cambridge for the Ph. D. degree.

Coombs, R. R. A., Coombs, A. M. and Ingram. D. G. (1961). The Serology of Conglutination and its Relation to Disease. Blackwell Scientific Publications, Oxford, England.

von Jettmar. H. M. (1923). Z. Immun. Forsch., 36, 148—201

Lackman, P. S. and Richards, C. B. (1964). Immunochemistry, 1, 37—41.

DISCUSSION

D. Schmid: Dr. Ebertus, have you examined the seasonal variation in the antibody production concerning anti-J in cattle?

R. Ebertus: The titre of anti-J is subject to seasonal fluctuations. That is known. Although we did not work in this direction, we got some information that complete antisera were not obtained before the middle of April.

D. Schmid: In co-operation with Dr. Mancic from Novi Sad, Yugoslavia, we had the possibility of testing 150 Podolic steppecattle from five villages of the Fruschka Gora mountains. The tests showed that the steppe-cattle cannot be considered as the only ancestor of the spotted breed or the Brown Swiss. On the other hand, there are references to a relationship between the primitive steppe-cattle and the spotted breed and Brown Swiss.

O. Böhm: The contribution on the Podolic breed of cattle in Vojvodina that was presented by Dr. V. Jovanović and Končar gives information on the programme of a rather broad study of cattle breeds in Yugoslavia. The study of the breeds in Bavaria, carried out at the Blood Group Institute in Munich, is of great interest to us, because the main breeds in the northern Yugoslavia belong to the same group of breeds. We have compared our results with the data published in Munich, and two conclusions can be drawn:

1. The occurrence of different blood groups and especially phenogroups in the B-system is very alike and we are looking forward to the possibility of comparing the gene frequencies we have calculated for some of our breeds.

2. There are some characteristic differences. We have found some B-phenogroups which have not been published for the Bavarian breeds so far. This was the reason, why we determined to study the two indigenous cattle breeds, the Podolic and the small brachycephalous breed Duša. Unfortunately, only small isles of pure populations of these breeds are found.

We know a few B-alleles in Podolic cattle, but we would like to extend our knowledge by blood-typing the animals from different sources and more numerous family material.

D. Schmid: Since 1961 we have a monovalent reagent, which is reacting only with the phenogroups $U_1H'M\ddot{u}_1$ and $H'M\ddot{u}_1$. This reagent does not react with the antigens U_1 and U_2. The factor $M\ddot{u}_1$, we call it U_x, belongs according to our serological, genetic and statistical tests, to the blood-group system S. The position of the factor U_x in the S-system, proposed by Grosclaude, should be tested.

F. Grosclaude: I didn't know about the existence of your anti-U_x reagent, it would be naturally interesting to compare this reagent with our anti-U_2.

J. Bouw: Dr. Grosclaude, are those so-called reagents, which you call mixtures of reagents, in fact, not monovalent?

We should take care that we come to a sound nomenclature about this system, since more laboratories will find opportunities to go into these studies.

F. Grosclaude: I believe that it is possible to admit, that mixtures of antibodies exist here; in fact, if we examine the case of the serum anti-S_1, anti-$(S_1 U_1')$ anti-$(S_1 U_1' U_1)$ then erythrocytes possessing the factor U_1 (and not the factors S_1 and U_1') absorb anti-$(S_1 U_1' U_1)$, from the mixture and leave anti-S_1 and anti-$(S_1 U_1')$; furthermore, erythrocytes, possessing U_1' factor, absorb anti-$(S_1 U_1')$ and leave anti-S_1. The picture of cross-absorptions, concerning the erythrocytes, carrying these different factors, shows well that many antibodies exist there.

Besides this, when we immunized one cow of the $U_2 H'$ phenotype with the erythrocytes of the $U_1 U'' H'$ phenotype, we obtained a monovalent anti-$(S_1 U'')$ antibody, which proved the existence of these secondary antibodies.

J. G. Hall: Let me illustrate the importance of a critical genetic analysis as Dr. Kraay describes:

In Britain my colleague Jamieson found a significant association between milk yield and a certain transferrin gene. This occurred when Friesian cattle were being selected because these had a higher milk yield than the cattle then in use. But Friesian cattle also had a higher frequency of this gene; so the association of a higher milk yield with the gene in the improved cattle was not necessarily caused, but may merely have reflected the general influx of Friesian characteristics.

P. Imlah: I would like to congratulate Mr. Kraay on his paper. There is one omission, however, which I think is very important and deserves some comment, that is the relationship between blood group factors and disease. I know we cannot find any physiological function for the red cell antigens, but with the biochemical polymorphs we can. They all appear to have some physiological function, for example, the transferrins carry iron, the haptoglobins bind haemoglobin; there are the copper-binding proteins and enzyme variants such as esterases, dehydrogenases and possibly other enzyme systems. Therefore, the question posed by Dr. Harboe earlier this morning is a very important one. There are at least 14 haemoglobin variants in man, and most appear to be involved in some form of anaemia. This could mean that all the different biochemical polymorphs may play some part in various disease conditions. Furthermore, I believe blood groups may help us to observe those animals, which have an inherited resistance to disease. Recent investigations on the identification of litters from mixed insemination experiments in pigs have shown that in one litter of eight, four animals belonging to one boar all died of E. coli infection, whereas the other four belonging to the other boar showed no signs of disease.

I. Granciu: The only question I want to raise is upon the scientific importance of work Dr. Gasparski had done and upon the need of finding possibilities to extend these studies in order to find out what are the phylogenetic relationship and differences between these two species. This can be done I think by using both blood groups and protein polymorphism studies.

As we have imported a certain number of wisents for a natural reservation, we'll be glad to find a possibility for co-operation in the future in this regard.

J. G. Hall: How did Dr. Ingram distinguish the Forssman antibody from anti-J and is he sure he was measuring the Forssman and not the anti-J antibody?

D. G. Ingram: In the titration of these sera for Forssman antibody, R-negative sheep red blood cells were used and these red cells do not react with the anti-J antibody. All of the cattle sera tested possessed antibodies against these R-negative red cells and thus the antibody being titrated could not have been anti-J.

In the titration of conglutinin, all sera were absorbed with sheep red blood cells before titration in order to remove any effect which might be caused by either the anti-J or the Forssman antibody.

BLOOD GROUPS OF PIGS

A STUDY OF BLOOD GROUPS IN PIGS

P. IMLAH

Blood Group Research Unit, Royal School of Veterinary Studies, Edinburgh

Autosomal linkage between genes for red cell groups and other marker genes has been widely investigated in man. None of these investigations have revealed linkage between serum protein markers and genes for the red cell groups. Linnet-Jepson et al. (1958) showed independent segregation of the genes controlling haptoglobin, Gm groups and red cell blood groups: ABO., MNS., Rh., P, Lewis, Lutheran and Duffy. Smithies and Hiller (1959) also found independent segregation of genes at the transferrin and haptoglobin loci, and the red cell blood groups; ABO., MNS., P and Rh.

In this paper I will present the results of an investigation carried out in pigs to detect the possibility of association and genetic linkage between 11 red cell factors and the genes controlling serum proteins at the transferrin, haptoglobin, haem-binding globulin, ceruloplasmin and amylase loci.

Red cell factors

The source and type of antibodies used to detect the 11 red cell factors are shown in table 1. These factors represent five known systems, which have

Table 1

Source and type of antibody

Antibody	Type of antibody	Source
A (No. 1)	haemolysin	cattle anti-J
E_b (No. 2)	agglutinin	isoimmune
K_b (No. 3)	haemolysin	heteroimmune (rabbit)
K_a (No. 4)	haemolysin	heteroimmune (rabbit)
E_a (No. 5)	agglutinin	isoimmune
F_a (No. 8)	agglutinin	isoimmune
E_e (No. 10)	agglutinin	isoimmune
E_f (No. 11)	agglutinin	isoimmune
X (No. 12)	agglutinin	isoimmune
G_a (No. 13)	agglutinin (incomplete)	isoimmune
G_b (No. 14)	agglutinin (incomplete)	isoimmune

already been established by Andresen (1962). Tests for association between the red cell factors in 2×2 contingency tables are shown in table 2. Factor X appears to be independent of the other factors, and may represent a sixth

Table 2

Tests for association between red cell factors in 2×2 contingency tables
X^2 values for 1 d. f. shown in parenthesis

No. 2 E_b + −	No. 3 K_b + −	No. 4 K_a + −	No. 5 E_a + −	No. 8 F_a + −	No. 10 E_e + −	No. 11 E_f + −	No. 13 G_a + −	No. 14 G_b + −	No. 12 X + −	Red cell factors
33 28 / 57 32 (1·5)	48 13 / 72 17 (0·1)	20 41 / 39 50 (1·5)	25 36 / 27 62 (1·4)	12 49 / 14 75 (0·4)	49 12 / 66 28 (1·6)	12 46 / 18 67 (0·005)	33 28 / 41 48 (0·9)	13 7 / 23 3 (2·4)	3 38 / 8 55 (0·08)	+ − No. 1 A
	69 21 / 51 9 (1·6)	36 54 / 23 37 (0·04)	25 65 / 27 33 (4·7)	15 75 / 11 49 (0·7)	55 35 / 60 0 (28·8)	7 77 / 23 36 (3·2)	39 51 / 35 25 (0·8)	15 2 / 21 8 (0·8)	6 67 / 5 26 (1·4)	+ − No. 2 E_b
		36 84 / 23 7 (21·8)	46 74 / 6 24 (3·6)	22 98 / 4 26 (0·4)	94 26 / 21 9 (0·9)	27 88 / 3 25 (2·2)	56 64 / 18 12 (1·1)	28 8 / 8 2 (0·08)	10 74 / 1 19 (0·8)	+ − No. 3 K_b
			14 45 / 38 53 (5·14)	9 50 / 17 74 (0·3)	42 17 / 73 18 (1·6)	12 46 / 18 67 (0·005)	31 28 / 43 48 (0·4)	11 2 / 25 8 (0·07)	1 45 / 10 48 (0·01)	+ − No. 4 K_a
				13 39 / 13 85 (3·3)	52 0 / 63 33 (21·4)	3 47 / 27 66 (10·4)	32 20 / 42 56 (4·7)	11 5 / 25 5 (0·6)	3 33 / 8 60 (0·3)	+ − No. 5 E_a
					21 5 / 94 30 (0·3)	7 17 / 23 96 (1·2)	15 11 / 59 65 (0·9)	3 3 / 33 7 (1·6)	0 20 / 11 73 (1·3)	+ − No. 8 F_a
						30 80 / (0 33 / 9·6)	58 57 / 16 19 (0·24)	31 9 / 5 1 (0·04)	10 65 / 1 28 (1·2)	+ − No. 10 E_e
							13 17 / 59 54 (0·75)	8 1 / 22 8 (0·27)	3 18 / 8 75 (1·2)	+ − No. 11 E_f
								16 10 / 20 0 (8·3)	2 46 / 9 47 (2·7)	+ − No. 13 G_a

Table 3

Factor 1 or A

Phenotype matings		No. of matings	No. of offspring		
			+	−	Total
+	+	9	47	3	50
+	−	8	37	12	49
−	+	6	15	10	25
−	−	25	0	114	114

Factor 2 or E_b

Phenotype matings		No. of matings	No. of offspring		
			+	−	Total
+	+	32	135	7	142
+	−	7	26	20	46
−	+	5	18	7	25
−	−	3	0	18	25

Factor 3 or K_b

Phenotype matings		No. of matings	No. of offspring		
			+	−	Total
+	+	29	130	0	130
+	−	6	20	10	30
−	+	11	55	11	66
−	−	1	0	4	4

Factor 4 or K_a

Phenotype matings		No. of matings	No. of offspring		
			+	−	Total
+	+	9	30	9	39
+	−	11	28	30	58
−	+	11	30	22	52
−	−	16	0	89	89

Factor 5 or E_a

Phenotype matings		No. of matings	No. of offspring		
			+	−	Total
+	+	5	30	0	30
+	−	6	13	26	39
−	+	11	33	28	61
−	−	26	0	108	108

Factor 8 or F_a

Phenotype matings		No. of matings	No. of offspring		
			+	−	Total
+	+	1	5	2	7
+	−	2	7	10	17
−	+	4	12	12	24
−	−	41	0	190	190

Factor 10 or E_e

Phenotype matings		No. of matings	No. of offspring		
			+	−	Total
+	+	15	77	11	88
+	−	3	11	4	15
−	+	17	56	22	78
−	−	12	0	57	57

Factor 11 or E_f

Phenotype matings		No. of matings	No. of offspring		
			+	−	Total
+	+	1	5	0	5
+	−	0	0	0	0
−	+	6	15	17	32
−	−	41	0	201	201

Factor 12 or X

Phenotype matings		No. of matings	No. of offspring		
			+	−	Total
+	+	2	6	2	8
+	−	3	9	4	13
−	+	4	10	8	18
−	−	38	0	194	194

Factor 13 or G_a

Phenotype matings		No. of matings	No. of offspring		
			+	−	Total
+	+	9	27	14	41
+	−	14	45	34	79
−	+	7	15	25	40
−	−	17	0	78	78

system. Ten of the antigenic factors decribed have been found within 50 pig families to be controlled by Mendelian genes dominant to their absence. The results are shown in table 3.

Serum proteins

The five serum proteins investigated are briefly described as follows:

(1) Transferrins: Applying the technique of starch gel electrophoresis first introduced by Smithies (1955), Ashton (1960) and Kristjansson (1960) working independently, reported the existence of two alleles at this locus. It would appear from the description given in both papers, that these alleles are similar to the Tf^a and Tf^b reported here. A further allele Tf^c first observed by King (1962) in the Landrace breed, and then by Imlah (1963) in the Wessex breed is also presented as a new allele at this locus. A photograph of the transferrin types described is shown in figure 1. The schematic drawing gives the interpretation of the patterns presented by the transferrin types. From the segregation studies of offspring from over 50 pig families of tested parents as shown in table 4, it is assumed that three bands represent the homozygous expression of the genes, and five or six bands the heterozygous condition. The type — c transferrin has not as yet been found in the homozygous state. The segregation of offspring are in accord with the theory of simple Mendelian inheritance with co-dominance. The transferrin locus in the pig therefore presents three alleles Tf^a, Tf^b, and Tf^c with six possible phenotypes. Gene frequencies and the observed and expected distribution of transferrin types in a random population of the Large White and Landrace breeds are presented in table 6. It has been established that all the bands representing the three alleles at this locus bind iron (Imlah, 1963).

(2) Haptoglobins: In 1961, Kristjansson using the starch gel technique presented data to show that six haptoglobin phenotypes occurred in pig sera, and were under the control of alleles called Hp^1, Hp^2 and Hp^3. Then in 1963 Hesselholt reported the existence of four alleles called 0, 1, 2 and 3, which appeared identical to Kristjansson's three types with the exception of type 0, which was a faster migrating component. Investigations carried out in 1961 with pig sera and fresh haemoglobin failed to reveal any types similar to those reported by Kristjansson and Hesselholt. However, as in man (Harris, 1962), it was found that pigs possessed a protein component which bound fresh haemoglobin, and the mobility of the haptoglobin-haemoglobin complex was different from the original unbound protein. It was also found, that some pigs like man did not possess this protein, and were unable to bind fresh haemoglobin (Imlah, 1963). Therefore, the true haptoglobin

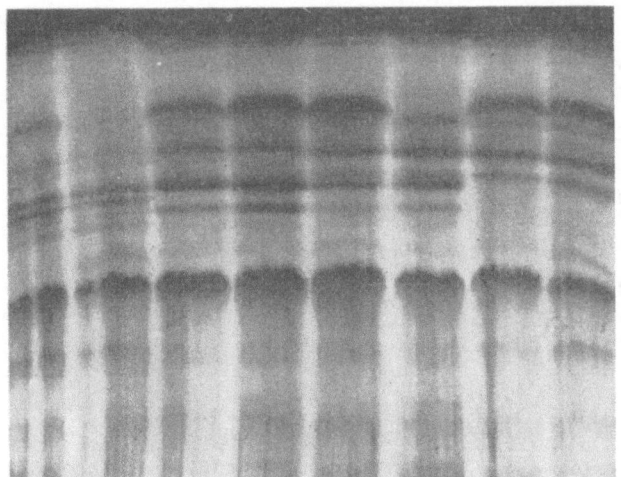

Fig: 1a

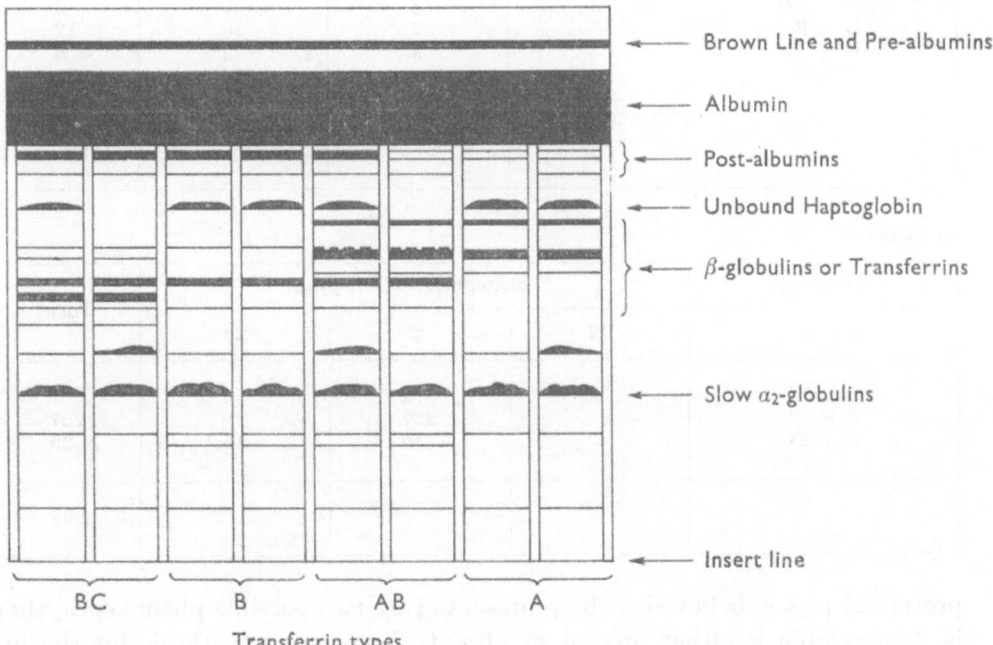

←— Brown Line and Pre-albumins

←— Albumin

}←— Post-albumins

←— Unbound Haptoglobin

}←— β-globulins or Transferrins

←— Slow α_2-globulins

←— Insert line

B.C.   B   AB   A
Transferrin types

Fig: 1b

Table 4

Segregation of serum protein types among offspring of different matings transferrins

Mating type ♂ × ♀	A	AB	AC	B	BC	Total
AB × A	3	6				9
AB × AB	2	3		3		9
AB × B		2		3		5
AB × BC		3	4	4	2	13
B × A		4				4
B × AB		18		22		40
B × B				172		172
B × BC				4	3	7
						259

Ceruloplasmins

Mating type ♂ × ♀	A	AB	B	Total
AB × AB	2	3	0	5
AB × B		7	10	17
B × AB		3	3	6
B × B			223	223
				251

Amylases

Mating type ♂ × ♀	12	2	23	Total
2 × 12	3	4		7
2 × 2		227		227
2 × 23		15	14	29
				263

protein of pigs is believed to be represented by two possible phenotypes, that is, haptoglobin is either present or absent. A genetic hypothesis for the inheritance of the anhaptoglobinaemic type has not been established yet.

(3) Haem-binding globulins: It was observed by King (1962), that the addition of old haemoglobin to pig sera revealed complexes similar to

those described by Kristjansson and Hesselholt. In view of the investigations carried out with fresh haemoglobin, it was therefore thought that the so-called haptoglobin types described in pigs were not true haptoglobins, but were complexes of breakdown products of haemoglobin. Fresh haemoglobin was

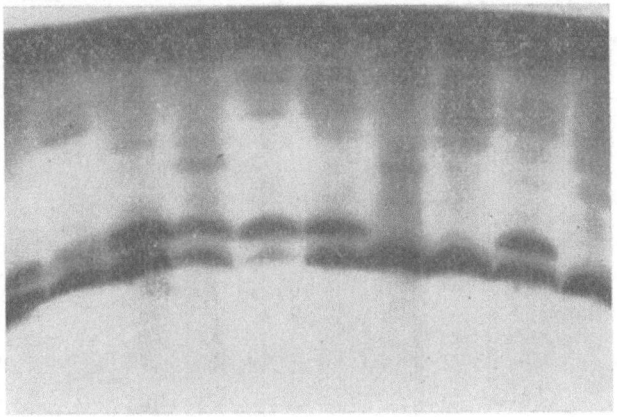

Fig. 2a

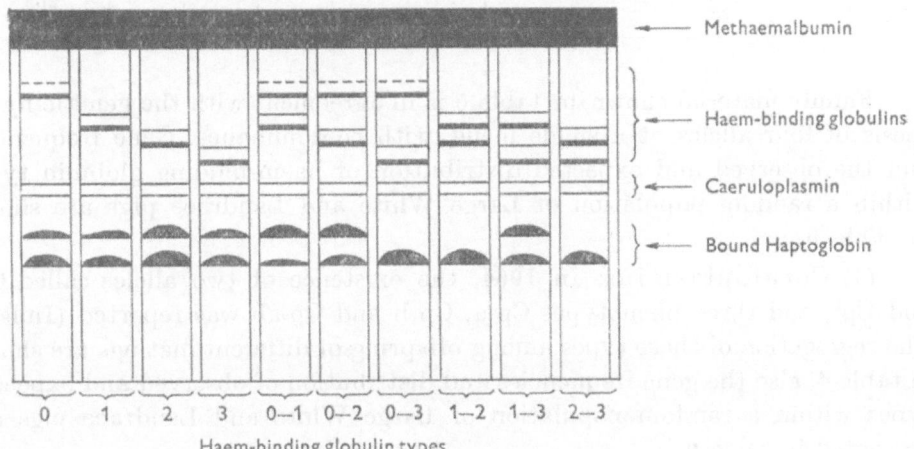

Fig. 2b

converted to alkaline haematin by boiling in alkali, and added to unhaemolysed pig sera. Staining gels with benzidine stain revealed complexes similar to those described by Kristjansson and Hesselholt, but there was no binding of the protein which normally attached itself to fresh haemoglobin. Four alleles called Hg⁰... Hg^0, Hg^1, Hg^2, and Hg^3 were found. The ten possible phenotypes resulting from a combination of all these types are shown in figure 2 with a schematic interpretation shown below.

Table 5

Segregation of serum protein types among offspring of different matings haem-binding globulins

Mating type $\male \times \female$	Segregation of offspring								Total
	0	01	03	1	12	13	23	3	
03 × 03	2		3					0	5
03 × 3			3					4	7
1 × 1				116					116
1 × 13				8		8			16
1 × 3						9			9
13 × 03		3	5			1		1	10
13 × 1				10		7			17
13 × 13				9		5		6	20
23 × 1					2	0			2
23 × 13					3	3	1	0	7
23 × 3							1	3	4
3 × 1						10			10
3 × 13						17		11	28
3 × 23							8	3	11
									262

Family material shown in table 5 is in agreement with the genetic hypothesis of four alleles at a single locus, with co-dominance. Gene frequencies and the observed and expected distribution of haem-binding globulin types within a random population of Large White and Landrace pigs are shown in table 6.

(4) Ceruloplasmins: In 1964, the existence of two alleles called Cp^a and Cp^b, and three phenotypes Cp-a, Cp-b and Cp-ab was reported (Imlah). The segregation of these types among offspring of different matings are shown in table 4, also the gene frequencies and distribution of observed and expected types within a random population of Large White and Landrace pigs are presented in table 6.

(5) Amylases: It was first observed by King (1963) in gels stained for ceruloplasmin with paraphenylenediamine, that clear zones appeared in gels which had stood overnight in the stain. These clear zones showed variations between pigs, and appeared in the gel around the slow $alpha_2$-globulins. Investigations showed that these clear areas appeared as a non-migratory component at the insert line, and as a migratory component in the slow $alpha_2$-globulin region. The migratory component showed genetic variation.

Table 6

Gene frequencies and distribution of serum protein types in two breeds of pig Large White

Transferrins	A	AB	B	BC	C						No. of animals
Gene frequency	0·36		0·64		0						161
Obs.	19	79	63								161
Exp.	(21)	(74)	(66)								161
Ceruloplasmins	A	AB	B								
Gene frequency			1.0								161
Amylases	1	12	2	23	3						
Gene frequency	0·05		0·95		0						161
Obs.	1	15	145								161
Exp.	(0·4)	(15·3)	145.3								161
Haem-binding globulins	0	01	02	03	1	12	13	2	23	3	
Gene frequency	0·05				0·84			0·04		0·06	85
Obs.	0	4	3	2	66	0	7	3	0	1	85
Exp.	(0·2)	(7·5)	(0·4)	(0·6)	(60)	(5·8)	(9·3)	(0·1)	(0·4)	(0·3)	85

Landrace

Transferrins	A	AB	B	BC	C						No. of animals
Gene frequency	0		0·98	0·02							95
Obs.		91	4								95
Exp.		(91)	(4)								95
Ceruloplasmins	A	AB	B								
Gene frequency	0·06		0·94								95
Obs.		12	83								95
Exp.		(11)	(84)								95
Amylases	1	12	2	23	3						
Gene frequency	0·2		0·8		0						95
Obs.	5	23	67								95
Exp.	(2·7)	(27)	(65·4)								95
Haem-binding globulines	0	01	02	03	1	12	13	2	23	3	
Gene frequency	0·06				0·51			0·05		0·38	55
Obs.	0	4	0	3	16	3	17	0	2	10	55
Exp.	(0)	(3·5)	(0·3)	(2·7)	(15)	(2·5)	(21)	(0·2)	(1·9)	(8)	55

Further investigations indicated that these components have an enzyme activity similar to the starch-splitting enzyme called amylase (Imlah, 1963). There are three criteria on which this hypothesis is based. They are:

(i) The digested areas reduced alkaline copper solution (ie. Benedict's reagent) indicating the presence of a reducing sugar.

(ii) The areas did not stain with iodine solution indicating that the starch had been hydrolysed.

(iii) The activity of this component was inhibited after 15 hours' incubation of a serum sample with a 5% solution of urea prior to electrophoresis.

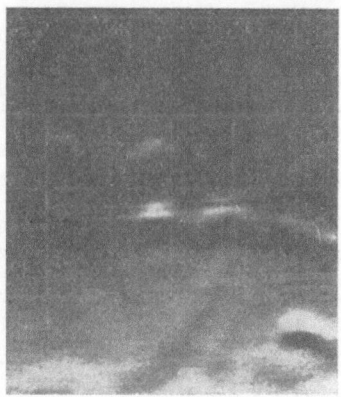

Fig. 3a

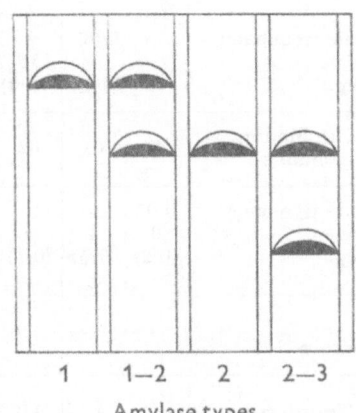

Amylase types

Fig. 3b

Three alleles called Am^1, Am^2 and Am^3 were identified in pig sera giving six possible phenotypes. A photograph and schematic drawing showing four amylase phenotypes are presented in figure 3. Family data presented in table 4 are in agreement with the hypothesis of three alleles at a single locus with co-dominance. The gene frequencies and distribution of amylase types within the Large White and Landrace breeds are shown in table 6.

Association and Linkage

Before establishing whether association or linkage exists between the red cell systems and the serum protein types described above, an account of the analyses carried out will be explained. The term linkage is used in the strictly genetic sense of two genes being on the same chromosome, but being sufficiently

Table 7

Test for association between red cell factors and haem-binding globulins in $2 \times n$ tables

Haem-binding globulins

		$Hg^{1/1}$	$Hg^{1/3}$	$Hg^{3/3}$	Total
Red cell factor K_a	+	67 (79·5)	37 (33)	21 (12·5)	125
	−	136 (123·5)	47 (51)	11 (19·5)	194
		203	84	32	319

$$X^2 \text{ for 2 d. f.} = 13·6$$

Haem-binding globulins

		$Hg^{1/1}$	$Hg^{1/3}$	$Hg^{3/3}$	Total
Red cell factor K_b	+	187 (170·5)	74 (70·5)	7 (26·8)	268
	−	16 (32·5)	10 (13·5)	25 (5·2)	51
		203	84	32	319

$$X^2 \text{ for 2 d. f} = 101·1$$

Haem-binding globulins

		$Hg^{1/1}$	$Hg^{1/3}$	$Hg^{3/3}$	Total
Red cell factor E_e	+	132 (130)	54 (50)	14 (20)	200
	−	69 (71)	24 (28)	17 (11)	110
		201	78	31	310

$$X^2 \text{ for 2 d. f.} = 6·05$$

far apart for crossing over to take place between them. Their linkage with one another being estimated as a recombination percentage, that is the proportion of times the two genes recombine as a result of crossing over between them. The necessary information is provided only by matings of a certain genetic type, that is the double backcross.

The term association on the other hand does not mean linkage in this sense, but describes the number of times two factors are found together. This may be due to linkage, but can occur for other reasons.

The data from over 300 pigs were therefore first examined in 2×2 tables to see what associations were present. Among the many comparisons which were made the only associations which were significant at the 1 in 20 level of probability were as follows: They were between the K_a, K_b and E_e factors and the haem-binding globulins Hg 1 and Hg 3. The results are shown in table 7.

Table 8

Double backcross matings involving the K^b allele and the haem-binding globulin alleles

Mating	Segregation of offspring			
♂ × ♀ $K^{bb}/Hg^{33} \times K^{Bb}/Hg^{23}$	K^{Bb}/Hg^{23}	K^{Bb}/Hg^{33}	K^{bb}/Hg^{23}	K^{bb}/Hg^{33}
1st family 2nd family	8 1	0 0	0 0	3 4
Total obs.	9	0	0	6
No. exp.	(3·75)	(3·75)	(3·75)	(3·75)
$K^{bb}/Hg^{33} \times K^{Bb}/Hg^{13}$	K^{Bb}/Hg^{13}	K^{Bb}/Hg^{33}	K^{bb}/Hg^{13}	K^{bb}/Hg^{23}
3rd family 4th family 5th family	3 6 2	1 0 0	0 0 0	2 3 3
Total obs.	11	1	0	8
No. exp.	(5)	(5)	(5)	(5)
$K^{Bb}/Hg^{03} \times K^{bb}Hg^{33}$	K^{Bb}/Hg^{03}	K^{Bb}/Hg^{33}	K^{bb}/Hg^{30}	K^{bb}/Hg^{13}
6th family 7th family	1 2	0 0	0 0	1 3
Total obs.	3	0	0	4
No. exp.	(1·75)	(1·75)	(1·75)	(1·75)
$K^{bb}/Hg^{11} \times K^{Bb}/Hg^{13}$	K^{Bb}/Hg^{13}	K^{Bb}/Hg^{11}	K^{bb}/Hg^{13}	K^{bb}/Hg^{11}
8th family	3	0	0	3
Total obs.	3	0	0	3
No. exp.	(1·5)	(1·5)	(1·5)	(1·5)

Double backcross matings involving the K^a allele and the haem-binding globulin alleles

$K^{aa}/Hg^{11} \times K^{Aa}/Hg^{13}$	K^{Aa}/Hg^{13}	K^{Aa}/Hg^{11}	K^{aa}/Hg^{13}	K^{aa}/Hg^{11}
1st family	1	1	1	1
2nd family	0	3	3	0
Total obs.	1	4	4	1
No. expr.	(2·5)	(2·5)	(2·5)	(2·5)

The possibility of the genes for these types being linked was therefore investigated in families of double backcross matings. Table 8 shows eight families of this type of mating involving all the haem-binding globulin genes and the K^b gene, also two families involving the haem-binding glubulin gene Hg^3 and the K^a gene. The two families involving K^a point to a linkage in repulsion, but the numbers are not sufficient, and applying "Yate's correction of continuity" for values less than 5 does not make the linkage significant ($X^2 = 1·6$ for 1 d. f.). However, the proportion of non-recombinants to recombinants in the eight families involving the K^b gene is statistically significant. Applying Finney's method (1940) the probability of linkage can be demonstrated at the 1 in 100 level:

$$S(\lambda) > 2·33 \sqrt{S(K)}$$

$$142 > 2·33 \sqrt{152}$$

$$142 > 28·7 \text{ (At the 1\% level.)}.$$

Also, an estimate of the recombination frequency can be calculated as follows:

$$\frac{1}{2}\left[1 - \sqrt{\frac{S(\lambda)}{S(K)}}\right]$$

$$\frac{1}{2}\left[1 - \sqrt{\frac{142}{152}}\right]$$

$$= 0·017 \text{ or } 1·7\% \text{ recombination frequency.}$$

In conclusion therefore, although the numbers involved in these families are small, one cannot ignore the posibility that there are two genes in the pig; one of which controls a type of red cell antigen, and the other which controls a biochemical polymorph, which are closely linked on the same chromosome.

References

Andresen, E. (1962). Annals of the New York Academy of Sciences 97, 205.

Ashton, G. C. (1960). Nature, Lond. 186, 991.

Finney, D. J. (1940). Ann. Eugen., Lond. 10, 171.

Harris, H. (1962). Human Biochemical Genetics, p. 204, Cambridge University Press.

Hesselholt, M. (1963). Acta Vet. Scand. 4, 238.

Imlah, P. (1963). Thesis, Edinburgh.

Imlah, P. (1964). Nature, Lond. 203, 658.

King, J. W. B. (1963). Personal communication.

Kristjansson, F. K. (1960). Canad. J. Genet. Cytology 2, 295.

Kristjansson, F. K. (1961). Genetics, 46, 907.

Linnet-Jepson, P., Galatius-Jensen, F., & Hauge, M. (1958). Acta genet. 8, 164.

Smithies, O. (1955). Biochem. J. 61, 629.

Smithies, O., and Hiller, O. (1959). Biochem. J. 72, 121.

INVESTIGATIONS ON PRODUCING TEST SERA FOR DETERMINATION OF BLOOD GROUPS IN PIGS*

S. ALEXANDROWICZ, A. KACZMAREK and I. WIATROSZAK

Agricultural College, Department of Animal Husbandry, Poznaň

Investigations on the production of test sera were initiated in Poland, in 1956, by A. Wróblewski at the Polish Academy of Sciences and continued by the authors at the Agricultural College in Poznaň.

We present here the results of these investigations.

1. Normal sera. Initially we investigated blood cell antigens in different pigs using normal sera. It was found that the young animals, at the age from 4 to 6 months, did not possess any antibodies in blood serum, or, if they did, the titre was very low, while on the contrary, the sera of adult pigs agglutinated blood cells pretty often.

Table 1

Appearance of antibodies resulting from successive immunizations on example of 20 donors and 20 recipients

Serum solution	before imm.		I		II		III		IV		V		VI		VII	
	aggl.	haem.	aggl.	haem.	aggl.	haem.	aggl.	haem.	aggl.	haem.	aggl.	haem.	aggl.	haem.	aggl.	haem.
1 : 1	20	2	20	1	20	10	20	9	20	9	20	11	20	7	20	9
1 : 2	17	2	19	1	18	10	19	9	20	9	17	11	18	7	18	9
1 : 4	10	2	14	1	14	9	19	9	16	8	16	11	16	7	17	9
1 : 8	7	2	12	1	12	9	15	8	12	8	13	10	17	7	15	9
1 : 16	—	—	8	1	11	7	12	8	11	8	10	10	10	7	13	9
1 : 32	—	—	3	1	8	8	11	8	8	6	9	10	11	7	12	8
1 : 64	—	—	2	—	8	—	8	4	7	5	6	8	10	7	11	7
1 : 128	—	—	—	—	2	—	4	4	2	5	4	9	8	6	9	6
1 : 256	—	—	—	—	—	—	—	1	—	5	1	7	1	5	7	3
1 : 512	—	—	—	—	—	—	—	—	—	—	—	—	—	—	2	—
1 : 1000	—	—	—	—	—	—	—	—	—	—	—	—	—	—	—	—
1 : 2000	—	—	—	—	—	—	—	—	—	—	—	—	—	—	—	—

Successive immunizations — Number of recipients with positive reaction

*) The work is partly financed by U. S. Department of Agriculture, Agriculture Research Service.

Tests, performed at monthly intervals, of piglets from ten litters of Zlotniki breed, from the age of 8 weeks to 6 months, revealed that the process of formation of normal antibodies in pigs lasted to the age of 6 months. Intensity of this process at different ages was an individual feature of the animal, but it also depended in the environmental conditions. This phenomenon has been previously observed in other breeds by Szent-Ivanyi and Szabo (1954), Goodwin and Coombs (1956), and Golders (cited after Coombs). They found that reactivity of the normal antibodies increased with the age of pigs.

2. Attempts to obtain isoimmune sera. As normal pig sera had low titre of antibodies, about 200 immunizations were carried out in order to

Table 2

Number of isoimmunizations with blood cells of known antigenic system for obtaining the desired antibodies

To obtain	No. of pigs imm. and reimm.	No. of reacting pigs	No. of sera unobtainable by absorption	No. of obtained test sera	Test titre	Sera No.
anti-A	—	—	—	—	—	—
D_1	2	2	—	—	—	—
E_a	8	8	6	2	1 : 64	7018
Eb	3	2	—	2	1 : 64	7539
Ee	9	5	4	3	1 : 16	7145
					1 : 16	7617
					1 : 64	7245
Ex	15	8	1	1	1 : 64	3340
Fa	4	4	3	1	1 : 128	7144
Ga	3	—	—	—	—	—
Gb	16	13	11	2	1 : 1	4659
					1 : 16	4659R
Ha	32	23	9	—	—	—
Hb	4	—	—	—	—	—
Xa	9	4	5	—	—	—
Ib	19	12	7	—	—	—
Ka	13	13	6	6	1 : 1	2885
					1 : 1	6648
					1 : 4	7018
					1 : 64	7539
					1 : 8	4098
					1 : 64	7018
Kb	3	1	2	—	—	—
La	6	2	4	—	—	—
Lb	3	1	2	—	—	—
Ld	10	5	5	—	—	—
Ma	28	27	1	—	—	—
Mb	—	—	—	—	—	—
Mc	18	17	1	—	—	—
O	12	6	6	—	—	—
S_1	10	2	2	—	—	—
S_2	1	1	—	—	—	—
X	1	1	—	—	—	—
Z_8	23	14	8	1 in work	—	—
Z_{a1}	4	1	3	2	—	—

obtain higher titre. Table 1 presents the process of antibody formation. It was found that the titre of antibodies increased up to the seventh immunization, though the titre was sufficiently high (1 : 128) after the third immunization.

In order to limit the number of different antibodies formed by the recipient, in a further study the donors and recipients were specially chosen. Immunizing such animals, the first selection was made to avoid formation of undesirable antibodies.

Looking at the data in table 2 it can be seen that some test sera were easy to obtain and some not. Easy obtainable sera in our investigations were: Anti-Eb, Ee, Ea, Fa, then successively Ed, Ka, Gb. The remaining sera, in spite of numerous immunizations, could not be obtained as complete antibodies. These were: anti-Ha (32 isoimmunizations), Ia (9 isoimmunizations), Ib (19 isoimmunizations), Ld (10 isoimmunizations), Ma (28 isoimmunizations), Mc (18 isoimmunizations), S_1 (10 isoimmunizations), Z_8 (23 isoimmunizations). The blood cell antigens were determined in the Laboratory of the Czechoslovak Academy of Sciences at Liběchov.

In our conditions, re-immunizations usually increased the titre after one or two treatments. But it did not mean that the titre of the desired antibody would always increase, as the injection of a single antigen caused formation of a wide range of antibodies in the recipient. As it appeared, the result of serological reaction depended on many factors. A further difficulty in obtaining pure immune sera is that, up to now, the known blood cell antigens are not numerous.

3. Absorption of sera. For absorption 1 part of dense blood cells was added to 1 part of serum. The suspension prepared in this way was mixed and absorbed for different periods of time: 30 minutes at room temperature, then 45 minutes at $+4°C$, and again 30 minutes at room temperature. Two, three and even six different blood cells were used for absorption. The same serum was sometimes absorbed seven times. Each absorption decreased the titre of the serum.

4. Species antibodies for differentiating blood cell antigens. Our research also included the normal antibodies in sera of other species of domestic animals. In 583 tested samples of bovine sera we found one antibody giving haemolytic reaction and corresponding to anti-O in pig. The bovine serum anti-J was used as pig serum anti-A.

By the heteroimmunization of a cow with pig blood cells the serum anti-Kb was obtained, while by immunizing a ram with pig blood cells we obtained the sera anti-Ed and anti-P_p 1.

Table 3 presents the list of all test sera produced in our Laboratory.

5. Some observations. Pigs are not a very easy material for serological studies, due to difficutlies in taking blood samples from them. At

125

first, we immunized the animals by intravenous injections, but as they caused shocks in some animals, we were forced to use intramuscular injections which also gave good results. A high titre of the serum enabled the repeated absorptions, but it did not guarantee obtaining a test serum. The generally accepted terms of reading are: after 1/2 hour, after 2 hours and after 4 hours. Each haemolyzing serum, however, has its specific way of reacting and the serum should not be regarded as bad if it does not give any result of haemolysis after half an hour.

Table 3

Test sera obtained

Blood systems	Blood group factors		Antibodies	Way of obtaining
A	1	A	haemolysing agglutinating	normal bovine serum anti-A normal pig serum
	2	O	haem.	normal bovine serum
E	3	Ea	aggl.	isoimmune
	4	Eb	aggl.	isoimmune
	5	Ec	aggl.	isoimmune
	6	Ed	aggl.	heteroimmune
	7	Eh	aggl.	isoimmune
F	8	Fa	aggl.	isoimmune
G	9	Gb	aggl.	isoimmune
K	10	Ka	aggl.	isoimmune
	11	Kb	haem.	heteroimmune
New sera of a not defined system				
	12	P_z1	aggl.	heteroimmune
	13	P_p2	aggl.	isoimmune
	14	P_p5	haem.	isoimmune

Analysis of blood cells, which were stored for a long time, gives weaker results in agglutination, while in haemolysis, especially after the third reading, the results are false.

Activity of test sera, namely the binding of the antibody to the antigen, causing agglutination or haemolysis, is specific for each serum.

Results

1. Appearance of antibodies in blood can be first observed in one-month-old piglets. In the following months the number of individuals possessing normal antibodies increases.

2. Immunization of individuals below the age of 6 months for obtaining immune serum gives rather poor results.

3. Shocks caused by intravenous injection can be avoided when immunization is carried out by intramuscular injections, and the results are equally good.

4. Absorption of isoimmune sera does not guarantee the content of one antibody if it is not absorbed and verified with blood cells of pigs from different environments and, if possible, of different breeds.

5. Titre of the serum was decreased by absorption as well as by multiple freezing and thawing of serum. It is specially true for haemolytic sera.

BLOOD GROUP STUDIES IN PIGS

H. BUSCHMANN

Institute for Animal Breeding Research, Munich

1. Blood Group Serological Studies

The following reagents are available for blood group studies (table 1)

Table 1

Source and nature of reagents used for blood-typing of pigs

Reagent	Nature of antibody	Source
A	lysin	cattle normal anti-J
E_a	agglutinin, lysin	isoimmune
E_b	agglutinin	isoimmune
E_e	agglutinin, incompl. aggl.	isoimmune
E_f	incompl. agglutinin	isoimmune
F_a	agglutinin	isoimmune
G_a	incompl. agglutinin	isoimmune
G_b	incompl. agglutinin	isoimmune
H_a	lysin	isoimmune
H_b	lysin	isoimmune
I_b	incompl. agglutinin	isoimmune
K_{a1}	lysin	isoimmune
$K_{a1} + K_{a2}$	lysin	isoimmune
K_b	lysin	isoimmune
L_a	incompl. agglutinin	isoimmune
$L_e + L_c$	incompl. agglutinin	isoimmune
M_a	lysin	isoimmune
Nf 1 (= B_a?)	agglutinin	isoimmune
Nf 3	lysin	isoimmune

The symbol Nf denotes a new factor, which could not be assigned to a system so far. However, the results of the first international comparison test for pig blood grouping reagents gave some evidence, that the Munich factor Nf 1 might correspond to the factor B_a. By means of a reagent, denoted $K_{a1} + K_{a2}$, it was possible to detect a linear subgroup relation in the K-system.

In our routine work, every serum sample is tested for naturally occurring antibodies, soluble A-substance and serum haptoglobin and transferrin types. The importance of the inhibition tests for the detection of soluble A substance, especially in parentage control, should be stressed. When a potent A reagent is available and the inhibition tests are performed, then parentage exclusions based on the A system need not be regarded with any reservation according to our experience. In our work, there was no case where an A-positive off-spring appeared and where both parents were neither A^{cs} nor A^s positive, thus being an exception to the theory of dominant inheritance (Andresen 1962).

In table 2, the A blood group system was analyzed in three groups of pigs:

Table 2

Distribution of A-phenotypes within three age-groups of pigs. Observed and expected values in the case of homogeneity

Group	Number of animals $A^{cs}+$, anti-A	Number of animals A^s+ anti-A—	Number of animals A— anti-A+	Number of animals A— anti-A—
I. Piglets 6—9 weeks N = 343	174 (153·15)	20 (18·20)	63 (142·17)	86 (29·47)
II. Pigs from slaughterhouse, 5—6 months of age N = 462	201 (206·28)	26 (24·52)	233 (191·49)	2 (39·70)
III. Sows and boars N = 382	155 (170·56)	17 (20·27)	116 (158·33)	14 (32·83)

It appears from table 2 that there are significant deviations from the expected values in the right two columns. Relatively less anti-A positive individuals and relatively more A negative, anti-A negative individuals appear to be among young piglets, than among adult pigs. The number of A negative, anti-A negative individuals decreases with increasing age and a further analysis of the 14 individuals found in the last group class (sows and boars) showed, that 11 of these animals were young boars under one year of age. In the remaining 3 animals possibly a weak anti-A titre escaped detection.

In table 3 our family material is summarized:

Table 3

Inheritance of the A substance in pigs

Mating type	Number	Offspring			
		$A^{cs}+$ anti-A —	A^s+ anti-A —	A — anti-A +	A — anti-A —
anti-A × anti-A	59	—	—	182	24
anti-A × A^{cs}	46	82	6	59	9
A^{cs} × anti-A	40	73	6	52	17
anti-A × A^s	1	—	—	3	1
A^s × anti-A	7	16	—	6	—
A^s × A^{cs}	3	10	—	1	1
A^{cs} × A^s	3	6	1	2	—
A^{cs} × A^{cs}	29	93	8	10	12
— × anti-A	5	—	—	9	4
A^{cs} × —	1	7	—	1	—

At the present state of our knowledge, it seems that two kinds of piglets exist: one group inheriting the possibility of forming A substance in their serum, which is absorbed on to their red cells later in development, and another group lacking this possibility and being exposed to immunization through external agents (food, microorganisms) and therefore forming anti-A in their serum. Pigs from one year of age upwards seem to possess either A (as A^{cs} or A^{s}) or anti-A. Animals lacking both A-substance in cells or serum and anti-A in their serum apparently are very rare and perhaps do only exist, because a weak anti-A titre cannot be detected in the serum.

2. Parentage Control of Pigs in Bavaria

The high percentage of erroneously determined parentages found in progeny testing groups of South Germany gave rise to the following measures which are in force in Bavaria since 1964:

1. 15% of all fattening groups at the testing stations have to be blood-typed and a parentage control is performed. Selection of the groups is done by the State authorities. If the specified descent has to be disputed, the group is eliminated from the testing station and the breeder must pay for all expenses.
2. From time to time the Bavarian Agricultural Ministry orders, that a parentage control must be performed in all young boars gathering on a breeding market.
3. The following rules for securing paternal descent were put in force by the breeders' associations:
   a) Within the same heat period a sow should be mated only with one boar. When more boars were used, a parentage control is required.
   b) A determined parentage is only acknowledged when the gestation period is between 109 and 121 days for the German landrace.
   c) Matings in succeeding heat periods should be done by the same sire.
4. All young boars joining the herd book should be blood-typed.

3. Blood Group Studies in Wild Boars (Sus scrofa ferus)

Through the courtesy of the German Hunter Associations, 238 blood samples from wild boars could be tested. The animals were shot in different regions of South Germany during the last winter.

By starch gel electrophoresis two different haemoglobin types could be demonstrated in wild boars. Their mode of inheritance is still unknown.

The occurrence of blood-group factors in wild boars was studied with our pig blood-group reagents. The results are summarized in table 3.

Table 4

Blood-groups and haemoglobin types in wild boars N = 238

Reagent	Number of positive animals	Number of negative animals
A	232	6
E_a	22	216
E_b	78	160
E_e	230	6
E_f	39	128
F_a	—	238
G_a	167	—
G_b	—	210
H_a	—	138
H_b	—	238
K_{a1}	2	236
K_{a2}	4	163
K_b	167	—
L_a	89	149
M_a	—	15
Nf 1(B_a?)	236	2
Nf 2($L_e + L_c$)	60	178
Nf 3	—	167
Haemoglobin type II:	88	
Haemoglobin type I:	70	

There are some blood group systems (F, G, H, M) without polymorphism in wild boars, as one allele has been fixed (G_a), or some blood group factors do not appear at all (F_a, H_a, H_b, M_a). In other systems (A, K), the frequency distribution is completely different from that found in domestic pigs. Now the question arises, whether the collected blood samples are representative of the species. In table 4 the origin of the samples is listed.

4. Multiple sire mating and insemination in swine

When sows are mated with two sires in the same heat period or inseminated by sperm mixed from two boars, then litters arising from such experiments will be composed of offspring from two boars. From the genetic point of view, there is a new situation, because the litter of one sow is composed of piglets being half-sibs within the same litter. The situation is demonstrated in table 5, further the co-variance and causal components are analyzed.

Following the argumentation of Falconer (1961), the total environmental variance V_E is divided in table 5 into a component V_{Ec} (= common environment), which is the cause of similarity between members of a group, and V_{Ew}, which is the environmental variance due to differences other than those resulting from whether the individuals are related or not. In the multiple mating system we succeeded in isolating the V_{Ec} component as shown in table 5. So far this has not been possible in population genetics.

Table 5

Origin of blood samples from wild boars

Name	District	Number of samples
Public Forest Board	Forstenried	2
Zoological Garden	Frankfurt	5
Thurn und Taxis	Donaustauf	154
Fürst Löwenstein	Wertheim	8
Wittelsbacher Fonds	Ingolstadt	70
Schloss Granheim	Granheim	1
Forest Board	Warndt Saar	1
Forest Board	Ehinger Danube	1

Furthermore, we get an improved method of estimating the breeding value of boars by eliminating the components of maternal environment. For this purpose, several sows are inseminated in the same heat period by a mixture of sperm from one boar which has been tested extensively and ascribed the index 100, and from another boar of unknown breeding value. By blood-group tests the offspring of each litter is ascribed to the two sires and the piglets are tested at a station. Then the results can be compared.

Table 6 shows blood-grouping of piglets from a litter produced by inseminating a sow with mixed sperm in the same heat period.

In our investigations several attempts were made to produce piglets from two different boars within the same litter.

From group 1 it can be seen that, when natural service was followed immediately by insemination, we did not succeed to produce mixed litters, except for one case. However, there is some advantage as to the number of piglets born when after the first service a second natural service follows after 10 hours. The best results were achieved by artificial insemination with mixed sperm. Our procedure was very simple: sperm from two boars were mixed immediately before insemination. It should be possible to get a better proportion of piglets approaching more the desired 1 : 1 ratio within the litters by adjusting the density of the two sperm samples by colorimetric methods.

133

Table 6

Design of double sire mating

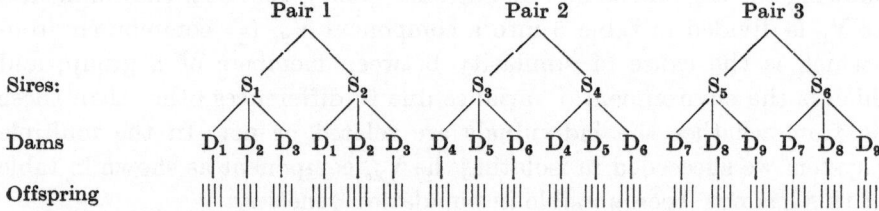

Genetic covariance of full sibs: $\frac{1}{2}V_A + \frac{1}{4}V_D + V_{Ec}(+\frac{1}{4}V_{AA} + \frac{1}{8}V_{AD} + \frac{1}{16}V_{DD})$

Genetic covariance of half sibs: $\frac{1}{4}V_A \quad (+\frac{1}{16}V_{AA})$

Genetic covariance of half sibs within the same litter: $\frac{1}{4}V_A + V_{Ec} \quad (+\frac{1}{16}V_{AA})$

In brackets: Components for two-factor gene interaction

V_A = additive variance
V_D = dominance cariance
V_{Ec} = variance due to common environment

Analysis of variance for the hierarchial structure of table 5:

$$MS_{Pairs} = \sigma_\omega^2 + 4_D{}^2 + 3 \cdot 4_s{}^2 + 2 \cdot 3 \cdot 4_P{}^2$$
$$M_{SS} \text{ (material half sibs)} = \sigma_\omega^2 + 4_D{}^2 + 3 \cdot 4_s{}^2$$
$$M_{SD} \text{ (paternal half sibs)} = \sigma_\omega^2 + 4_D{}^2$$
$$MS \text{ (full sibs)} = \sigma_\omega^2$$

$$\sigma_s{}^2 = MS_s - MS_D/3 \cdot 4$$
$$\sigma_s{}^2 = \text{cov HS}(♀) - \text{cov HS}(♂)$$
$$\sigma_s{}^2 = \frac{1}{4}VA + V_{Ec} - \frac{1}{4}V_A$$
$$\sigma_s{}^2 = V_{Ec}$$

Table 7

Blood Group System	A	E	F	G	K	L	M	Nf	Serum antibodies
Boar 1: Verdi	—/—	ae/e	—/—	a/a	a₂ /b	a/	—/—	1, 2, —	anti-A
Boar 2: Triton	—/—	b/b	—/—	a/a	/b	a/	—/—	1, 2, —	anti-A
Sow: 936	A^cs/—	ae/ef	—/—	a/b	/b	—/—	—/—	1, 2, —	—
Piglet 1:	A^cs/—	ae/**b**	—/—	a/a	/b	—/—	—/—	1, 2, —	—
Piglet 2:	A^cs/—	ef/**b**	—/—	a/b	/b	a/	—/—	1, —, —	—
Piglet 3:	A^cs/—	ae/ef	—/—	a/b	a₂/b	—/—	—/—	1, 2, —	—
Piglet 4:	—/—	ae/**b**	—/—	a/b	/b	a/	—/—	1, 2, —	—
Piglet 5:	A^cs/—	ae/**b**	—/—	a/b	/b	—/—	—/—	1, 2, —	—
Piglet 6:	A^cs/—	ef/e	—/—	a/b	a₂/b	—/—	—/—	1, 2, —	—
Piglet 7:	—/—	**ae**/ef	—/—	a/b	/b	a/	—/—	1, —, —	—
Piglet 8:	A^cs/—	ae/	—/—	a/a	a₂/b	a/	—/—	1, 2, —	—
Piglet 9:	—/—	ae/	—/—	a/b	a₂/b	a/—	—/—	1, 2, —	—

Result: 4 piglets from Triton, 5 piglets from Verdi.

Group I
Artificial insemination with sperm from one boar, immediately after natural service

Number of sows	Piglets from insemination	Piglets from natural service	Unknown descent	Total
12	1	71	11	83

Group II
First natural service with one boar, immediately followed by natural service with another boar

Number of sows	Piglets from first service	Piglets from second service	Unknown descent	Total
2	1	4	13	18

Group III
First natural service with one boar, natural service with another boar after 10 hours

Number of sows	Piglets from first service	Piglets from second service	Unknown descent	Total
12	25	9	34	68

Group IV
Insemination with mixed sperm from two boars

Number of	Piglets with determined descent	Piglets with not determined descent	Total
12	62	18	80

Proportions of piglets within the 12 litters:

1. 3 : 7	3. 5 : 1	5. 8 : 2	7. 3 : 0	9. 4 : 5	11. 2 : 2
2. 3 : 0	4. 2 : 1	6. 3 : 2	8. 3 : 0	10. 2 : 0	12. 4 : 0

Group V
At first natural service, insemination 10 hours thereafter

Number of sows	Piglets from natural service	Piglets from insemination	Unknown descent	Total
5	27	1	8	36

Furthermore, it should be mentioned that we got the sperm from an insemination station where only sperm from one boar was taken every day. So in our experiments the sperm mixture contained sperm from one boar that had been stored longer than the sperm from the other boar. But all these disadvantages can be avoided.

By this method of multiple insemination, a chance is given to improve the evaluation of the breeding worth of boars by comparing simply the performance of the offspring of a standard boar with that of a test boar, thereby eliminating maternal environmental influences. Furthermore, some knowledge of fertilization and artificial insemination in pigs can be obtained.

Summary

1. The high percentage of erroneous parentage cases found in progeny testing groups of South Germany gave rise to official measures for the parentage control in Bavaria.
2. Investigations on insemination of sows with mixed semen in the same heat period were carried out. The theoretical and practical aspects of the problem are discussed.
3. Blood group genetic studies in wild boar populations of Germany revealed the absence of genetic polymorphism in many blood group systems as compared with the domestic pig.
   A new haemoglobin type was detected.

A NEW APPROACH TO BOAR PROGENY TESTING

R. W. WIDDOWSON and T. A. NEWTON

BOCM Pig Advisory Department, Stoke Mandeville, Aylesbury, Bucks

Introduction

The progeny testing of boars was pionneered in Denmark in 1926. Farmers submit four teams of 2 hogs and 2 gilts, each sired by the same boar but by unrelated dams to stations controlled by a Central Testing Authority. The pigs are fed individually from 20—90 kilos liveweight, and slaughtered. Records of growth rate, food conversion and carcase quality are kept. Differences in the average performance of the 12 pigs sired by each boar are then taken as reflecting genetic differences between boars. This system of testing, with minor variations, has been adopted by other countries. In the United Kingdom, for example, Progeny Testing was pioneered by the British Oil & Cake Mills in 1953. The system of testing originally adopted was a modification of the standard Danish System. Four half litters of three hogs and one gilt all sired by the same boar and from four unrelated sows were taken to a testing station at between 50 and 60 lb. liveweight. At 70 lb. liveweight the test was

Table 1

Comparison of results of 4 years boar testing

	United Kingdom				Denmark			
Year	1954	1955	1956	1957	1954	1955	1956	1957
Daily liveweight gain*)	1·10	1·07	1·08	1·10	675	678	680	681
Food conversion†)	3·50	3·57	3·49	3·40	3·03	3·03	3·01	2·97
Length mms.	790	792	790	789	937	938	941	944
Thickness of back fat mms.	49	48	49	48	33	33	32	31
% Grade A or better	76	77	76	81	90	91	92	91

(Danish Report on Boar Progeny Testing, 1962—63; N. P. B. A. Report on Progeny Testing, 1958, Vol. 4.)

*) lbs/day (U. K.), gms/day (Denmark).

†) lbs. of food per lb. liveweight gain (U. K.). Fodder units per kilo liveweight gain (Denmark).

commenced. The pigs were fed ad libitum to 200 lb. liveweight, and slaughtered. Various carcase measurements were made and results for boars evaluated. This system with slight modification was used by the Pig Industry Development Authority who look after boar progeny testing. Table 1 show comparative results in Denmark and U. K.

These figures show that progress in the improvement of economic factors in the U. K. has been more erratic than in Denmark. This has led to an examination of the methods of testing used to ascertain whether or not improvements could be instituted which would lead to more accurate assessment of the boar's genetic worth so enabling greater progress to be made.

The accuracy of Progeny Testing depends on the assumption that the genetic value of the various dams is the same, and on this assumption, the performance of the offspring can represent the boar's genetic worth. It is therefore usually stipulated that the dams should be of average breed performance. Even with this stipulation the dams are bound to vary one to another, and it is usual to eliminate this variation by testing the boar over four matings.

In Great Britain no control is exercised over the selection of dams, and there is a tendency by the farmer, unaware of the genetic implications, to use the better females in his herd. The use of these females means that the boar's value given by the offsprings' performance is too high.

This therefore is the first source of error in British Boar Progeny Testing.

The second weakness occurs from the fact that under the British system of testing the pigs for each boar to be tested are born at or about the same time of year, and more importantly are always reared on the same farm, and therefore subject to the same pre-test environment (disease, management and feeding). Environmental effects for each boar will be different, and therefore the comparison of results will be less accurate. As an example, age of pigs at start of test at 70 lb. have differed between 61 days and 113 days, and it is difficult, if not impossible, to know whether or not these differences, which certainly effect performance in later life and carcase assessment, are genetic or environmental.

A third source of weakness in the United Kingdom system of testing lies in the fact that only half litters are tested. Theoretically then, pigs should be the average for the litter, but due to human nature, farmers tend to select the better pigs for testing. Practically, this means that the boar is being judged by the performance of his best pigs, no account being made for the normal distribution to be expected in any mating.

These three weaknesses are not applicable to Danish Testing methods, as care is exercised by the testing authority to ensure that only impartially selected females are used as progeny test group dams; the pigs for the test also being impartially selected. Additionally, the widespread use of skim milk/barley diets, the similarity of pig housing, all tend to even out pre-test environmental differences.

As the actual test procedures in the two countries are almost identical, differences in the rate of genetic improvement in pigs can be attributed to the pre-test management and selection practised in the United Kingdom. This paper describes an improved progeny test which eliminates as far as possible differences due to the genetic make-up of dams, and the pre-test environment. The test involves several new techniques which are described in detail and preliminary results are given.

Experimental Techniques

1. General Principle

The basis of this test depends on the artificial insemination of four sows with semen which contains sperm from two boars. The paternal parentage of the resultant litter is ascertained as soon after birth as possible by blood grouping techniques. Litters are split according to sire and the two part litters are fed and managed identically. As each boar is mated to the same female the genetic effects of the dam are reduced if not eliminated, and the environment (due to dams, climate, feeding and management) is identical from conception to slaughter; therefore, differences must be due to boars. Embryonic placentation may have some effect, but it is felt that this must be near zero when the boars are tested over four litters.

In practice, a herd of 36 Large White sows of average performance and known blood group are maintained at the testing station for use as dams in the Progeny Test. There are also three Large White Boars of known blood groups. Each boar has been performance tested and progeny tested over at least ten litters. These boars act as standards.

Each test boar is blood grouped as it comes to the testing station, and from this knowledge it is possible to select four sows and one of the standard boars according to their blood group variance with the test boars so that paternal identification of the resultant litter will be 100%. Four sows are mated with the standard boar and a test boar and the litters are tested. This procedure is repeated with other test boars. Comparison of the performance of test boars can then be made by reference to their respective performance to the standard boar.

2. A. I. Techniques

As one boar's semen may die off more quickly than the others, semen is collected from the two boars as quickly as possible. It is collected in three fractions — pre-sperm, sperm rich and post-sperm — each fraction being held in Thermos jars at a temperature of 37°C. Densities for all fractions for both boars is determined by the method of Madden. Samples stained with Nigrosin are microscopically examined and an estimation of healthy live sperm from each boar is made. For the purpose of this test, sperm with twisted tails, decaps, or drops is regarded as useless. Calculations are then made so that each sow is inseminated with the same numbers of healthy sperm from each boar. A total dosage rate of 60 cc. of mixed semen, with a minimum of 15 cc. from one boar is aimed at, but if the two boars differ greatly in the number of healthy sperm per cc. then dosage rates can exceed 60 cc. Mixing of the semen needs care, vigorous agitation being very likely to cause severe damage to the sperm. As soon as the semen has been mixed it is used in a fresh, undiluted form.

3. Blood Grouping

The pigs are grouped by the red cell antigen method. Samples from the litter are taken from anterior vena cava when the pigs are between 7 and 14 days. All blood samples are collected into bottles containing a normal isotonic sodium citrate chloride solution. From the result of the blood grouping determination it is possible to make up two teams of pigs, according to the sire. (Appendix I.)

4. Feeding and Management

After service, the sows are penned together in yards and fed on the following scale: 4 lbs. per day for the first three months of pregnancy: 6 lbs. per day for the last months of pregnancy, and after farrowing they receive 8 lbs. per day. The diet which is fed wet is a normal sow ration. Water is available ad libitum.

The sows are farrowed in wooden huts fitted with sleeping creeps. At three days all pigs receive an injection of iron/dextrose compounds as an anaemia preventative. All male pigs are castrated at this age. Early weaning diets are offered to the pigs from 7 days of age.

Table 2

Litter No.	Boars	No. of sperm/cc. $\times 10^9$	% Live Sperm	Dosage Rate cc.	No. Born
1	Test Boar A	2·8	75	34	15
	Standard Boar X	3·4	85	24	—
2	Test Boar A	12·2	74	20	5
	Standard Boar X	3·3	86	66	4
3	Test Boar A	5·0	69	15	3(7)*)
	Standard Boar X	2·0	43	71	—
4	Test Boar A	5·0	69	15	10
	Standard Boar X	2·0	43	71	5
1	Test Boar B	3·2	80	30	7
	Standard Boar X	2·0	84	47	—
2	Test Boar B	2·6	74	25	7(3)*)
	Standard Boar X	2·0	43	56	—
3	Test Boar B	4·6	95	20	5
	Standard Boar X	4·5	75	26	—
4	Test Boar B	7·2	76	25	Not yet farrowed
	Standard Boar X	5·2	66	40	
1	Test Boar C	8·8	77	31	2
	Standard Boar X	7·1	82	36	4
2	Test Boar C	8·6	81	15	3
	Standard Boar X	2·6	88	45	3
3	Test Boar C	2·3	84	22	5
	Standard Boar X	2·3	74	25	6
4	Test Boar C	4·3	85	30	5
	Standard Boar X	4·3	82	31	2

*) In these cases positive identification of the number of pigs given in brachets was not possible.

141

Litter No.	Boars	No. of sperm/cc. $\times 10^9$	% Live Sperm	Dosage Rate cc.	No. Born
1	Test Boar D	2·0	62	34	6
	Standard Boar Y	2·5	38	45	6
2	Test Boar D	8·6	65	45	4
	Standard Boar Y	15·2	86	20	10
3	Test Boar D	11·2	67	26	2
	Standard Boar Y	11·2	70	25	7
4	Test Boar D	15·2	81	25	3
	Standard Boar Y	15·1	77	26	5
1	Test Boar E	7·2	60	29	4
	Standard Boar Y	5·9	70	25	6
2	Test Boar E	12·8	87	24	7
	Standard Boar Y	15·8	67	26	4
3	Test Boar E	12·8	87	24 }11	Not yet
	Standard Boar Y	15·8	67	26	typed
4	Test Boar E	11·2	75	25	Sow
	Standard Boar Y	12·5	50	33	returned
1	Test Boar F	15·2	69	25	Not yet
	Standard Boar Y	15·2	52	33	farrowed
2	Test Boar F	4·8	88	50	Not yet
	Standard Boar Y	7·8	88	25	farrowed
3	Test Boar F	3·6	87	30	Not yet
	Standard Boar Y	4·4	89	25	farrowed
4	Test Boar F	9·5	58	25	Not yet
	Standard Boar Y	8·6	58	31	farrowed

When the pigs weigh 12 lbs. liveweight the litters are removed from the sow and split according to male parentage. Each team is given the same sex ratio, and is made as large as possible. If, for instance, there were 10 pigs born, 6 by the standard and 4 by the test boar, and if the latter consisted of 3 hogs and a gilt then 3 hogs and a gilt are selected to constitute the standard boar's test team. The part litters are reared out of doors on a normal early weaning system (Bellis). Food consumption and growth rate for all pigs is recorded. The pigs remain on this system until they reach 50 lbs. liveweight when they are transferred to fattening accommodation. They are kept in this accommodation until they are 200 lbs. liveweight when they are slaughtered. During this fattening period they receive a normal fattening ration and water on an ad libitum basis. Full records of growth rate and feed conversion are recorded. Carcase measurements are made at slaughter, and then all carcases are dissected to ascertain percentage of lean, its distribution, and eye muscle (longissimus dorsi) area.

Interim Results and Discussion

1. Artificial Insemination

Table 2 gives all the relevant information concerning the insemination of all sows used in this new test.

Examination of this table shows that where standard Boar X was used only six mixed litters were obtained. Of the remaining six litters, in three, only pigs by the test boars were born, in two, positive identification of all pigs was not possible, and in the remaining case the sow has not farrowed. After the boar had been used it was discovered that when used by natural service methods he was intermittently sterile, and he has therefore been withdrawn from use for progeny test purposes. There is no apparent reason for this fault although, as can be seen, the percentage of live sperm in some ejaculates was very low.

All litters where standard boar Y has been used have split into two teams. One team, C. l., has been rejected as it gave insufficient pigs for the Progeny Test. Of the 24 sows so far inseminated only one has returned to service some three weeks later. This gives a conception rate of over 95%. This is, of course, a very high rate, and it would not be expected to get a rate better than 75%. The use of fresh undiluted semen may have been a factor in attaining such high rates.

2. Blood Grouping

Originally, it was considered that this test could be conducted using colour marking as a method of identifying the piglets at birth. Unfortunately,

Table 3

Boar	Litter no.	Nos. and sex of pigs in each team	Birth weight lb.	Daily liveweight gain			Food conversion		Health notes
				Birth to 12 lb.	12 lb. to 50 lb.	50 lb. to 120 lb.	12 lb. to 50 lb.	50 lb. to 120 lb.	
Test A	2	2 H	2·62	0·35	0·66	2·00	2·10	2·28	
Standard X		2 G	2·75	0·34	0·88	1·43	2·08	2·28	
Test A	4	3 H	2·77	0·29	0·86	1·79	2·03	2·27	
Standard X		2 G	2·66	0·29	1·15	1·32	1·61	2·34	
Test C	2	1 H	3·25	0·33					All pigs died
Standard X		1 G	2·87	0·35					
Test C	3	1 H	2·87	0·26					Pigs in test team died
Standard X		1 G	3·25	0·29					
Test C	4	2 H	2·80	0·35					1 pig in test team died
Standard X			2·81	0·39					

Table 3 (cont.)

Boar	Litter no.	Nos. and sex of pigs in each team	Birth weight lb.	Daily liveweight gain Birth to 12 lb.	12 lb. to 50 lb.	50 lb. to 120 lb.	Food conversion 12 lb. to 50 lb.	50 lb. to 120 lb.	Health Notes
Test D	1	1 H	2·87	0·42					2 pigs in each team died
Standard Y		3 G	2·71	0·42					
Test D	2	1 H	2·25	0·44					2 pigs in each team died
Standard Y		1 G	2·11	0·46					
Test D	3	1 H	3·21	0·42					2 pigs in each team died
Standard Y		1 G	3·33	0·43					
Test D	4								All pigs died within 24 hours of birth
Standard Y									
Test E	1	2 G	3·44	0·51					All pigs healthy
Standard Y			3·25	0·46					
Test E	2	1 H	3·29	0·42					All pigs healthy
Standard Y		1 G	3·44	0·41					
Test E	3		3·25						All pigs healthy Not yet typed
Standard Y									

this would not have been sound genetically as the comparison would have been between crossbred and pure bred offspring.

Starch Gel electrophoresis of the β globulins was also considered but this was discarded as there were not enough phenotypes available to make identification easy.

Red cell antigen techniques were eventually used as the large number of systems and factors give a higher rate of exclusion of sires.

Standard Boar Y has a rare (for British Large Whites) homozygosity in the E system namely E edf/edf and the use of this boar has so far made it possible to identify the sires with 100% accuracy.

It is considered too early to give any information on the gene frequencies but it is hoped that this can be reported in a final paper.

3. Performance of Teams to Date

Table 3 gives the performance of the test and standard teams.

A severe outbreak of scouring and death of piglets has been experienced at the farm, and the performance of all animals privately owned, test and standard, on the station has been affected. The causative organism has been identified by the International Reference Laboratory, Copenhagen, as E. coli 0141 K85 ac (B) H4. The organism although not uncommon is highly pathogenic and losses have been severe.

Nevertheless, preliminary examination of the results gives confirmation that environmental effects have been almost wholly eliminated, as although the performance of the test boar and standard boar offspring vary (as would be expected) from litter to litter, the superiority of one boar over the other in any one phase is manifest in each of the four teams. Average birth weights of the team follow a haphazard pattern, thus giving confirmation of the low heritability of the factor. However, the varying birth weights do not appear to alter the subsequent performance of the pig.

Early weaning appears to have overcome the environmental differences due to different rates of milking of the dam's teats, as the progress of the pigs from 12 lb. to 50 lb. follow a definite pattern.

The disease outbreak which has been experienced has set back the test, but nevertheless it has confirmed the view that for valid comparisons between boars to be made, all pigs should be subjected to the same disease environment. Under traditional progeny testing the poor performance of these pigs would be explained as genetic, whereas it is almost certain that it is environment which has caused the poor growth. Additionally, it is interesting to note that losses have been heaviest in the test teams. Whether this fact has any

Additional material from *Blood Groups of Animals,*

ISBN 978-94-017-5834-5, is available at http://extras.springer.com

genetic significance must await further investigation. However, Smith and King gives a heritability ot 0·25 for scouring in pigs between 50 lb. and 200 lb. liveweight, and it is therefore possible that resistance to E. coli infections may be under genetic control.

Full results giving details of lean meat and other economic factors will be given in a later report.

Acknowledgments

Thanks are expressed to the Directors of BOCM Limited for allowing facilities to develop the test; to Professor J. Moustgaard and his colleagues at the Royal Veterinary College, Copenhagen, for the invaluable help, assistance and encouragement which they have given in the setting up of the Immunogenetic Laboratory; to M. J. Bell and N. L. Handscombe for taking charge of all the artificial insemination work; and lastly, thanks are due to the Manager and staff of the Testing Station for their daily supervision and recording of the stock in this test.

Summary

(1) The paper describes a new and improved method of progeny testing boars.
(2) By inseminating sows with mixed semen from boars, the test eliminates dam and environmental differences which influence present day boar progeny tests. Parentage of the litter is determined at 10—14 days by red cell antigen techniques. Litters are then split according to sire and reared under standard feeding and management routines.
(3) Fresh undiluted semen has been used and conception rates have been over 95%.
(4) Apart from three litters where one boar was later proved to be periodically sterile, all litters have split into satisfactory teams.
(5) To date six boars have been brought forward for tests, and preliminary results are discussed.

References

Madden, D. H. L., Hampshire Cattle Breeders Association, Lyndhurst. Verbal Communication.

Bellis, D. B., Animal Production, Vol. VI, Part. II, Page 253. "Outdoor Rearing of the Early Weaned Pig" — abstract of paper given at the Society of Animal Production.

Smith, C. and King, J. W. B. Animal Production, Vol. IV, Part I, Pages 128—143. Genetic Parameter of British Large White Bacon Pigs.

STUDY OF PIG BLOOD GROUPS IN VOJVODINA

V. JOVANOVIČ and Z. STOJANOVIČ

Faculty of Agriculture, Novi Sad

The study of pig blood groups in Vojvodina was started in 1962. In order to produce monospecific reagents, a lot of isoimmunizations were done on experimental pigs. We are indebted to The Royal Veterinary and Agricultural College, Copenhagen, Denmark, and The Veterinary Institute, Ljubljana, Slovenia for help so that we were able to use 40 blood typed pigs, tested with 16 monospecific reagents from Denmark, for isoimmunization.

A. Production of monospecific reagent YU-V-1

From the group of "pairs" for isoimmunization, we picked the following "pair" of pigs: Donor 1947 (Holland Landrace) and Recipient 6589 (Large White) (table 1).

Table 1

Blood types of donor and recipient

Tested on	System Factor	A E F G H I J K L M A a b e f a a b a b a b a a a b a a
Donor Recipient		. . . e . . a . . . . a . b a a A . . e . . a b a . . a . b . .
Expected antibodies		La Ma

We expected to get the immune sera with the antibodies anti-La and anti-Ma at the end of isoimmunization.

Information tests of the red cells of our tested pigs gave the reaction which did not show the existence of anti-La and anti-Ma. However, there was a series of reactions with red cells of our standard, which indicated the existence of some antibodies different from those expected.

By means of test-selective absorptions from the crude immune sera we separated a monospecific reagent which in the test with red cells of our tested pigs showed the reaction different from all 16 reagents from Denmark, used in blood typing of our 40 tested pigs. This reagent was designated YU-V-1.

Reagent anti-YU-V-1 contains haemolytic antibodies in ratio 1 : 32. A rather great number of animals react with it. At 40 tested absorptions on monospecificity it was always possible to separate antibodies for the absorbant and all the other red cells which had previously reacted. Monospecificity was thus verified.

We also studied the heredity of antigen-YU-V-1 by observing the progeny from different mating combinations (table 2).

Table 2

The observed distribution of factor YU−V−1 in the progeny from different types of mating

Mating type	No. of litter	No. of animals	+	−	X^2
YU-V-1 × YU-V-1	15	119	107	12	
YU-V-1 × −	49	392	200	192	
− × −	32	276	−	276	
YU-V-1 (− × −) −	46	369	177	192	0·609

Finding the value of X^2 and comparing it with the critical value 3·841 as corresponding to the 5 per cent level of significance of X^2 with one degree of freedom we see that the segregation ratio of the respective factors in progeny is in good agreement with its expectation 1 : 1.

The relation between factor YU-V-1 and other factors discovered by our reagents was tested by means of contingency tables (table 3).

The results of these investigations can be seen in contingency tables (2×2 tables) (table 3). The X^2 values were computed, and in the cases where one of the groups were less than 5, Fisher's exact method was applied (Race and Sanger 1958).

On the basis of the test of dependence between the factors discovered by our reagents, the factor YU-V-1 was found to have an extremely significant correlation with the factor A (table 3).

With our reagent anti-YU-V-1 we took part in the International comparative test which was organized by the Laboratory ol Physiology and Genetics in Libĕchov. The results show that our reagent YU-V-1 is in fact monospecific reagent anti-O.

Table 3

Investigation of a possible relationship between YU-V-1 and other known group factors

System	Factor	Reaction	Reaction with YU-V-1		
			+	−	
A	A	+	0	102	
		−	101	18	P < 0·00001
E	Ea	+	17	27	
		−	19	16	X² = 1·92
	Eb	+	42	36	
		−	28	38	X² = 1·86
	Ee	+	17	17	
		−	21	23	X² = 0·03
F	Fa	+	36	18	
		−	28	26	X² = 2·45
G	Ga	+	22	26	
		−	29	24	X² = 0·79
	Gb	+	21	30	
		−	21	25	X² = 0·19
H	Ha	+	34	43	
		−	43	44	X² = 0·45
K	Ka	+	33	32	
		−	31	41	X² = 0·81
	Kb	+	27	30	
		−	21	28	X² = 0·21
M	Ma	+	3	7	
		−	4	4	P = 0·352

B. Determination of pig blood types of native autochthonous primitive pig races "Lasasta mangulica" and "Crna-slavonska" and of type "Berkshire"

We studied the proportion of blood factors and the gene frequencies for blood group antigen factors with two types of our native autochthonous pigs: "Lasasta mangulica", "Crna-slavonska svinja" and with the type "Berkshire" which was involved in developing the race "Crna-slavonska" pig.

"Lasasta mangulica" developed from "Šumadinka". "Šumadinka" is a pure representative of the tamed Europe wild pig (Sus scrofa ferus). "Lasasta mangulica" developed from "Šumadinka" as a result of systematically bringing it up under better conditions of feeding and tending, withour crossing with other races.

Table 4

The amount of blood factors in percent for three above mentioned pig types

	LM	CS	B
A	25·0	37·5	32·0
Ea	33·0	39·4	61·1
Eb	80·0	47·1	32·0
Ee	67·0	93·2	97·0
Ef	81·0	93·2	100·0
Ed	94·0	90·3	92·2
Fa	6·0	50·9	89·3
Ga	100·0	94·2	71·4
Gb	2·0	42·3	75·7
Ha	1·0	16·3	30·0
Hb	2·0	3·8	3·8
Ka	3·0	15·3	6·7
Kb	97·0	72·1	91·2
Ma	3·0	0·0	5·8
YU-V-1	62·0	53·8	37·8
YU-V-2	51·0	46·1	66·0

LM = "Lasasta mangulica".   CS = "Crna-slavonska".   B = "Berkshire".

"Crna-slavonska" pig (also called "Pfajfer" pig) is a result of combined mixing of "Lasasta mangulica" and "Berkshire" by adding pig blood of "Polandchine" type.

Blood samples for our test were taken from 16 herds of "Lasasta mangulica" pigs in the surrounding area of the Sava in Srem, 14 herds of "Crna-slavonska" pigs in the vicinity of Ruma, Stara Pazova and Slankamen and 19 herds of the "Berkshire" race pigs in the surroundings of Sombor. In all cases, the pigs were chosen which were not related through at least three generations. Blood samples from 100 or more pigs of each race were used.

Table 5

Gene frequencies for blood factors

	LM		CS		B	
	f	s	f	s	f	s
A	0·134	0·025	0·210	0·030	0·176	0·029
Ea′	0·130	0·027	0·221	0·028	0·325	0·032
Eb′	0·565	0·035	0·269	0·031	0·174	0·026
Ef′	0·435	0·035	0·730	0·031	0·825	0·026
Ed′	0·820	0·027	0·778	0·028	0·674	0·032
Fa	0·031	0·044	0·300	0·035	0·674	0·047
Ca	0·990	0·007	0·750	0·029	0·485	0·035
Gb	0·010	0·007	0·250	0·029	0·514	0·035
H−	0·985	0·001	0·893	0·048	0·815	0·048
Ha	0·005	0·005	0·086	0·019	0·164	0·017
Hb	0·010	0·007	0·020	0·009	0·009	0·000
K−	0·001	0·047	0·437	0·028	0·234	0·049
Ka	0·021	0·002	0·082	0·006	0·036	0·003
Kb	0·867	0·460	0·479	0·069	0·727	0·236
Ma	0·016	0·008	0·000	0·000	0·030	0·011
YU-V-1	0·384	0·039	0·321	0·036	0·212	0·030
YU-V-2	0·300	0·036	0·267	0·033	0·417	0·040

LM = "Lasasta mangulica". CS = "Crna-slavonska". B = "Berkshire".
f = gene frequency. s = standard error.

Table 6

Comparison of gene frequencies between the pig races

	CS		LM			B		
	f. g.	s. g.	f. g.	s. g.	s. s.	f. g.	s. g.	s. s.
A	0·210	0·030	0·134	0·025	1.94^{ns}	0·176	0·029	0.8^{ns}
Ea	0·221	0·028	0·180	0·027	1.1^{ns}	0·325	0·032	2.4^{+}
Eb	0·269	0·031	0·565	0·035	6.4^{++}	0·174	0·026	2.3^{+}
Ee	0·730	0·031	0·435	0·035	6.4^{++}	0·825	0·026	2.3^{+}
Ef	0·741	0·028	0·564	0·045	3.3^{++}	1·000	0·000	9.2^{++}
Ed	0·778	0·028	0·820	0·027	1.1^{ns}	0·674	0·032	2.4^{+}
Fa	0·300	0·035	0·031	0·044	3.9^{++}	0·674	0·047	6.2^{++}
Ga	0·750	0·029	0·990	0·007	8.2^{++}	0·485	0·035	5.8^{++}
Gb	0·250	0·029	0·010	0·007	8.2^{++}	0·514	0·035	5.8^{++}
H−	0·893	0·048	0·985	0·001	1.91^{ns}	0·815	0·048	1.1^{ns}
Ha	0·086	0·019	0·005	0·005	4.2^{++}	0·164	0·017	3.1^{++}
Hb	0·020	0·009	0·010	0·007	0.9^{ns}	0·020	0·009	0.0^{ns}
K−	0·437	0·028	0·001	0·047	8.0^{ns}	0·234	0·049	3.6^{++}
Ka	0·082	0·006	0·021	0·002	10.1^{++}	0·036	0·003	7.6^{++}
Kb	0·479	0·068	0·867	0·460	0.8^{ns}	0·727	0·236	1.0^{ns}
Ma	0·000	0·000	0·016	0·008	2.8^{+}	0·030	0·011	2.7^{++}
YU-V-1	0·321	0·036	0·384	0·039	1.1^{ns}	0·212	0·030	2.3^{+}
YU-V-2	0·267	0·033	0·300	0·036	0.6^{ns}	0·417	0·040	2.9^{+-}

CS = Crna-slavonska, LM = Lasasta mangulica, B = Berkshire.
f. g. = gene frequencies, s. g. = standard error, s. d. = statistical significant.

Table 4 shows the proportion of blood factors in the percentage for the three above pig types.

Tables 5 and 6 show the gene frequencies for blood factors and the comparison of gene frequencies between the pig races.

Observing the proportion of blood factors and gene frequencies for blood factors of the tested pig races, and determining the ratio by the statistical analysis, we can conclude that the values for ,,Crna-slavonska" pig type are mainly in the middle between the extremes of "Lasasta mangulica" and "Berkshire".

A CONTRIBUTION TO THE STUDY OF THE BLOOD GROUP SYSTEM A IN PIGS

J. HOJNÝ and K. HÁLA

Laboratory of Physiology and Genetics of Animals, Czechoslovak Academy of Sciences, Liběchov

The A system of pig blood groups, the first factor of which has been first described by Szymanowski, Stetkiewicz and Wachler 1926, remains of particular interest for some of its special features. It is interesting not only serologically, but also from the genetic point of view. The majority of older authors (Kaempffer 1932, Schott 1932, and others) demonstrated the dominant inheritance of the blood factor A. On the other hand, Andersen et al. (1959, Nielsen 1960) found positive offspring from negative parents in their litters. The same observation was made later by Rasmusen (1963) studying at the same time the O factor, which is recessive for the A antigen.

As Saison and Ingram (1962), we also observed differences in the number of positive animals at a parallel testing of the cattle antibody anti-J and the pig sera anti-A (Hojný, Hála 1963). Therefore, we began to look for a more suitable method for a reliable identification of A-positive animals and to concentrate on the study of inheritance of the A blood factor in the progeny.

Material and Methods

To determine blood factor A we selected 4 sera: cattle anti-J (17/119), pig serum (D_2), the incomplete pig serum (167), both obtained from the slaughterhouse, and finally the heteroimmune serum, prepared by immunization of 7 rabbits with human A_1-positive erythrocytes (Bednekoff et al. 1963). We compared the antisera with each other in the current haemolytic test, using rabbit complement, in a direct agglutination test and in an indirect Coombs test (Hála, Hojný 1962), and in a dextran-test (Hojný, Hála 1964).

The most suitable sera and technique were used for testing (see Table 3) of 1094 offspring, older than 21 days originating from 161 families, mainly of the Large White breed and further of the Black and White - Přeštice and some of the Landrace breeds.

By means of the inhibition test (Podliachouk and Eyquem 1956) we also observed the presence of A substance in the serum of 548 piglets from all types of mating and in 305 older pigs and their parents from the progeny testing stations.

A preliminary phenotype frequency has been determined in 222 pigs of the Black and White breed by using blood from the slaughtergouse in Plzeň and from some breeding farms.

Results

All sera, including the haemolytic reagent anti-J, reacted with the majority of A-positive erythrocytes in a high titre. We have designated these erythrocytes Ac. Some sera, however, and only in certain tests, showed reactions with erythrocytes, designated before A_2 or A_x (Hojný, Hála 1963). Since pig sera were the most suitable to demonstrate these erythrocytes, we designated them Ap. The differences in titre in all sera used with regard to Ac and Ap red cells, were, according to the sensitivity of the method, 2 to 8 dilutions. The greatest difference was observed in a heteroimmune serum which hardly reacted with Ap erythrocytes, whereas it completely agglutinated Ac erythrocytes up to a titre of 1/128 in the direct agglutination test and up to 1,024 in the dextran-test. The heteroimmune serum reacted identically with the cattle anti-J. The best reactions with Ap erythrocytes were obtained with the pig serum D_2 in the dextran-test (Table 1).

Table 1

Sera used for the detection of A-substance on pig erythrocytes

Reagents designation origin		Titre in							
		haemolysis		agglutination		Coombs-test		dextran-test	
		Ac	Ap	Ac	Ap	Ac	Ap	Ac	Ap
89/117	anti-J of cattle	32—64	0—(2)	0	0	16—32	0	(4—16)	(0—2)
167	anti-A of pigs	0	0	0	0	256—512	(0—8)	1—2	0
D 2	anti-A of pigs	4—8	0	4—8	**1**	32—64	(0—4)	128—256	**4—16**
R-mixture	anti-A of rabbits	256—512	0	**64—128**	0	—	—	512—1024	(0—8)

[1]) Anti-J and heteroimmune rabbit serum was absorbed by pig A-negative erythrocytes.
[2]) Values in parentheses represent weak, independent reactions.
[3]) In the indirect Coombs test only a rabbit antiglobulin serum was used.

Serologically, Ac and Ap have a subgroup character. In absorption tests, Ac red cells absorbed the reactions against all erythrocytes (Ac and Ap), whereas Ap red cells absorbed the reactions only against Ap erythrocytes, when absorptions were repeated several times.

The subgroup character of the Ac and Ap phenotypes was refuted by the observation of 1,094 offspring from 161 families of all types of mating (Table 2). For A-negative animals we chose the designation "a". From matings of the type Ac × Ac, Ac × Ap, Ac × a and Ap × a offspring of all three groups (Ac, Ap and a) were born. The most interesting is the mating Ap × a. Of the 141 offspring, 60 were A-negative, 23 belonged to the group Ap and 58 animals were Ac which would mean, if only the haemolytic reagent were used, that 58 A-positive offspring were born from negative parents. An example of the determination of A factor in one family is given in Table 3. To prove the theory of the influence of another gene pair on the phenotypic expression

Table 2

Distribution of 1·094 offspring from 161 families for Ac, Ap and a in different types of mating

Type of mating	Number of		Offspring		
	families	offspring	Ac	Ap	a
Ac × Ac	33	249	177	18	54
Ac × Ap	15	77	49	16	12
Ap × Ap	1	6	0	3	3
Ac × a	55	382	195	15	172
Ap × a	22	141	58	23	60
a × a	35	239	0	0	239
Total	161	1094	479	75	540

of A substance, the mating Ap × Ap has been most suitable, but is represented by only one family. The last type of mating a × a is represented by 35 families from which, as expected, 239 exclusively A-negative offspring were born.

By means of the inhibition test we found in a set of 548 offspring, at the age of 1—2 months, that all sera inhibited the agglutinating anti-A serum from 1/4 to 1/256, Ap sera had a low inhibition activity or lacked it completely. The sera of A-negative animals never inhibited the anti-A serum. Among animals of the Ac group, belonging to the Large White and Black and White breeds, a difference was found in the inhibition titre, in the Large White breed it was usually 1/16—1/32, in the Black and White breed 1/64 (Table 4). In older pigs, detection of A-substance was more difficult because of the presence of natural antibodies. Of the 110 A-negative animals, only 15 could be

Table 3

An example of determination of A-factor in 1 family from the mating Ap × a

Reagents	Parents			No. of offspring									
	Dil	65	539	11	12	13	14	15	16	17	18	20	21
Anti-Ac(R-mix.)	1/8	—	—	—	4	—	—	—	—	4	4	—	—
1) Anti-Ap (D₂)	N	3	—	2	4	3	—	3	—	4	4	—	—
2) Anti-Ap (D₂)	1/2	4	—	3	4	3	—	4	—	4	4	—	—

1) direct agglutination test,
2) dextran test.

evaluated. No A-substance has been found in their serum. A weak inhibition activity was detected only in 8 out of 35 Ap animals, the remaining 160 pigs of the Ac group has a soluble A-substance in their serum with titres ranging from 1/16 to 1/32.

Table 4

The A substance in the serum of 548 piglets at the age of 1—2 months

Number of piglets	Groups	Distribution of animals according to the inhibition capacity of the serum										
		non-inhib.	Inhibition in the titre									
			N	2	4	8	16	32	64	128	256	512
246	Ac	0	—	—	10	15	46	59	83	31	2	—
183	Ap	5	15	6	1	—	—	—	—	—	—	—
275	a	275	—	—	—	—	—	—	—	—	—	—
(Group Ac divided according to breeds)												
LW 153	Ac	0	—	—	10	15	45	49	25	10	—	—
BW 93	Ac	0	—	—	—	—	1	10	58	21	2	—

The serum of tested animals inhibited the anti-A serum at a titre of 1/4 in a direct agglutination test.
N = undiluted serum, LW = Large White, BW = Black and White breed.

Contrary to the Large White pigs, where the phenotype frequency of the Ac group was 37·81 % and the Ap group 5·37 % (Hojný, Hála 1963), we found 64·86 % of the Ac group and only 2·7 % of Ap animals among 222 pigs of the Black and White breed.

Discussion

In our work we found it possible to classify A-positive animals into Ac and Ap groups. In their classification, a special strong antiserum must be used, mainly for the determination of Ap specificity. The method used is of the same importance. The dextran-test appeared to be most suitable for the determination of the Ap specificity, as it increased the titre and strength of the reactions.

Contrary to Munk-Andresen (1956) who used the dextran-test for the detection of antibodies of the human ABO system with a final dextran concentration of 3 %, we used a lower concentration of 1·5 % in order to decrease further the danger of unspecific agglutinations.

Saison (1964) also observed a slight haemolysis of Ap erythrocytes, when using anti-J after a prolonged incubation period.

The determination of soluble A substance in the serum has no essential significance in the classification of pigs into Ac and Ap groups. In exceptional cases, we found the same inhibition activity of the weakly inhibiting Ac sera and strongly inhibiting sera of the Ap group. Podliachouk and Eyquem (1956) found a strong inhibition in 40 out of 81 A-positive animals, a weak in 15 and no inhibition activity of all animals of the Ac group, in 26 sera, whereas the sera of Ap animals inhibited very weakly or not at all. We do not assume, however, that the A substance is not present in the serum, but we rather think that the detecting system is not sensitive enough. In agreement with the cited authors, we have never detected the A substance in the serum of A-negative animals.

While determining the blood groups of parents and their progeny, we observed frequently that the offspring with the same blood groups as their parents were A-positive, although the A antigen could not be detected in their parents by means of cattle anti-J. Therefore we observed families with regard to the A system and its Ac and Ap groups.

The results from the studied matings prove the assumption of Andresen et al. (1959a) and the results of Saison and Ingram (1962) and Rasmusen (1963). The presence of A substance on erythrocytes and its level in the serum was found to depend on another pair of alleles which, in a homozygous state, had an influence on A substance production. It probably does not exert its influence on the erythrocytes directly, but primarily on the A substance level in the serum. We found that the inhibition activity of the A substance in Ap animals is much lower than in Ac animals. According to the theory of influence of two gene pairs on the expression of A substance on erythrocytes, it is possible to assume the following genotypes:

Phenotypes	Ac		Ap		a	
Genotypes	A/A	i/—	A/A	i/i	—/—	i/i
	A/A	—/—	A/—	i/i	—/—	i/—
	A/—	i/—			—/—	—/—
	A/—	—/—				

This theory is confirmed especially by the mating Ap × a, where the amount of A antigen is suppressed due to the i/i homozygous state, but in the i-heterozygous offspring it appears normally as Ac.

An equally suitable mating type is Ap × Ap where the animals born must be i/i homozygous and if they are A-positive, they must show the phenotype Ap.

Unfortunately, we had no cattle anti-sheep O serum at hand, therefore we have no information about the relation of Ap red cells to the O antigen. We suppose, however, that Ap erythrocytes, as Ac red cells, will not react with the anti-O serum, as Ac and Ap are, in our opinion, 2 different phenotypes of the A factor of pig blood groups.

Comparison of the phenotypic frequency in two of our breeds results not only in a considerable difference in the number of A-positive animals, but also in the difference in the ratio of Ac : Ap animals. The presence of Ap erythrocytes in Large White pigs which is relatively high, enabled us to study this problem.

Summary

Pig A-positive erythrocytes were broken down into Ac red cells, giving strong reactions with all anti-A sera (including cattle anti-J) and Ap cells which gave a weak reaction with pig antibodies anti-A, as best demonstrated in the dextran-test.

The inheritance of the blood factor A (Ac and Ap) was observed in 1,094 offspring from 161 families. If we count, not only Ac but also Ap among A-positive erythrocytes, the inheritance of the A factor is dominant. At the same time the results prove the theory of the suppressive effect of another gene pair on the A phenotype. In i/i homozygous individuals the A antigen occurs phenotypically only as Ap.

In the serum of all Ac-positive pigs we demonstrated the presence of a soluble A substance with an inhibition titre of 1/4—1/256, Ap sera had a low inhibition capacity or none at all, in A-negative animals the A substance was not identified in the serum.

The phenotypic frequency of Ap red cells is considerably higher in Large White pigs (5·73 %) than in Black and White pigs (2·70 %).

On the basis of the results obtained the A system can be used for progeny control in pigs.

References

Andresen, E., Larsen B., Neimann-Sörensen A. (1959). XVI. Congress Mundial de Veterinaria, I. A. 2 : 71—89.

Andresen, E., Højgaard N., Jylling B., Larsen B., Möller F., Moustgaard J. and Neimann-Sörensen (1959). Report of the VI. int. bloodgroups congress in Munich.

Bednekoff, A. G., Tolle A., Datta S. P., Friedman J. and Stone W. H. (1963). J. immunol. 91 : 369—373.

Bräuner-Nielsen, P. (1960). Report of the VII. int. bloodgroup congress in Edinburgh.

Hála, K. and Hojný, J. (1962). Živočišná výroba 7 : 319—325.

Hojný, J. and Hála, K. (1963). Monatshefte f. Vet. Med. (Supp.) 19 : 10—15.

Hojný, J. and Hála, K. (1964). Report of the IX. European conference of animal blood groups, Prague.

Kaempffer A. (1932). Z. indukt. Abstamm. u. Vererb Lehre 61 : 261—300.

Munk-Andresen, G. (1956). Acta path. microbiol. scand. 38 : 259—272.

Podliachouk, L. and Eyquem A. (1956). Ann. Inst. Pasteur 91 : 751—758.

Rasmusen, B. A. (1963). In: Genetics today. Proc. XIth int. Congr. Gent (The Hague).

Saison, R. and Ingram, D. G. (1962). Ann. N. Y. Acad. Sci. 97 : 226—232.

Saison, R. (1964). Personal communication.

Schott, A. (1932). Wiss. Arch. f. Landw. 7 : 68—108.

Szymanowski, Z., Stetkiewicz, St. and Wachler, B. (1962). C. R. Soc. Biol. (Paris) 94 : 204—205.

BLOOD GROUP SYSTEM O IN PIGS

J. HOJNÝ and K. HÁLA

Laboratory of Physiology and Genetics of Animals,
Czechoslovak Academy of Sciences, Liběchov

By systematic studies of pig blood groups new blood factors on pig erythrocytes could be detected by means of immunizations and classified into 13 genetic systems (Andresen 1957, Bräuner-Nielsen 1961, Andresen 1962, Andresen and Baker 1964, Hála and Hojný 1964).

In the present work we submit a report on the inheritance of the blood factor Cz_1, independent of any so far known system, and its classification into a new genetic system.

Material and Methods

Sows of production breeds in Řepy and Uhřiněves were immunized (Hojný and Hála 1963). In addition to serological tests, described in the same report, the dextran test was used. One drop of 6% dextran (Spofa) was added to each tube containing 2 drops of serum and 1 drop of a 2% erythrocyte suspension. After vigorous shaking and incubation in the thermostat at 37°C a macroscopic evaluation was made after 2 and 4 hours. The relationship of other blood factors to Cz_1 was calculated from a set of 312 unrelated animals and the inheritance observed in 1,424 of offspring from 199 families, from which suitable types of double backcross matings were chosen. The frequency of the blood factor was calculated according to the Hardy-Weinberger law.

We are indebted to Mrs Glázrová, Miss Hyková and Mrs Sulovská for their technical assistance and to Mr Ling for his help with immunizations and bleeding of animals.

Results

By reimmunizing the sow Ř 10 with the blood of the same donor as in the previous immunization (Ř 3), a serum was obtained which reacted in the Coombs test at a titre of 1/8 to 1/16. In addition to Ea, another antibody was detected in the serum. For working purposes we designated it S_1 (Hála, Hojný 1962), and after the introduction of the uniform nomenclature Cz_1 (Hojný, Hála 1963). 6 months later, the sow Ř 10 was immunized for the third time. The serum gave a weak reaction in the direct agglutination test. A serum of high quality was obtained after a fourth immunization in the same combination of donor-recipient (Table 1). This time two antibodies must have been absorbed from the raw serum, anti-Ea (1/16) and anti-Ka (1/32). The serum anti-Cz_1 reacted in the Coombs test at a titre of 1/64, in the dextran-test at 1/128. Although 24 further immunizations and reimmunizations with the intention of obtaining the Cz_1 factor were carried out, no other animals was found to produce this antibody.

Table 1

Blood groups of recipient and donor

Sow	Blood group systems													
	A	B	C	E	F	G	H	I	J	K	L	M	N	O
Recipient Ř 10	A/–	a/a	–/–	bdg/edgh	–/–	a/b	a/	b/b	a/	b/	c/df	–/–	a/	–/–
Donor Ř 3	A/–	a/a	–/–	bdg/edgh	–/–	a/b	a/	a/b	a/	b/	c/df	–/–	a/	a/

The blood groups of both animals were completed according to later typing.

After verifying its specificity, we used anti-Cz_1 to determine the corresponding blood factor in the progeny. We found that the antigenic factor Cz_1 was inherited dominantly by one gene pair. The ratio between Cz_1-positive and Cz_1-negative offspring from all types of mating was balanced and the differences

Table 2

Distribution of the factor CZ_1 in 199 families with 1·424 offspring

Type of mating	Number of				Offspring				x^2
	familes		offspring		CZ_1 +		CZ_1 –		
	obs.	exp.	obs.	exp.	obs.	exp.	obs.	epx.	
$CZ_1 \times CZ_1$	17	17·75	122	127·16	92	96·70	30	25·30	1·09
$CZ_1 \times$ –/–	85	83·40	609	596·80	338	331·48	271	277·52	0·27
–/– × –/–	97	97·75	693	699·47	0	0	693	693·00	–

The gene frequency has been calculated from 398 parents.

between observed and theoretically expected numbers of animals were of no statistical importance. (Table 2.) For the theoretically expected values the gene frequency was chosen, which was calculated from the population of the parents.

Table 3

Determination of posible relations between CZ_1 and other blood factors by means of a contingency table 2×2 in 312 pigs

		A		Bb		Ea		Ed		Eb		Ee	
		+	−	+	−	+	−	+	−	+	−	+	−
CZ_1	+	47	49	9	87	12	84	92	4	37	59	92	4
	−	95	121	17	199	38	178	212	4	86	130	197	19
x^2		0·682		0·196		1·280		P = 0·2059		0·045		P = 0·1101	

		Ef		Fa		Ga		Gb		Ha		Hb	
		+	−	+	−	+	−	+	−	+	−	+	−
CZ_1	+	27	67	6	90	86	10	49	47	65	31	13	83
	−	66	145	14	202	176	40	127	89	154	62	25	191
x^2		0·20		0·005		3·241		1·625		0·408		0·240	

		Ia		Ib		Ja		Ka		Kb		La	
		+	−	+	−	+	−	+	−	+	−	+	−
CZ_1	+	85	11	56	40	82	14	42	54	82	14	28	68
	−	183	33	137	79	171	45	78	138	189	27	65	151
x^2		0·800		0·730		1·693		1·638		0·252		0·027	

		Lc		Ld		Ma		Mc		Na		Ca	
		+	−	+	−	+	−	+	−	+	−	+	−
CZ_1	+	74	22	44	31	14	82	10	86	53	43	2	94
	−	172	44	102	58	26	190	18	198	106	110	12	204
x^2		0·258		0·560		0·385		0·353		1·00		P = 0·1405	

Furthermore, we investigated whether the factor Cz_1 was inherited independently or in dependence on alleles of already known genetic systems. In a set of 312 unrelated animals, using a contingency table 2×2, no relation to other antigenic factors of all 13 genetic systems has been found (Table 3). Attempts were made to confirm this result by segregation in double backcross matings (Table 4). The segregation results obtained come near to the expected ratio $1 : 1 : 1 : 1$ and give a convincing proof of non-allelism. Only for the factor Ca (Cz_5) no suitable families were available.

On the basis of the results obtained, we classified the new blood factor Cz_1 into a new genetic system designated O and the factor Cz_1 was called Oa. The blood factor Oa can often be found in the Large White breed (Hojný,

Table 4

Segregation results of double backcross matings

Reagents	Number of families	Offspring			
		CZ_1			
		+		−	
		Other reagents			
		+	−	+	−
A	8	10	12	13	11
B	6	9	9	9	11
C	0	−	−	−	−
E	14	23	26	22	28
F	4	9	9	10	12
G	11	18	16	20	18
H	6	8	8	7	12
I	9	19	14	14	18
J	2	5	3	8	4
K	13	17	16	25	18
L	8	15	16	14	16
M	12	20	16	20	20
N	6	9	10	6	18

Table 5

Comparison of O^a and O allele frequency in large White and Black and White (Přeštice) Pigs

Breed	Number of pigs			Frequency	t
	Total	Oa +	Oa −		
Large White	558	189	369	$qO^a = 0 \cdot 1868 \pm 0 \cdot 0123$ $qO = 0 \cdot 8132$	5·6 $P = > 0 \cdot 01$
Black and White	222	36	186	$qO^a = 0 \cdot 0847 \pm 0 \cdot 0135$ $qO = 0 \cdot 9153$	

Hála 1963). In comparison with a set of 222 Black and White pigs, chosen at random, a highly important statistical difference between the two breeds was found as to the frequency of the O^a allele. In the Landrace and Cornwall breeds the blood factor Oa has not been detected so far.

Discussion

Our results from the last immunizations confirmed the weak antigenicity of the blood factor $Oa(Cz_1)$ (Hojný, Hála 1963), or the incapacity of animals to produce antibodies against this factor. A serum of high quality was obtained because of the chance to repeat reimmunizations, which shows the advantage of preparing specific test sera from animals which are kept for our further experiments.

The antibodies anti-Ea and -Ka detected in the raw serum of the sow Ř 10, have been probably formed as a result of vaccination. Anti-Ka may have also appeared due to passed farrowing, since the mating boar was Ka-positive. Andresen and Wroblewski (1961) also found the antibody anti-Ka in an immune serum, which could not have been produced against donor erythrocytes. The antibody Ea is frequently found in normal sera (Hojný, Hála 1962).

The dextran-test is often mentioned among the methods, suitable for the demonstration of antigenic factors in pigs, but it is rarely used because of the danger of unspecific agglutinations of the red cells. The dextran-test as described by Bräuner-Nielsen (1960) did not give satisfactory results. The main difficulties were encountered in the evaluation of the reactions. Only by using a lower concentration of dextran was it possible to ensure reliable reading without centrifugation. No unspecific reactions were observed, if the serum was used at the dilution 1/32 and the reactions were carefully evaluated twice; this was proved by testing the antibody, anti-Oa in parallel dextran and Coombs tests.

The classification of blood factors into genetic systems is much easier in pigs than in cattle, because the double backcross matings in a large number of offspring constitute a reliable criterion of allelism or non-allelism. In Table 4 these matings are summarized in groups according to the systems (Andresen and Baker 1964). The values of x^2 or P (Table 3) could be confirmed in all systems, except for C. For the proof of non-allelism between O^a and C^a it was impossible to find any litter among 199 families, owing to a rare occurrence of blood factor Ca in the populations of our pigs. For the same reason, we were not able, e. g. to classify the blood factor $Ca(Cz_5)$ into a genetic system before, although we had anti-Cz_5 at hand for a long-time (Matoušek et al. 1962).

The present case is based only on the probability values (P = 0·1405) which suggest non-allelism. For decisive proof of non-allelism it will be necessary to use a special form of mating in animals of the Black and White breed where the occurrence of the Ca factor is more favourable.

The importance of blood factors for genetic studies is in close relation with their frequency. From this point of view, the blood factor Oa can be very useful in Large White pigs.

Summary

We described a new O system of pig blood groups with one antigenic factor Oa. For the detection of this factor the dextran-test has been used with success.

The frequency of the O^a allele in Large White pigs amounts to 0·1868, in the Black and White (Přeštice) breed to 0·0847. This allele has not been detected so far in the Landrace and Cornwall breeds.

References

Andresen, E. (1957). Nord. vet. med. 9 : 274—284.

Andresen, E. (1962). Ann. N. Y. Acad. Sci. 97 : 205—225.

Andresen, E. and Baker L. N. (1964). Genetics 49 : 379—386.

Andresen, E. and Wroblewski A. (1961). Acta vet. scand. 2 : 267—280.

Bräuner-Nielsen, P. (1960). Report of the 7 International Bloodgroup congress in Edinbourgh.

Bräuner-Nielsen, P. (1961). Acta vet. scand. 2 : 246—253.

Hála, K. and Hojný J. (1962). Živočišná výroba 7 : 319—326.

Hála, K. and Hojný J. (1964). Folia biologica (Prague) 10 : 239—244.

Hojný, J. and Hála, K. (1963). Monatshefte f. Vet. Med. (Suppl.) : 10—15.

Matoušek, J., Hála, K., Hojný, J. and Schröffel, J. (1962). Report of the VIII. int. bloodgroup Congress in Ljubljana.

DISCUSSION

H. Buschmann: According to our investigations with a potent cattle anti-J reagent, it seems that pigs are either A^{CS}, A^S or possess anti-A in their serum. Among pigs under one year of age some animals have neither A^{CS}, A^S nor anti-A. The assumption of epistatic gene interaction, suppressing the A-character is justified only, when an exception to dominant inheritance is really established, i. e. A-positive piglets will arise from a mating type −/− × −/−. But in this case, only the inhibition test of sera from both parents will provide definite evidence.

K. Hála: By means of the inhibition test we detected the A substance only in the serum of Ac- or Ap-positive animals, but not in the A-negative animals. We accepted the hypothesis concerning the effect of the gene pair of another locus on the phenotypic expression of the A-substance only when a large number of offspring were studied from the families of all mating-types. This hypothesis, suggested also by other authors, is appropriate for the explanation of the phenotype Ap, which we described here.

O. Böhm: I want to stress the interesting results of the extensive study of the pig blood groups in Munich that has been presented by dr. Buschmann. According to these results, there is no indication of an epistatic gene action in A-system. Here I would like to report the results of a study of A-system carried out in the laboratories in Ljubljana and Novi Sad, Yugoslavia. In these laboratories genetic analyses have been performed on a material of some 100 families with approximately 1,000 offspring. Dr. Stojanović from Novi Sad produced so-called YU-V-I reagent which could be classified as anti-O reagent according to the last comparison test. The factor YU-V-I is really contrasting to A factor and the phenotypes can be classified within the A-system: A, O, or negative animals. There has been only one exception: Three animals from one litter have reacted with both sera, anti-A and anti-O. Unfortunately, the animals had been slaughtered before the nature of YU-V-I factor was recognized. Therefore, no retesting and absorption test have been performed.

In the two Yugoslavian laboratories, the bovine anti-J and porcine anti-A were used for the detection of A factor. Some more positive reactions were obtained with the bovine anti-J. Nevertheless, there were some 5—6 cases in our family studies, where A-positive animals occurred in the progeny of A-negative parents. Parentage is not dubious at all at least for one of the mentioned families. The conclusion can be drawn therefore that our results suport the hypothesis of the epistatic action of secretor gene.

C. Stormont: I am particularly interested in dr. Harboe's comments regarding variation in the bands or zones in the region of the post-albumins in the diagrams shown by dr. Imlah. In studies in our laboratory on serum protein polymorphisms in horses, we observed variation in the proteins which migrate (on starch gels) in the region of the albumins. (Three phenotypes were seen and they were readily accounted for on the basis of two alleles.) We also noted that this variation could be diagnosed by examining the bands or zones which appeared in the post-albumin region. Thus, the same pair of alleles seemed to be controlling variation in two distinct regions of the gels and the question has arisen whether this variation involves different proteins. The observation that the Gc group in man can be detected on starch gels as well as by immuno-electrophoretic methods leads me to ask dr. Harboe whether he believes this variation in post-albumins in swine and horses may actually be Gc proteins.

BLOOD GROUPS IN CHICKENS, DUCKS, RABBITS, RATS AND MINK

THE EFFECT OF BLOOD GROUP GENOTYPES OF THE B SYSTEM ON THE PERFORMANCE OF HYBRID CHICKENS

E. M. McDERMID

Thornber Bros. Ltd., Mytholmroyd, Halifax, Yorkshire

Introduction

Briles (1956) and Gilmour (1954) have shown that chickens of inbred lines heterozygous for genes of the B blood group system are superior in performance to homozygotes in a number of traits. Briles and Allen (1961), Allen (1962) and Allen and Gilmour (1962) studied the effects of B system blood group genotypes on performance in commercial inbred lines and in crosses established between such lines. All concluded that an effect of blood group on performance was demonstrated in their data. The experiment reported here extends these observations to cover crosses between non-inbred closed flocks.

Materials and Methods

Reciprocal matings between birds of two White Leghorn non-inbred closed flocks established as control lines (Gowe, Robertson and Latter 1959) were made so that in each cross the same four blood group genotypes of the B system were represented in the chicks. The only selection exercised was that all parents were heterozygous for B blood group antigens. This experimental design ensured that the effect of the genotype of the parent birds, other than that of B blood group, would have a random effect on the progeny.

Single pens of approximately 40 pullets and 5 cockerels were used. The matings were doubly heterozygous, B19/B21 birds of strain X and B2/B14 birds of strain Y were used. The four possible chick genotypes are shown in table 1. Up to 100 pullet chicks of each cross were placed on each of eleven locations. When put into the laying house at 18—22 weeks of age, according to the location, they were randomised into single bird laying cages.

The pullets were blood typed at the B system at housing time using iso-immune antisera prepared in the two strains. Tests were carried out in 50×10 mm test tubes one drop of suitably diluted antiserum and one drop of an approximately 4% suspension of red blood cells in 0·95% saline solution being mixed together and allowed to stand for 2 hours before being examined macroscopically. Reactions were classified as positive or negative on the basis of the sedimented appearance of the cells (Gilmour 1959). Doubtful reactions were repeated until a clear-cut positive or negative results was obtained. All tests were adequately controlled with suitable positive and negative cell suspension (McDermid 1963).

Table 1

Progeny genotypes at B blood group system

Cross	Genotype			
X♂♂ × Y♀♀	19/2	21/2	19/14	21/14
Y♂♂ × X♀♀	2/19	2/21	14/19	14/21

Records were taken, on each location, of the mortality up to housing time (rearing mortality) and the mortality in the laying house (adult mortality); of the body weight, in ounces per bird, at 23 weeks and at 64 weeks of age; of the egg weight, in grams per egg, at 32 weeks of age; and of egg numbers for a period of up to 300 days in the laying house. The data for each genotype in each cross were summed over all locations and the totals were compared in analysis.

Two methods of statistical analysis of the results were used. The mortality records were tested for "goodness of fit" of the numbers of pullets of each genotype to the expected 1: 1: 1: 1 ratio by X_2 tests. The other data were subjected to analysis of variance, using the autocode routine for the Elliot 803 computer, developed for a 2-way cross classification for unequal numbers in subclasses. My appreciation and grateful thanks are due to Dr. J. C. Bowman and Mr. J. C. Powell who performed the statistical analysis and computation.

Results

Mortality. The number of pullets of each genotype reared successfully is shown in table 2. In the YX cross the distribution of progeny amongst the 4 genotype classes differs significantly from the expected 1: 1: 1: 1 ratio. There was a significant excess of B2/B19 pullets compared with B2/B21 pullets. Because blood typing was not performed until housing time it is not possible to say whether departure from expectation is due to maternal effects (Allen 1962), to differential fertilisation, to differential hatchability, or to differential survival during the rearing period.

Table 2

Numbers of pullet progeny of each B blood group system genotype successfully reared and housed in laying house

Cross	Genotype				X^2	D. F.
	2/19	2/21	14/19	14/21		
XY	226	233	209	206	2·348	3
YX	167	120	141	135	8·190*	3

N. B. The X^2 given is that for the combined pullet rearing figures over all farm locations. The summing of farms was tested for heterogeneity with 30 degrees of freedom (D. F.) and none was found (X^2 32·113 and 29·248). Combining is thus a valid procedure.
*) P = <0·05.

Distribution of deaths in the laying house amongst the different genotypes is shown in Table 3. The dead pullets of each genotype were compared as proportions of the total number of pullets housed. These proportions did not differ from a 1: 1: 1: 1 ratio.

Body Weight. Table 4 shows the body weights of the pullets of each genotypic class at 23 weeks of age. There were no differences between the classes.

Table 3

Number of pullets of each B blood group system genotype dying in the laying house

Cross	Genotype				X^2	D. F.
	2/19	2/21	14/19	14/21		
XY	17	16	24	19	0·021	3
YX	4	3	19	6	0·336	3

The body weights at 64 weeks of age are shown in Table 5. In cross YX there was a significant interaction between genotype and farm (P = < 0·01) and in cross XY 21/2 pullets were significantly heavier than either 19/14 or 21/14 pullets (P = < 0·05.) Pullets carrying the B2 antigen were found to be significantly heavier than birds carrying the B14 antigen (P = < 0·05).

Egg Weight. Birds of genotype B19/B2 laid eggs which were heavier (larger) than those of all the other 3 genotypes in the XY cross. This can be seen in Table 6. The heaviest birds in this cross did not lay the largest eggs.

Table 4

Body weight of pullets of each B blood group system genotype at 23 weeks of age ozs. /bird

Cross	Genotype			
	2/19	2/21	14/19	14/21
XY	52·0	51·6	51·1	51·4
YX	54·8	54·6	55·1	53·6

Table 6

Egg weight of pullets of each B blood group system genotype at 32 weeks of age. gms./egg

Cross	Genotype			
	2/19	2/21	14/19	14/21
XY	53·8 *	52·5	52·6	52·2
YX	54·4	53·9	53·6	53·4

*) P = <0·001.

Table 5

Body weight of pullets of each B blood group system genotype at 64 weeks of age. ozs/bird

Cross	Genotype			
	2/19	2/21	14/19	14/21
XY	66·3	67·9	65·4	67·1
YX	58·9	60·1*	58·2	58·1
	2	14	19	21
XY	67·1	66·3	65·9	67·5
YX	59·5*	58·2	58·6	59·1

*) P = <0·05.

Table 7

Numbers of eggs laid by pullets of each B blood group system genotype whilst in laying house Approximately 300 days of production

Cross	Genotype			
	2/19	2/21	14/19	14/21
XY	192·4	192·1	187·0	191·0
YX	183·0	181·0	190·4	188·9
	2	14	19	21
XY	192·3	189·0	189·7	191·6
YX	182·0	189·7*	186·7	185·0

*) P = <0·05.

Egg Numbers. The numbers of eggs laid by pullets of each genotype are shown in Table 7. Statistical difference can be demonstrated between pullets carrying the B2 and B14 antigens in cross XY (P = < 0·05). B14 pullets laid more eggs than B2 pullets.

Discussion

Real differences between the genotypic classes were found in rearing mortality, in egg weight, in body weight at the end of the laying year and in egg production. The differences due to the effect of different farm environments were always significant (P usually = <0·001) and in the case of 64 week body weight there was an interaction between the blood group and the farm effect.

The deviation of class numbers from that expected was found in successfully reared pullets in one only of the reciprocal crosses and suggests that a parental effect was operating. Allen (1962) demonstrated a maternal effect in his experiments. Differences between pullets inheriting the B2 and B14 antigens and those inheriting the B19 and B21 antigens were not significant. However, if substantiated in further experiments at present in progress such differences would also suggest a maternal effect as they were apparent only when the particular groups were inherited from the dam (see Table 2). Delay in performing the blood grouping tests until housing time has prevented any conclusion on the mechanism operating to cause the observed disturbance in the segregation ratio and it may well be that all the alternatives suggested have combined to produce the observed effect. The existence of this disturbance in the ratio underlines the need to carry out blood grouping tests of the progeny and suggests that the suppositions made by Briles and Allen (1961), Allen (1962) and Allen and Gilmour (1962) that the progeny in their experiments segregated into B blood group genotypic classes according to a Mendelian ratio may not always be justified. Blood group did not affect survival in the laying house.

The characteristic interaction between egg weight and body weight was not fully apparent in this experiment. The largest, heaviest bird did not lay the largest, heaviest egg. There appeared to be a maternal effect on body weight. In cross XY pullets descended from B2 dams were heavier than those descended from B14 dams. In cross XY the interaction of farm environment and blood group on body weight suggests that the pullets from B21 dams would similarly be heavier than those from B19 dams but this has not been tested for.

The better egg production, in terms of number of eggs laid, of B14 pullets in cross XY means that lighter, smaller birds laid more but smaller eggs.

Conclusions

The effects of B system blood groups in this experiment have been on mortality in one cross and on egg production (numbers, size and body size) in the reciprocal cross. This indicates that blood groups can affect the performance of hybrid chickens, but this effect may vary for each cross and can depend upon the way in which the cross progeny are produced possibly because of maternal influence on progeny performance.

References

Allen C. P. (1962). Ann. N. Y. Acad. Sci. 97 : 184—193.
Allen, C. P. and Gilmour D. G. (1962). Genetics 47 : 1711—1718.

Briles W. E. (1956). Poultry Sci. 35 : 1134—1135.

Briles, W. E. and Allen, C. P. (1961). Genetics 46 : 1273—1293.

Gilmour, D. G. (1954). Heredity 8 : 291.

Gilmour, D. G. (1959). Intern. Blood Group Congr. 6 : 50—79.

Gowe, R. S., Robertson, A. and Latter, B. D. H. (1959). Poultry Sci. 38 : 463—471.

McDermid, E. M. (1964). Vox Sang. 9 : 249—267.

RED CELL ANTIGENIC POLYMORPHISM IN A STRAIN OF THE WYANDOTTE HEN (M 11)

A. PERRAMON

Central Station of Animal Genetics, Jouy-en-Josas

Because of the intra-breed specificity of the immune antibodies in poultry it is necessary to prepare antibodies specific for the particular breed, if samples from a given breed are to be analyzed.

In the primary studies of blood groups in poultry, various breeds have been used, such as Barred Plymouth Rock and New Hampshire (Briles, McGibbon and Irwin, 1950) or Cockin, Brahma and Plymouth Rock (Gilmour, 1949). However, so far, White Leghorns have been the most widely investigated breed.

This breed is now the best known of all the breeds with regard to red cell antigenic polymorphism.

So far as we know, the Wyandotte breed has not yet been submitted to such investigations. In France, this breed is considered to be one of the most interesting, because of the food results obtained by crossing it with the Rhode Island Red breed. For this reason, we decided to choose it for an immuno-genetic investigation.

Material and Methods

Our birds belong to the M 11 strain, which has been kept for 12 years at the Poultry Breeding Station in Le Magneraud. The strain carries the R (rose comb), a (recessive white) and w (yellow legs) genes and is inbred in approximately 60% of cases.

Isoimmune sera were prepared after an intramuscular injection of a 50% suspension of red cells in physiological saline. The cells were previously washed at least three times. The injections were made alternately on each side in the "bréchet" (breast) muscle every third day, so as to obtain a sufficient antibody titre, but without ever exceeding 15 injections.

The females were mostly treated, because of their decidedly lower haematocrit index (35% instead of 45% in the males) which makes it possible to collect a higher amount of plasma from each bleeding. At the beginning, however, a few cocks were immunized. As in other laboratories we often observed that it was preferable to collect plasma instead of serum because of the imperfect retraction of the coagulum (Gilmour, 1959). The use of the plasma instead of serum offers no disadvantage.

Samples of immune plasma were stored at $-28°C$ without any addition of preserving substances.

The specific reagents for each antigenic factor were obtained by submitting "polyvalent" plasma to selective absorption by red cells reacting positively with such a plasma. The red cells, after three washings in physiological saline, were mixed to an equal volume of plasma. Occasionally, no absorption was necessary, the immune plasma being "monovalent".

The purity of each reagent was checked by a classical purity test.

Furthermore, the results of matings between heterozygous individuals and individuals double negative for a given antigenic factor were analyzed statistically. Such matings should theoretically give 50 % negative and 50 % positive (heterozygous) offspring (Stone and Irwin, 1963). The results of these analyses are given along with the description of each reagent.

Agglutination reactions were made in glass tubes, 0·8 cm in diameter, where a drop of reagent in the chosen concentration was mixed with a drop of a 2 % red cell suspension. Readings were made under a magnifying glass after 30 minutes at room temperature. No microscopical examination was considered necessary.

Results

Eight "monovalent" reagents were obtained and numbered in the order of their discovery. Thus, the first corresponding to the W_1 antigen (W for Wyandotte) is called anti-W1, and the gene responsible for the biosynthesis

Table 1

Reagent	Antigenic factor	Type of mating	Number of matings	Total number of offspring	Offspring reacting with the reagent		x^2
					positively	negatively	
anti-W1	W_1	$W^1/- \times -/-$	24	127	64	63	0·008
anti-W2	W_2	$W^2/- \times -/-$	31	676	314	362	3·408
anti-W3	W_3	$W^3/- \times -/-$	27	242	125	117	0·264
anti-W5	W_5	$W^5 - \times -/-$	20	455	241	214	1·602

of the W_1 antigen will be called W^1. This system of notation was proposed by Briles (1958).

The purity of the first four reagents, obtained by the imumnization of three cocks, is demonstrated in table 1.

The first bird ♂ Y 3059 was injected with blood from ♀ J 9088 and gave a polyvalent plasma called Y 3059 (J 9088) which contained four fractions

Table 2

Type of mating	Number of matings	Total number of matings	Phenotypic type of the offspring (total number)			
$W^1/W^2 \times -/-$	2	8*)	W_1	W_2	W_1, W_2	0
			1	6	0	0
$W^1/W^3 \times -/-$	2	8*)	W_1	W_3	W_1, W_3	0
			4	4	0	0
$W^1/W^5 \times -/-$	4	26	W_1	W_5	W_1, W_5	0
			14	12	0	0
$W^2/W^2 \times -/-$	9	73	W_2	W_3	W_2, W_3	0
			29	44	0	0
$W^2/W^5 \times -/-$	7	56	W_2	W_5	W_2, W_5	0
			23	33	0	0
$W^3/W^5 \times -/-$	3	47	W_3	W_5	W_3, W_5	0
			29	18	0	0

*) The low number of these matings is partly due to the low frequency of the W^1 allele.

Two of these had too low a titre to permit them to be studied and were discarded. The two remaining ones were designated anti-W1 and anti-W2, respectively.

The second immune plasma Y 3056 (J 9086) contained a single antibody called anti-W3 and similarly the Y 3061 (J 9085) plasma gave a reagent called anti-W5 (anti-W4 was obtained from the cock Y 3049 but our supply is exhausted because the cock was soon lost).

When the purity of each reagent had been proved, it was possible to study the relations between the genes determining the corresponding antigenic factors.

Six types of matings, involving a bird carrying two antigens on the one hand and a bird carrying none on the other are presented in table 2.

The absence of heterozygotes and recessives in the progeny (see last two columns in table 2) proves that the genes are alleles. Moreover, among 3,200 tested birds (the total number bred in 1964 at le Magneraud for the M 11 strain), none was found to be negative with regard to the reagents at the same time. This again confirms the existence of a tetraallelic system.

The four antigenic factors described above also have similar serological properties. Their rather high antigenicity makes it possible to reach a titre of 1/32 to 1/64 after five injections and, 15 days after the first injection at the latest. These antigens can be detected on both red cells and lymphocytes (the method of the buffy-coat, Schierman and Nordskog, 1961).

Table 3

Reagent	Antigenic factor	Number of matings of the type $W^6/- \times -/-$	Total number of matings	Offspring reacting with the reagent		x^2
				positively	negatively	
anti-W6	W_6	29	653	306	347	2·574

Table 4

Matings of the type $W^2/-, W^6/- \times -/-, -/-$

Phenotypes of the possible offspring	W_2, W_6	W_2, O	O, W_6	O, O	x^2 for the ratio $1:1:1:1$
Observed number of offspring for each phenotype and each mating	6	12	5	11	4·352
	2	1	4	5	3·332
	3	3	6	2	2·571
	3	11	6	8	4·858
	7	11	8	10	1·110
	4	14	10	4	9·000 (P < 0·005)
	8	7	11	4	3·332
	2	8	7	10	5·149
Total					33·704

The total x^2, with 24 degrees of freedom, is not significant.

The next reagent, called anti-W6, corresponds to an antigenic factor (W_6) with widely differing properties. Its antigenicity is very low. Immunization must be continued sometimes for 6 weeks, with a high number of injections and the titre obtained never exceeds 1/8. The unit character of the corresponding factor is obvious in table 3. So far, we have not found any other antibody related to the W^6 locus. This factor does not belong to the preceding system, because it has been found to be associated with the heterozygotes, W^1/W^2, W^1/W^3, W^1/W^5, W^2/W^3, W^2/W^5 and W^3/W^5. The data in table 4 show the independence of the W_6 locus from the locus of the first system.

Table 5

Reagent	Antigenic factor	Matings of the type	Total number of matings	Offspring reacting with the reagent		x^2
				positively	negatively	
anti-W7	W_7	$W^7/- \times -/-$ ――――― 21	370	204	166	3·903
anti-W8	W_8	$W^8/- \times -/-$ ――――― 18	382	181	201	1·047

Two further factors W_7 and W_8, with still higher antigenicities than those of the first system, produce antibodies between 1/128 and 1/256 in titre. Table 5 shows the purity of corresponding reagents. Over 3,000 analyses were made using these two reagents. All samples reacted with at least one of them. This suggests that the corresponding genes are alleles. The data in tables 6 and 7 confirm this hypothesis.

Table 6

Matings of parents supposed W^7/W^8 (negative with anti-W8) $\times$ parents W^7/W^8

Number of matings	Total number of offspring (all reacting positively with anti-W7)	Number of offspring possessing W_8	
		observed	expected
18	431	219	212

Table 7

Matings of parents supposed W^7/W^8 (negative with anti-W7) × parents W^7/W^8

Number of matings	Total number of offspring (all reacting positively with anti-W8)	Number of offspring possessing W_7	
		observed	expected
11	224	113	111

Table 8

Matings of the type $W^5/-, W^8/- \times -/-, -/-$

Phenotypes of the possible offspring	W_5, W_8	W_5, O	O, W_8	O, O	x^2 for the ratio $1:1:1:1$
Observed number of off-spring for each pheno-type and each mating	5	4	6	6	0·527
	7	8	10	6	1·129
	5	2	4	4	1·268
	4	7	9	4	2·998
	6	7	3	3	2·686

Table 9

Matings of the type $W^6/-, W^8/- \times -/-, -/-$

Phenotypes of the possible offspring	W_6, W_8	W_6, O	O, W_8	O, O	x^2 for the ratio $1:1:1:1$
Observed number of off-spring for each phenotype and each mating	7	4	4	3	1·999
	18	8	5	23	15·778
	7	2	0	5	5·786
	6	4	0	6	6·000
	11	6	4	9	3·866
	8	6	7	7	0·286
Total					33·751

Table 10

Reagent	Antigenic factor	Matings of the type $W^9/- \times -/-$	Total number of matings	Offspring reacting with the reagent		x^2
				positively	negatively	
anti-W9	W_9	18	155	76	79	0·058

184

Thus, a third system exists, which is biallelic. The data in table 8 show that the first and third system are independent. The data in table 9 are insufficient to establish independence between W^6 and W^8. Further data are required to test the linkage hypothesis.

The last reagent anti-W9 is described in table 10. Factor W_9 probably belongs to a fourth system, because it was found in birds heterozygous for the first and third system. Moreover, with regard to W^6, animals having both W_6 and W_9 gave entirely negative offspring. This excludes allelism between W^6 and W^9.

However, the possibility of a very tight linkage to one of the preceding loci cannot be excluded.

At the present time, the preparation of our anti-W9 is too recent to enable us to elucidate the relations between W^9 and the other genes described.

Discussion and Conclusion

The four loci considered above are autosomal. With regard to the first and third system, the existence of heterozygous hens (heterogametic sex) excludes the sex-linkage hypothesis. Neither gene W^6 nor gene W^9 can be borne by the sex chromosome because it can be shown, by mating hens carrying these antigens to negative cocks, that the hens transmitted these antigens to their daughters.

So far, twelve systems have been described by Gilmour and Briles (1962) and designated as follows: A, B, C, D, E, H, I, J, K, L, N, and P.

When comparing our four systems with the above twelve systems, our first system would possibly be identical with the B system studied abroad. The B system presents the following characteristics which make it the most interesting system:

1. it is highly polymorphic and remains polymorphic (at least 2 alleles) even in highly inbred strains (99·4 % in the CH strain of Gilmour):

2. its factors are highly antigenic and the antigens are found on both lymphocytes and red cells.

Our first system shows essentially the same characteristics.

However, valid comparisons can only be made through an exchange of reagents between laboratories.

Furthermore, we made 40 immunizations on Faverolles cocks, a breed probably of a very different origin from the Wyandotte and more inbred than our M 11. We only obtained 8 immune sera, 7 of which were identical and contained a single antibody called anti-F2.

By analyzing over 300 blood samples from Faverolles the 5 reagents anti-F2, anti-W1, anti-W2, anti-W3 and anti-W5 gave exactly the same results.

We are thus led to suppose that the factor F_2 belongs to the same system as W_1, W_2, W_3 and W_5, this system probably being the B system.

Summary

The immunogenetic study of the antigenic polymorphism of red cells has been undertaken in chickens of the Wyandotte breed (M 11); the reasons for the choice of this material are given.

8 reagents were produced. Their unit character was controlled by serological and statistical tests. Antibodies were designated by the letter W (Wyandotte) followed by a number, in the chronological order of production.

These antibodies detect antigenic factors controlled by 4 genetic systems determined by autosomal loci: one tetraallelic and one biallelic, which are both "closed", and two biallelic which are "open". In order to identify these systems with those already described by other investigators using different breeds, comparisons were made on the basis of serological and genetic criteria.

References

Briles, W. E. (1958). Proc. of the X Intern. Congress of Genetics, Montreal 2 : 33—34.

Briles, W. E. (1962): VIII European Conference on Blood Groups in Animals. Ljubljana, Yugoslavia.

Briles, W. E., McGibbon, W. H. and Irwin, M. R. (1950). Genetics 35 : 633—652.

Gilmour, D. G. (1949). The identification of red cell antigens in fowls. Ph. D. dissertation, Cambridge.

Gilmour, D. G. (1959). Genetics 44 : 14—33.

Schierman, L. W. and Nordskog, A. W. (1961). Science 137 : 620—621.

Stone, W. H. and Irwin, M. R. (1963). In Advances in Immunology 3 : 316—345.

THE BLOOD GROUPS OF DUCKS

L. PODLIACHOUK

Pasteur Institute, Paris

This study, part of a research work on genetic markers of the duck, was made by a team headed by Professor J. Benoit.

The domestic ducks examined during this study — the Khaki Campbel (K) and Peking and their hybrids and the Peking ducks altered by the injection of Khaki DNA (1) — were all taken from the Gif-sur-Yvette breeding.

The technique used was the agglutination test on plates of plexiglass. Isoimmune agglutinins were used as reference antibodies.

Results

Natural isoagglutinins are rare in the duck. Almost all of those disclosed by us had a low titre and an equal specificity. We named them anti-A (present in about 5% of the sera examined), and the corresponding red cell antigen was called A.

The study of the blood from these ducks by means of anti-A gave us certain information about the choice of the fowl for isoimmunization.

We immunized altogether 82 ducks, using red cells of 27 donors, and obtained a certain number of complete isoimmune agglutinins, the titre of which varied from 1 to 1/2,000. We selected the sera which titrated 1/16 and more. Three of them were identical with natural anti-A. After numerous differential absorptions by means of selected red cells, we finally isolated 5 antibodies of different specificities. These antibodies determined 5 red cell antigens which we designated, in the chronological order of their discovery, by the capital letters A, B, C, D, and E.

There is no relationship whatsoever between this nomenclature and those used in man or other animal species.

We possess 3 or 4 different samples of some of these reference antibodies (anti-A, anti-D, and anti-E) coming from Khaki or Peking ducks, and only one sample of the other 2 antibodies (anti-B and anti-C).

In the strain examined (table 1), we found the antigens A and B in the Khaki duck and absent from the Peking. The other 3 antigens C, D and E were found in both races.

Table 1

Isoimmune antibodies			
Anti-factor	Number	Recipient	Donor
A	3	P*)	K**)
B	1	K	K
C	1	K	K
D	4	3 K 1 P	P P
E	4	2 K 1 K 1 P	K P K

*) Peking.
**) Khaki.

The hereditary transmission of these antigens has been studied on 65 families with a total of some 650 offspring. The blood groups of the parents, of the F_1 and F_2 hybrids of the matings P × P, K × K, P × K, K/BN × K* and others were determined.

The hereditary transmission of these antigens agrees with the Mendelian laws. They are transmitted as dominant characters. They are found in the offspring only if they exist in the parents.

Table 2

Race	Blood factors of ducks				
Khaki Peking	A	B	C C	D D	E E

*) K/BN × K: backcross of the hybrid male (Khaki with modified Peking) to his female Khaki relative.

The 88 ducklings of 3 families descending from Peking matings, negative for the factors A, B, D and E, also were all negative for these factors.

We observed that when the parents possessed a certain antigen, some of their offspring also possessed it and some not. We therefore came to the conclusion that the parents were heterozygous for most of the red cell antigens found in them. Only very few could be supposed to be homozygous for certain antigens. For instance, one female D+ mated to a male D— gave 26 offspring, all of them being D+.

All ducks heterozygous for a certain antigen, when mated to ducks negative for the same, produce the progeny which are in about 50% of cases positive and in 50% negative for that antigen.

Table 3

Factors	Mating types	Distribution of progeny phenotype "O" in testcrosses on allelic factors	
		Progeny	
		Number	Phenotype "O"
AB	AB × O, A, B, AB	224	0
CD	no sample		
CE	CE × O, E	35	0
DE	DE × D, E, DE	113	0

The observation of hereditary transmission of red cell antigens enables us to suppose the existence of a series of 3, 4 or more alleles, which explains better than any other the observed transmission.

We can admit that the antigens A and B are determined by a series of 3 alleles: A, B and O, where A and B are codominant with regard to O. Therefore if one of the parents is AB, the offspring cannot be O. Of the 224 ducklings, with one parent AB, none belonged to the O group. (Table 3.) Similar observation was made on 35 offspring of matings CE with O or with E, and on 113 offspring of matings DE with D, E or DE.

These genetic observations enable us to suppose that the antigens C, D and E are all located at the same locus, distinct from that of the factors A and B.

By mating a duck, having 2 antigens located at 2 different loci, to a duck with none of them, 25% of offspring negative for both are theoretically produced.

Of the 247 offspring produced from matings of ducks AD, AE, BC, BD and BE with ducks negative for the 2 corresponding antigens, 59 are negative

Table 4

		Distribution of progeny in testcrosses of hypothesis 1 : 1 : 1 : 1 on the factors A, B, C, D and E			
Types of mating	Number of progeny	Observed distribution of progeny phenotype			
		+ +	+ −	− +	− −
AB × −−	85	1 ?	40	44	0
AC × −−	0				
AD × −−	85	29	31	12	13
AE × −−	52	11	9	16	16
BC × −−	18	4	5	4	5
BD × −−	74	16	19	18	21
BE × −−	18	5	4	5	4
CD × −−	0				
CE × −−	18	0	10	8	0
DE × −−	0				

? error in pedigree.

for both antigens present in their parents. This number is neighbouring the theoretical one, which is 61·7.

The mating of ducks one parent of which possesses 2 antigens, located at 2 different loci, must result in a certain number of offspring negative for both these antigens.

We observed that in the matings where one of the parents had the factors AC, AD, AE, BD or BE, 90 ducklings out of 709 were negative for both factors of their parents.

This genetic analysis shows that the hypothesis, according to which the genes determining the antigens A and B are located at one locus and those

Table 5

		Distribution of progeny phenotype "O" in testcrosses on independent factors	
Factors	Mating types	Progeny	
		Number	Phenotype "O"
AC	AC × A	17	2
AD	AD × O, A, D, AD	200	26
AE	AE × O, A, E, AE	150	22
BC	BC × O, B	35	6
BD	BD × O, B, D, BD	187	27
BE	BE × O, B, E, BE	120	7

determining the C, D and E antigens on another, is not inconsistent with the observation. The statistical study, not presented here, does not refute our hypothesis either.

In the examined ducks, the antigens A and B of the first locus exist but in the Khaki race and are absent from the Peking. The antigens C, D and E of the second locus are present in the two races, Khaki and Peking, but their frequency is considerably lower in the latter.

Table 6

Blood groups of ducks					
Actual nomenclature				Former nomenclature	
Loci	Blood factors	Genes	Reference sera	Loci	Blood factors
A	A_1	A^1	A1	1	A
A	A_2	A^2	A2	1	B
B	B_1	B^1	B1	2	C
B	B_2	B^2	B2	2	D
B	B_3	B^3	B3	2	E

Now, as we are aware of the genetic structure of these red cell factors, we are going to establish a new nomenclature consistent with the genetic rules.

Henceforth the first locus will be the locus A; the 2 allelic factors A and B located at this locus will be designated A_1 and A_2. As to the corresponding genes, they will be named A^1 and A^2.

The second locus becomes the locus B; the 3 allelic factors C, D and E will be named B_1, B_2 and B_3; their corresponding genes — B^1, B^2 and B^3. The reference sera are designated A1, A2, B1, B2 and B3.

References

J. Benoit et al. (1957). C. R. Ac. Sc. 244 : 2320.

BLOOD GROUPS IN RABBITS

M. VARGA, M. TOLAROVÁ, M. TOLAR

Department for Research in Animal Genetics, Hungarian Academy of Sciences, Gödöllö,
Institute of Experimental Biology and Genetics, Czechoslovak Academy of Sciences, Prague,
and Institute of General Biology, Faculty of General Medicine, Charles University, Prague

The literature on the blood groups in rabbits has been recently reviewed by Cohen (1962) and Iványi and Tomášková (1960). Although almost 100 papers concerned with blood groups in rabbits or their utilization in various experiments were published, most of them deal with only one blood group system within which 3 antigens denoted by us Š ,N, O antigens of the system H_c (and 2 antigens as products of the "gene interaction") (see Cohen 1962) can be determined by means of isoimmune sera. This system can be designated as the "major" blood group system of rabbits. A number of authors refer even to other groups not falling in the main system or not assigned, but none of them was examined in detail. A review of nomenclatures employed for the major blood group system is presented in table 1.

In the present communication we would like to report of some results connected with the question of whether selection without any knowledge of the existence of blood groups in rabbits made by commercial breeders or other selective factors can lead to a change in the distribution of blood groups. This possibility was mentioned by Cohen (1962) in rabbits. He observed

1. a higher frequency of heterozygotes in random populations than expected. He calculated that six of the nine unselected populations reported in the literature by different authors deviated significantly form the expected. The source of deviation appeared to be the higher frequency of heterozygotes than expected. Eight out of the nine unselected populations showed a higher frequency of heterozygotes than expected.

2. There was a progressive decrease in the frequency of Š(A)-type during the course of an inbreeding experiment ($F_2 - F_4$). From studies of sizes and numbers of litters it appeared that there was a competition among sperm of zygotes with the non-Š(A) to become the more frequent winner.

3. It was found that a wild population of rabbits of the same species as the domesticated ones on the Farallon Island was made only of animals type (O/D), N(F), or ON(DF). Cohen supposed that selection against Š(A) antigen could occur in physically very unfavourable environments such as rocky islands.

Table 1

Nomenclature of blood group antigens system H_0* in rabbits.** Other or not yet analyzed blood groups: $K_{3,4,5}$ (Fischer); H_6 (Knopfmacher); H, J (Kellner); $R_{2,3}$ (Dahr); X (Heard); B, C, E, H (Cohen); C, D (Anderson); $H_{a,b,d}$ (Ivanyi); A, B, C, D, E, F, G, H, J; (Wright); J—X (Helmbold); a, b, c, d, e, (Grodecka); X (Nelken); A, B, C, D, (Matsumoto). (For literature see Ivanyi and Tomašková 1960).

Castle and Keeler 1933	Fischer 1935	Marcussen 1936	Kellner and Hedal 1953	Dahr and Fischer 1955	Anderson 1955	Heard 1955	Joysey 1955	Cohen 1955	Iványi and Tomášková 1958
H_2	K_1	K_b	g	r_1	A	Y	B	A	Š
H_1	K_2	K_a	G	R_1	B	Z	A	F	N
H_0	O			O	O	W	C	D	O

* Hg according to Cohen.
** Partly according to Cohen.

Table 2

Frequency of phenotypes in 1 145 rabbits from 10 random rabbit populations selected by commercial breeders.

Population of rabbits	No. tested	Observed frequencies (%)				p	q	r	Expected frequencies (%)				X^2
		S	SN	N	O				SS(SO)	SN	NN(NO)	OO	
Csincsilla-Gödöllö	344	11·92	34·59	53·49	—	0·292	0·708	—	8·50	41·40	50·10	—	2·730
Csincsilla-ČSSR	214	14·95	51·86	33·18	—	0·408	0·592	—	16·65	48·31	35·04	—	0·520
Csincsilla-SSSR	164	6·70	31·10	62·20	—	0·223	0·777	—	4·97	34·66	60·37	—	1·010
Csincsilla-DDR	26	42·31	38·46	19·23	—	0·615	0·385	—	37·82	47·36	14·82	—	3·510
Nagy-Ezüst	100	18·00	58·00	24·00	—	0·470	0·530	—	22·09	49·82	28·09	—	2·690
Kosorrú	69	33·34	40·58	26·08	—	0·536	0·464	—	28·73	49·74	21·53	—	3·380
Szürke Orias	49	34·70	44·90	20·40	—	0·572	0·428	—	32·72	48·96	18·32	—	0·780
Albino	17	23·53	58·82	17·65	—	0·529	0·471	—	27·98	49·83	22·19	—	3·260
G. Magyar Vadas	60	13·33	46·66	40·00	—	0·366	0·633	—	13·46	46·46	40·08	—	0·002
New Zealand	102	63·72	25·49	8·83	1·960	0·672	0·190	0·138	83·71	25·54	5·85	1·9	0·001

We have examined some points connected with these assumptions:

1. The blood groups were determined in 10 populations of rabbits bred in Hungary (see table 2) and intensively selected for production properties. Three of them were `lines imported specially for breeding high productive animals. Nine of them were European populations, one from American import; the blood group O was found only in the last group. A total of 1,145 rabbits were examined. From the observed frequencies of the blood groups the expected frequencies of various phenotypes were derived by calculations using the Hardy-Weinberg law. From these comparisons the X^2 value was derived.

Table 3

Frequency of heterozygotes in 1 145 rabbits from 10 random rabbit populations selected by commercial breeders.

| Population of rabbits | No. tested | Frequency of SN Rabbits | | X^2 |
		Observed	Expected	
ॽ Csincsilla-Gödöllö	344	34·59	41·40	1·12
Csincsilla-ČSSR	214	51·86	48·31	0·26
Csincsilla-SSSR	164	31·10	34·66	0·36
Csincsilla-DDR	26	38·46	47·36	1·67
Nagy Ezüst	100	58·00	49·82	1·34
Kosorrú	69	40·58	49·74	1·68
Szürke Orias	49	44·90	48·96	0·34
Albino	17	58·82	49·83	1·62
G. Magyar Vadas	60	46·66	46·46	0·0008
New Zealand	102	25·49	25·54	0·001

In none of the populations a significant deviation between the observed and expected values was found. The frequencies of observed heterozygotes were also practivally the same as expected (table 3). However, it must be noted that even Cohen claimed that eight out of the nine calculated populations showed a higher observed frequency of heterozygotes than expected, in fact only three of them differed significantly.

2. The blood groups of three semiinbred rabbit populations were examined. The rabbits under study were derived from the $4^{th}-10^{th}$ generation of intrafamilial matings (Fabián et al. 1963). Skin grafts exchanged between members of the semiinbred strains survived in the same way as those transplanted in rabbits in the $8^{th}-9^{th}$ generation of sib mating (Chai 1964). Table 4 shows the results of blood group examinations. In two populations the observed frequency was very similar to the expected one, in the third population it was considerably different but in the opposite sense than in the experiments of Cohen, the group Š significantly prevailed. In this population the number of observed heterozygotes was significantly higher than expected. The findings in our semiinbred rabbits cannot be discussed in detail, because the blood group

of the starting generation and the generations preceding the examined one was not known to us. The number of now available animals from individual generations was too small for a separate evaluation. It can hardly be explained why no animal of the group N was found in the semiinbred population Chinchilla. However, the differences between the observed and expected values in two semiinbred populations show clearly that inbreeding does not increase the frequency of heterozygotes or the group N in all instances.

Table 4

Blood groups in semiinbred rabbits.

Population of rabbits	No. tested	Observed frequency in %			p	q	Expected frequency in %			X^2
		S	SN	N			S	SN	N	
G. Magyar Vadas	101	2·3	28·2	69·3	0·165	0·835	3·8	25·3	70·7	0·892
Tihanyi Oroszos	33	28·0	42·8	29·0	0·494	0·506	24·6	49·4	25·8	1·768
Csincsilla	38	59·7	40·2	0·0	0·798	0·202	65·1	29·5	5·3	9·674

3. Wild rabbits and hares were also examined. The hares did not react specifically with anti-Š or anti-N sera. In wild rabbits, the blood groups Š, N and ŠN were found. Although this population is also living in physically unfavourable environments, the individuals Š were not eliminated. However, it is possible that these animals are already better adapted to these conditions than wild rabbits on the Farallon Island derived from the domesticated ones.

In conclusion, it can be said that our findings do not support the assumption of Cohen (1962) that the blood groups of the system H_c would have a selective value in rabbits. Even the findings in the semiinbred populations and wild rabbits remain under discussion; it was clearly demonstrated that the observed frequency of heterozygotes is practically the same as the expected one in random-bred rabbits selected by commercial breeders for production properties.

References

Cohen, C. (1962). Blood groups in rabbits. Ann. N. Y. Acad. Sci. 97 : 26.

Chai, C. K. (1964). Skin grafts between inbred rabbits. Transplantation 2 : 436.

Fabián, Gy., Iványi, P., Széky, P. (1963). Skin transplantation in partially inbred rabbits. Folia Biol. (Praha) 9 : 440.

Iványi, P., Tomášková, M. (1960). O značenii identifikacii grup krovi u krolikov. (The significance of blood group identification in rabbits; a review). Probl. gematol. pereliv. krovi 5/7 : 39.

ERYTHROCYTE B1 ANTIGEN IN INBRED RAT STRAINS

B. FRENZL, R. BRDIČKA, V. KŘEN and O. ŠTARK

Department of Medical Biology, Faculty of General Medicine, Charles University, Prague

Recognition of the antigenic mosaic in rats necessary for immunogenetic and immunobiological experiments is only in the beginning in comparison with our knowledge of the antigenic mosaic in mice. It is difficult to compare earlier findings because neither rat strains nor antisera used for the determination of erythrocyte antigens are available.

The first findings of erythrocyte antigenicity in the rat were carried out by means of normal isohaemagglutinins. In this connection, Eyquem in his book "Les groupes sanguins chez les animaux" (1953) mentions in the first place Rosenberg (1920) and Hibino. Later, there were found four types of normal isohaemagglutinins in wild rats by Friedberger and Taslokwa (1928). Burhoe (1932—1947) discovered antigen A by means of normal isoantibodies and a weaker antigen M, proved by immune isoantibodies. Both antigens were inherited independently and formed four blood groups: AM, A, M, O. Linkage with other known loci has not been proved. Owen (1948) selected a strain of rats which possessed normal isohaemagglutinins alpha reacting with erythrocyte antigen A of other rats. The presence of antigen A was dominant over the occurrence of agglutinin alpha. It is considered by Owen that the independent locus A-alpha is a possible analogy of Burhoe's antigen A. Further independent loci controlling erythrocyte antigens in rats were discovered by Owen (1948, 1962) by means of rabbit anti-rat absorbed sera. Allelic antigens C—D identified by Owen are codominant; the C—D locus was found in all tested rat strains. Further antigens E and F are also controlled by solitary loci. It was possible to obtain antibodies by isoimmunization only against antigen E. All four independent loci were autosomal; but Owen failed to prove linkage with other genes. Eyquem (1953) reports that Jaemeri and co-workers (1952) found a rat erythrocyte antigen in 56·4% of the 250 rats tested by means of absorbed rabbit sera.

The first indication of the existence of a histocompatibility locus in rats was Lumsden's discovery (1938) of immune isohaemagglutinins and cytotoxins against susceptible rats in those animals which had rejected Jensen's sarcoma or had been immunized with cells of susceptible rats. Bogden's and Aptekman's (1960, 1961, 1962) R — 1 locus is the first detected and examined histocompatibility locus which controls the occurrence of antigen B. Random-bred Wistar rats resistant to ascitic tumour AA form antibodies with cyto-toxic and isohaemagglutinic activity against cells of susceptible Wistar rats. The authors detected the erythrocyte antigen G which had no histocompa-tibility character. No relation between B and G antigens and Owen's antigens A, C, D, F was found. Further blood antigens were found by the absorption of isoimmune sera by Dr. Palm (1962). The antigenic specificities 1, 2, 3 described by her do not agree with any of the known antigens. Specificities 1 and 3 are allelic and according to the personal communication of Dr. Palm (1964) they are part of the histocompatibility system related to the Bogden R-1 locus.

In our laboratory we contributed to the solution of the problem of the antigenic mosaic in rats by experiments with tumour Walker 256. After in-oculation of this tumour into random-bred Wistar, Hooded and Black rats, we observed the production of isohaemagglutinins (Frenzl and Štark 1956). The formation of antibodies was not in direct relationship to the fate of in-oculum, but it drew our attention to the existence of certain haemagglutin-ating types. By the use of isoimmune sera and absorbed rabbit sera, we proved the presence of three antigens: W 1, B 1, and B 2 (Frenzl et al. 1960).

Our own inbred breeds were used for further immunogenetic studies and immunobiological experimental models. Among the antigens studied, antigen B 1 was found to be most valuable and its presence and absence was the basic selective criterion. From the sixth generation of the inbred animals, the sur-vival time of skin grafts exchanged between brothers and sisters served as the further criterion (Brdička et al. 1962).

Now there are two inbred rat lines available: albino Wistar strain, denoted WP, without antigen B 1, in the 18th generation of inbreeding; and the strain Black, designated BP, antigen B 1 is present, in the 13th generation of inbreed-ing. After the intralineal transplantation, the skin grafts survive in both strains practically throughout the life in 100% of cases.

After the administration of the antigenic material from the line BP to the WP rats, isohaemagglutinins anti-B 1 are formed which react with B 1 positive red blood cells in saline in titres 1/256—1/512. Cytotoxic antibodies are also produced in high titres. The source of the B 1 antigen are not only blood cells but also spleen cells, epidermal cells or tumour cells of the BP strain. B 1 antigen can also be proved by the rabbit anti-rat serum absorbed by B 1 negative RBC. Heterologous test serum reacts with B 1 positive RBC in saline in the titre of 1/64. It has the same specificity as the isoimmune serum. Im-munization of rats BP with the antigenic material WP evokes the production of saline antibodies anti-Wp; their titres are, however, low (1/8—1/64).

Both inbred lines were used for the induction of experimental foetal erythroblastosis, the inhibition of runting syndrome by means of isoantisera and the study of the state of immunological tolerance.

Experimental foetal erythroblastosis (Frenzl et al. 1960) was evoked in B 1 negative female rats by active immunization with BP strain blood or passively by the injection of anti-B 1 isologous sera. (Tab. 1.) The injury to newborn rats is manifested in both morphological changes in blood indicating haemolytic anaemia, and birth of dystrophic or dead foetuses. In two cases, a hydropic foetus was found. When female rats WP mated with a homozygous BP male were immunized, the injury to newborn animals in litters was almost uniform. In backcrosses of homozygous WP females to heterozygous male rats, the injury to offspring occurs in agreement with segregation of B 1 antigen. Newborn rats recognized as B 1 positive were injured, while those determined as B 1 negative were normal.

Table 1

Mating female male	Number of newborn animals			
	Total	B 1 positive	Strong changes in blood cells	Dystrophic young, stillbirths
W. P. × B. P.	58	58	41	17
W. P. × BP. WP.	22	11	6	5

Results of experimental erythroblastosis foetalis, induced by immunization of W. P. females with B. P. blood or by injection of anti-B 1 isoserum.

The runting syndrome develops after the injection of spleen cells of the adult BP rat to newborn WP rats (Křen et al. 1960). As a result of immune reaction of inoculated immunologically competent cells against antigens of a nonreactive host WP, 95% of animals die. It is possible to cure the runting syndrome by the injection of immune anti-B 1 isosera (Křen et al. 1960, 1962). The recovery depends on the dose and rate of development of runt disease. (Tab. 2). The percentage of surviving animals decreases in relation to the

Table 2

Dose of anti-B 1 isoserum	Number of survivors/total number of animals treated								
	Day on which administration of antiserum was started								
	1	2	4	6	8	10	12	14	16
0·5 ml.	36/36	5/5	7/9	8/10	6/11	7/14	5/13	2/12	
0·6—1 ml.					44/47	5/6			1/19
1·5—2 ml.					23/23	6/6	9/10	7/20	

Results of the treatment of runting syndrome with anti-B 1 isoserum against blood cells.

day when the treatment was begun. The greater the dose used, the larger number of animals survive. Antisera against blood, spleen cells, skin grafts or benzpyrene-induced tumours of the BP strain are effective in treatment. The serum against isolated erythrocytes is also, in part, effective. It proves the cytotoxic effect of anti-B 1 antibodies on inoculated lymphoid cells.

Table 3

Group of rats	Days from transplantation of B 1 positive skin							
	Before	5—6	8	10	13	17	26	35
Spontaneously survived the runting syndrome	0	0	2	4	1	0	0	0
Cured with anti-B 1 serum from $10^{th}-12^{th}$ day	0	—	1	—	2	4	1	0
Control	1	1	64	512	—	512	64	16

Average titres of anti-B 1 isohaemagglutinins in split-tolerant and control rats after transplantation of skin graft from B 1 positive donor.

We found almost complete suppression of anti-B 1 antibody formation in rats WP which recovered spontaneously from runt disease or which had been cured with isoantiserum applied on the tenth or twelfth day. (Tab. 3.) After immunization with BP blood or transplantation with BP skin, antibodies were produced only in low titres. In these animals skin grafts were destroyed within the same time as in controls. We evaluated this finding as a specific form of split-tolerance, in other words as a dissociation of antibody production and cellular mechanism of immunity (Křenová-Peclová et al. 1963). This residual tolerance indicated that the amount of antibodies necessary for recovery from the runting syndrome does not obviously destroy the whole inoculum.

Thus the findings exist that antigen B 1 is an important component of tissue cells. The strong cytotoxic activity of anti-B 1 isoserum was proved by Dr. Iványi in vitro. We found not only suppression of isohaemagglutinin formation but also cytotoxic antibodies in split-tolerant rats. The sera of split-tolerant rats, which had rejected their skin grafts, were not effective in the treatment of the runt syndrome. These data indirectly indicate the role of B 1 antigen in tissue incompatibility reactions.

After crossing BP and WP strains, B 1 antigen behaves as a simple dominant character. (Tab. 4.) All F_1 hybrids are positive with isologous and heterologous anti-B 1 sera. Of the 114 F_2 hybrids tested, 87 were B 1 positive and 27 B 1 negative. Back-crosses to WP line gave B 1 positivity in 65 cases and B 1 negativity in 48 cases. All 22 back-crosses to the parental BP strain were B 1 positive. These results support the idea of a simple allelic system controlling the occurrence of B 1 antigen. And just as in the case of Burhoe and Owen, we also failed to prove a linkage of B 1 antigen with another locus.

Table 4

Hydrid group	Number of animals	
	B 1 positive	B 1 negative
F_1 WP × BP	108	0
F_2 WP . BP × WP . BP	87	27
backcross WP × WP . BP	65	48
backcross BP × WP . BP	22	0

Results of testing hybrid rat RBC by isologous and heterologous anti-B 1 sera.

Table 5

Rat strain	Origin	Total number	Average titres of anti-B 1 sera
BP-Black	Prague, Med. Fac.	234	512
AVN	Prague, ČSAV	21	512
Druckrey I.—IX.	Berlin	9	256
Lewis	Wistar Institute	8	256
BN	Wistar Institute	4	256
LE-P, Long Evans	Prague, Med. Fac.	30	61
Capturowe	Poland	4	32
CN	Prague, ČSAV	5	16
WP-Wistar	Prague, Med. Fac.	116	0
Le-K, Long Evans	Prague, ČSAV	6	0

The presence of B 1 antigen on RBC of 10 inbred rat strains.

Table 6

Serum anti-B 1 absorbed with RBC	Rat strains, their genotype		
	PA B+ G+	RN B− G+	BGN B− G−
—	+	+	+
PA	−	−	−
RN	−	−	−
BGN	+	+	−

Dr. Bogden's results of testing RBC by unabsorbed and absorbed anti-B 1 serum.

We tested the presence of B 1 antigen in blood cells of members of further inbred strains by means of isologous and heterologous sera anti-B 1. (Tab. 5.) Positive results were obtained in all strains, except for our WP line and the

Long Evans line -K- from the ČSAV. Erythrocytes of S positive strains were agglutinated in different intensity of positivities and different rate of titres.

Dr. Bogden (personal communication 1962) found in his laboratory that our anti-B 1 isoserum agglutinated blood cells of PA strain (genotype B^+G^+) and BGN (genotype B^-G^-. (Tab. 6.) Complete loss of agglutinating activity was made possible by the absorption with PA and RN erythrocytes. The absorption with erythrocytes BGN did not remove antibodies against PA and RN strain. From these results Dr. Bogden concluded that our anti-B 1 serum contained minimally 2 components: anti-G^+ and another not defined clearly.

Palm (1962) classifies the two antigens differently and denotes Bogden's B antigen as B 2.

We tested a sample of anti R-l serum we were sent by Dr Bogden using RBC of five rat strains: WP, LEP, BP, AVN and Lewis. The last one obtained from the Wistar Institute was ascertained as R-l negative by both Dr Bogden and Dr. Palm. We obtained negative results with saline as a diluent of anti R-l serum with all five types of RBC. Using normal rat serum as a diluent we ascertained positive results with RBC of four our strains: WP, LEP, BP and AVN. The reaction with Lewis RBC was expectedly negative. When anti R-l serum was absorbed with RBC of the AVN strain, positive reaction remained with RBC of WP and LEP only and disappeared with AVN and BP RBC. By the absorption with RBC of the WP strain the positive reaction was removed with WP and LEP erythrocytes but remained with AVN and BP RBC.

We suppose our WP and LEP strains to be certainly R-l positive and AVN and BP strains to have an additional specificity being present in the original Dr Bogden's rat strain. This assumption seems to be proved by the following facts: 1. Dr Bogden did not find anti R-l specificity in our Wp anti BP serum. 2. Immunization of the Lewis rats with antigenic material of the WP strain leads to production of antibodies agglutinating WP, LEP, AVN and BP RBC similarly to Dr Bogden's anti R-l serum. 3. F_1 hybrid rats (Lw × BP) immunized with WP antigenic material produce antibodies similar to those of anti R-l serum after absorption with RBC of AVN strain, i.e. agglutinating RBC of WP and LEP strains only.

Dr Palm's recent work indicates the relation between 1 and 3 antigenic specifities and Dr Bogden's R-l locus. We take pains to find the relation, if any, of our B 1 antigenic specifity to these antigens in further experiments which are in progress in our laboratory.

References

Bogden, A. E., Aptekman, P. M. (1960). Cancer Research 20 : 1372.

Bogden, A. E., Aptekman, P. M. (1961). J. Nat. Cancer Inst. 26 : 641.

Bogden, A. E., Aptekman, P. M. (1962). Ann. N. Y. Acad. Sci. 97 : 43.

Brdička, R., Křen, V., Frenzl, B., Štark, O. (1962). Fol. Biol. (Praha) 8 : 352.

Burhoe, S. O. (1947). Proc. Nat. Acad. Sci. (U. S. A.) 33 : 102.

De La Rivière, R. D., Eyquem, A. (1953). Les Groupes Sanguins chez les Animaux. Editions Médicales Flammarion. Paris, France.

Frenzl, B., Štark, O. (1956). Frekvence isohemaglutininů krys ve vztahu k resistenci proti nádoru Walker 256. (The frequency of isohaemagglutinins in rats in relation to resistance to Walker tumour 256.) Universitas Carolina, Prague, Suppl. 2 : 358.

Frenzl, B., Křen, V., Štark, O. (1960). Fol. Biol. (Praha) 6 : 121.

Frenzl, B., Křen, V., Štark, O., Smetana, K., Kraus, R. (1960). Fol. Biol. (Praha) 6 : 135.

Friedberger, E., Taslokwa, T. (1928). Zschr. Immunitätsforsch. 59 : 271.

Křen, V., Veselý, P., Frenzl, B., Štark, O. (1960). Fol. Biol. (Praha) 6 : 333.

Křen, V., Braun, A., Štark, O., Kraus, R., Frenzl, B., Brdička, R. (1962). Fol. Biol. (Praha) 8 : 341.

Křenová-Peclová, D., Král, J., Baborovská, J., Křen, V. (1963). Fol. Biol. (Praha) 9 : 258.

Lumsden, T. (1938). Am. J. Cancer 32 : 395.

Owen, R. D. (1948). Genetics 33 : 623.

Owen, R. D. (1962). Ann. N. Y. Acad. Sci. 97 : 37.

Palm, J. (1962). Ann. N. Y. Acad. Sci. 97 : 57.

STUDIES ON ERYTHROCYTIC FACTORS IN RATS

M. SPITERI and A. EYQUEM
Pasteur Institute, Paris

The immunological relationship between different strains of rats has been studied by various workers using different techniques.

Grafting of tumours combined with serological studies led to the discovery of the locus "R", which determines the antigens common to grafts and red cells. The runt disease has shown the relationship between different strains of rats (Billingham 1960). This relationship has been confirmed by skin grafting. Frenzl and co-workers in Prague studied the relationship between 3 strains of rats: Black, Long-Evans, Wistar. They concluded that successful skin transplantation was the best test for homogeneity of histocompatibility.

Blood group factors were studied by different workers. In 1947, Burhoe discovered two antigens: The "A"-antigen was detected with a normal agglutinin discovered in one of the 10 groups of rats. — The "M"-antigen was detected with an immune agglutinin.

Owen (1948) studied more than 5,000 rats and detected one "A"-antigen by using the only normal isoagglutinin which was found in all these rats.

Other factors "C", "D", "E", "F" were detected by using rabbit immune serum against rat red cells.

The "C" and "D" factors were not detected by isoimmune sera (Palm). Inbreeding by brother × sister mating gave two homozygous strains:
— one albino homozygous for the "C"-factor
— another one non-agouti homozygous for the "D" factor.
In Prague, Frenzl, Křen, Štark (1956) observed that rats belonging to the same strain lacked normal isoagglutinin.

After isoimmunization of rats or immunization of rabbits to rat red cells, they were able to detect 5 factors: W, W_1, W_2 or W_k, Bl_1 and Bl_2. They observed a linkage between blood group factors and the colour of the fur.

Frenzl, Štark and co-workers (1960) detected a strong factor "B" in Black rats; Wistar rats are lacking this factor.

Skin or tumour transplantation from B^+ rats to B^- rats can produce isoimmune agglutinin anti-B. This B antigen of Frenzl is analogous to the B factor of Bogden and Aptekman which was found in 50% of the Wistar rats. It can be detected with sera diluted in saline without using a high concentration of protein or dextran.

Joy Palm (1962) studied the strains Lewis, B. N. (Brown-Norway) and Wistar I. F.

In addition to the "C" and "D" antigens, 6 antigenic specificities

$$\text{"B}_1\text{", "B}_2\text{", "G", "1", "2", "3"}$$

can be detected. All the rats possessed either C or D, or both. Antigens 1 and 3 are the product of allelic genes. Lewis strain has the antigen 1, the Brown-Norway strain has the antigen 3, and the antigen 2 is shared by the 3 strains:

Lewis, P. A. and Wistar I. F.

The G factor of Bogden is absent from Wistar I. F,. but is shared by the 3 strains: Lewis, Brown-Norway and P. A.

Material and Methods

Rats from 8 inbred strains obtained from the "Centre de Sélection des Animaux de laboratoire" were studied:

— 3 albino strains with pink eyes:
  — Wistar AC
  — Wistar C F (Carworth Farm)
  — Sherman (Columbia University)
— 1 Black with black eyes: — London -Black
(was obtained at the University College of London between an agouti wild male and a Wistar female.)
— 2 Long-Evans strains:
  — one black and white with black eyes
  — one black and white agouti with black eyes.
— 1 August, beige (pink-eyed)
— 1 P. V. G., black-eyes and black and white fur (Glaxo Laboratories).

Some rats were immunized with intraperitoneal injections of 1 ml. of a 20 % suspension of red cells in saline. They received 2 weekly injections during 2 months.

Heteroimmune sera were prepared in rabbits by a total of 8 injections of 20 % erythrocyte or liver and spleen suspensions in saline, administered twice a week. At the same time they received Freund adjuvant subcutaneously. They were immunized for 5 weeks.

Rats belonging to different strains were joined in parabiosis. They were usually separated after 12 days of parabiotic union. In a few cases, one of the parabionts died before.

In some cases, parabiosis could be accepted during 3 weeks by the animals. Very good results were obtained by separation of the 2 parabionts after a 2-week-period of parabiosis, and then a new parabiosis of 10 to 15 days. Parabiosis was made between:

Wistar C. F. and August or Long-Evans
or Wistar A. G.

and between: Wistar A. G. and Long-Evans.

The agglutination test used to analyze the heteroimmune sera involved red cells collected in sodium citrate and washed 3 times in 10 volumes of saline. Titration of the serum was done in saline. Isoimmune sera were tested by the same technique. Incomplete antibodies were detected with a Coombs indirect test using immune rabbit serum against rat serum, and also with red cells treated by papain during one hour, then diluted in saline.

They were also studied by means of a 20% human albumin solution and 2% Dextran or 1% Polyvinyl-Pyrrolidone solution.

Heteroimmune sera obtained from rabbits, injected with various strains of Salmonella and Escherichia coli, were also studied.

Results

Of the 200 adult rats, 60 belonged to the Wistar strain, the others belonged to the following strains: Wistar AG, Wistar CF, Sherman, London-Black, Long-Evans, Long-Evans California, August, P. V. G.

No natural isoagglutinins were found either at 37°C or at 4°C during the examination of groups of 10 rats belonging to the different strains or even by cross-examination of a group of 60 rats belonging to the Wistar albino-strain.

Natural isoagglutinins active in rats of the different strains were found only in a few animals of the Long-Evans California strain. This natural iso-agglutinin is active between 0 and 10°C against the red cells of the rats belonging to the 7 other strains, but not against the red cells of the rats of the Long-Evans California strain. Antigens shared by different strains of rats were detected after selective absorptions of rabbit heteroimmune sera.

After absorption with Wistar C. F. or Long-Evans red cells, there is still an antibody against Wistar A. G., London-Black and some August rats in the serum.

Absorption with Wistar A. G. red cells leads to the disappearance of all the antibodies. After absorption with some of the London Black rats, there is still an antibody against Wistar A. G. After absorption with Sherman or August red cells, there is still an antibody against Wistar A. G. and London-Black.

This is an indication of an antigen "C" shared by Wistar A. G. and London-Black. There is an antigen "B" found in London Black and another one "W" in Wistar A. G.

The presence of an antigen in both London-Black and Wistar A. G. can be explained: London-Black are produced by hybridization (1953) of a wild-agouti male and a Wistar albino female; this antigenic relationship is also shown with rabbit heteroimmune serum to London-Black.

Isoimmune agglutinins are very rare. One was obtained in a Wistar C. F. immunized to August. This agglutinin was active in saline between 37°C and 4°C. The titre with August, Long-Evans and Wistar C. F. was 1/256 using papainized red cells.

Isoimmune agglutinins are also detected after parabiosis usually in the serum of the strongest animals.

One agglutinin was found in a serum of a Wistar C. F. which was joined in parabiosis with a Wistar A. G.

Isoimmune agglutinins were also found in the serum of 2 Wistar C. F. in parabiosis with August and in the serum of Wistar A. G. in parabiosis with Long-Evans California.

Parabiosis seems to be a good way to obtain isoimmune antibodies. These isoimmune agglutinins are active in saline at a titre of 1/64 at 37°C, against August, London-Black and at a lower titre against Long-Evans California and P. V. G.

When studying rabbit heteroimmune sera to Salmonella and Escherichia coli, heterophile antigens were searched for. No significant results were obtained by testing:

— 11 sera of Rabbit anti-Coli 86 B_7

— 12 sera Rabbit anti-Coli 26 (1 excepted)

— 11 sera of Rabbit anti-Coli 111 B_4.

In the study of the anti-Salmonella sera we found that the serum anti-S. Weslaco is specifically active against Wistar A. G.; an anti-S. artis is also active at a titre of 1/40 to 1/60 against the cells of half of the Wistar A. G. and not active against the red cells of the other rats. The anti-S. Poona serum is active at a titre of 1/40 against rats of different strains.

Negative results were obtained in the study of the rabbits immunized against:

Salmonella-Milwaukee
  -Bulawayo
  -Johannesburg
  -Riogrande
  -Betjoky
  -Bukauv
  -Durham
  -Locarno
  -Worthington
  -Basel

The phyto-agglutinin extracted from Phaseolus vulgaris also shows the special character of Wistar A. G. The red cells of this strain are agglutinated at a titre of 1/16 and on the contrary the cells of the rats belonging to the 7 other strains are agglutinated at a titre of 1/512 or higher.

Summary

Rats belonging to 8 inbred strains were studied:
  (Wistar A. C., Wistar C. F,. Sherman,
  London-Black, Long-Evans black, L. E. Agouti,
  August, P. V. G.).

There is no natural isoagglutinin in rats belonging to the same strain. Rats belonging to the Long-Evans California may possess agglutinins against the red cells of the rats of other strains. Rabbit heteroimmune sera can be used to detect an antigen shared by Wistar A. G. and London-Black.

An antigen specific for August can be detected by isoimmune sera of Wistar C. F. anti-August cells.

Parabiosis seems a good method to obtain isoimmunization. Wistar C. F. seems to be easily immunized against August or Wistar A. G.

An antigen shared by Wistar A. G. and S. artis is detected by a rabbit heteroimmune serum against S. artis.

BLOOD GROUP STUDIES IN THE DOMESTIC MINK

J. RAPACZ, R. M. SHACKELFORD, J. JAKÓBIEC

Department of Cattle Breeding, College of Agriculture, Kraków,
Division of Genetics and Department of Meat and Animal Science, University of Wisconsin

Numerous blood group studies in many species of domestic animals have been made during the past twenty years, but reports of immunogenetic investigation in the domestic mink (Mustela vison) appeared only within the past two years (Rapacz and Shackelford 1962, 1963a, b, Saison 1962). A report presented here refers to blood group studies of the domestic mink and is a review of investigations carried out at the University of Wisconsin during the past two years.

Over 2,200 mink blood samples were studied. Agglutinins or haemolysins or both for mink red blood cells were found in the normal sera of the chicken, marten, rabbit, guinea pig, cattle and man; of the seven species tested, only the ferret failed to show crossreacting antibodies.

Four species were used to produce heteroimmune sera; no reagents from guinea pig sera were separated; three different reagents from rabbit: anti-A, B and E; two from chicken: anti-A and B; and three — anti-A, B and B_2 — reagents were obtained from ferret sera. All these reagents functioned as agglutinins. As heteroimmune sera, all isoimmune sera and normal mink sera also contained complete agglutinins, except for one which showed incomplete antibodies.

Using the reagents from iso-, hetero-, or normal sera, at least 12 blood antigenic specificities were found. Since mink is so small and most sera exhibit low titres, heteroimmune sera required at least a fourfold absorption, only seven reagents called anti-A, -B, -B_2, -C, -D, -E and anti-G were available for an extensive study of the seven blood factors and their mode of inheritance.

The data obtained gave evidence that the genes for these specificities were at four different loci and the segregation patterns suggested that they were probably located on four different chromosomes.

System A

Of the 2,200 mink classified in the A system, 810 were tested for the D system; 484 for the two preceding and the E system and 359 for all four systems: the A, D, E and G system.

Four specificities — the A, B, B_2 and C — belong to the A system. Six phenotypes A, ABB_2, ACB_2, BB_2, BCB_2, CB_2 were found in all tested mink and each individual was identified as belonging to one or another of these groups. It was established by planned matings resulting in more than 300 litters of mink with more than 1,100 offspring, that blood types A, B and C result from genes of the triple allelic series and a 1 : 1 ratio between genotype and phenotype can be demonstrated (table 1). The fourth blood factor is an over-lapping specificity produced by genes A^b and A^c; consequently, no distiction can be made between these two genes by the use of this reagent (table 1).

Table 1

The A system in mink

Red Cells			Reagents			
Phenotype		Genotype	Anti-A	Anti-B	Anti-B_2 (BC)	Anti-C
1	A	A^a/A^a	+	0	0	0
2	AB	A^a/A^b	+	+	+	0
3	AC	A^a/A^c	+	0	+	+
4	B	A^b/A^b	0	+	+	0
5	BC	A^b/A^c	0	+	+	+
6	C	A^c/A^c	0	0	+	+

Systems D, E and G

In each of the next three systems (D, E, G), only a single blood factor was found, and two phenotypes classified as presence or absence of the speci-ficity were observed. The frequency of D, E and G factors was 70%, 26%, and 8%, respectively. The frequency of blood type D and especially the E va-ried considerably between colour phase groups and in one it was entirely absent.

Matings of mink negative for D produced only negative offspring, and mink positive for D may produce offspring of both types of response; it is concluded that the presence of one gene, named D^d is responsible for this type. The same is true for the system E and G (8).

Genes at the A and D loci segregated independently and the numbers involved seem sufficient evidence for the assumption that they are located on different chromosomes. Smaller numbers were available for thee E and G loci, but it is clear that E segregates independetly of the genes at the A and

D loci; the linkage of G with the other three loci cannot be determined with a small number available, although it was observed in the combination with each of the other genes.

Normal antibodies

The sera of more than 1,500 adult mink were examined for normally occurring antibodies against mink red blood cells by means of agglutination test and in some cases haemolytic and antiglobulin tests were used. Normal antibodies were found in several mink sera following protective vaccination to virus enteritis in which mink tissue was used; it was supposed that antibodies had been produced as a side effect of virus enteritis vaccination. With this exception and one family with strong autoantibodies, others were found only in the sera of females after pregnancy and most of them in the mink which lost young or did not bear them. No antibodies were detected in sera of the males at that time. Usually antibodies were against the blood antigens of the male to which females had been mated and disappeared in 95% of cases two months after pregnancy.

Antibodies against five blood types: anti-A, B, B_2, C and E were detected; anti-A, and B_2 were most frequent. Numerous sera with unidentified antibodies, usually of very low titre, were found. It seems probable that, except for the appropriate genetic background, there are at least two sources of stimulation for antibody production in this species; pregnancy and virus enteritis vaccination.

It seemed appropriate to investigate the possibility that haemolytic disease resulting from incompatible matings might contribute to a high mortality of mink kits within a few days of birth and a large number of "misses" especially with females at their first pregnancy, a condition generally observed on commercial ranches.

Since "normal" antibodies against factors A and B_2 were found most frequently and in highest titre, incompatible matings within the A system suggested that the most likely cause of early deaths was haemolytic disease. However, anti-B and C, without anti-B_2 or A, were found only in those normal sera collected after pregnancy, never in the isoimmune sera; this may mean that selectivity is involved.

If haemolytic disease develops in mink, additional mechanism may be at work, since 17 females with antibodies from the research herd gave birth to their young and did not have them. The situation in mink may be comparable to that in humans where more than the Rh+(D)factor are involved as discussed by Levine (1962).

Erythrocyte antigen chimaerism in mink

To our knowledge, the chimaerism presumed to have arisen from an interchange of primordial haematopoietic tissue has been reported in three species of mammals; the classical work of Owen (4) with cattle where genetically different individuals from multiple births usually had the same blood types; the ewe of male-female pair in sheep (Stormont et al. 1959): four pairs of human twins (Cotterman 1958, Dunsford et al. 1953).

Some exceptional individuals were observed among the first juveniles of tested mink. It was noted that two females out of seven born in litter showed incomplete $(++)$ reaction for anti-A and B with a strong one in anti-B_2, which was a reverse for the genotype $A^a A^b$. During a second re-test of the entire breeding herd nine months later, the female No. 87 gave a weak anti-A and -B response and most surprisingly an anti-C $(++++)$ reaction as well, which might suggest some irregularity at the A locus. Records were checked and it was found that the No. 87 was one of young tested mink. She was then tested with anti-A from additional preparations and anti-B from four; with both types of reagents she persisted in giving an incomplete reaction which made it certain that she was an erythrocyte chimaera of blood types A and B, one or the other being descended from the genetically foreign source. The second erythrocyte chimaera-female was discarded in the meantime.

At that time the most interesting test was under way; female No. 87 was pregnant. In order to show what she was genotypically, it was necessary to test the offspring. None of the five offspring of No. 87 had gene A^a but received from her either B or C blood factor (her mate was of the $A^b A^c$ genotype), thus she must have been genetically $A^b A^c$.

Previous workers concerned with the phenomenon of erythrocyte chimaerism considered various ways by which the condition might have arisen. We have chosen to interpret the evidence presented here in the same manner as it has been done for cattle, sheep and humans, i.e. female No. 87 and her sister exchanged red blood cell precursors during the foetal life by anastomosis of placental blood vessels.

Erythrocyte chimaerism may occur with greater frequency than is generally realized; these two cases in the mink were accidentally discovered, but with the few blood type systems already found in this species it is easy to see how the condition could be missed.

So far as we are aware, this is the first case of blood chimaerism reported in a mammal which bears its young in litters although in sheep twins occur frequently.

References

Cotterman, C. W. (1958). J. Cell. and Comp. Physiol, 52 (supp. 1): 69—96.

Dunsford, I., C. C. Bowley, A. M. Hutchison, J. S. Thompson, Ruth Sanger and R. R. Race (1953). Br. Med. Jour. 2 : 81.

Levine, P. (1962). Proc. of Conference on Immunoreproduction, La Jolla, California, 133—141.

Owen, R. D. (1945). Science 102 : 400—401.

Rapacz, J. and R. M. Shackelford (1962). Nature 196 : 1340—1341.

Rapacz, J. and R. M. Shackelford (1963a). Immunogenetics Let. 3(1) : 55—62.

Rapacz, J. and R. M. Shackelford (1963). Immunogenetics Let. 3(2) : 77—83.

Rapacz, J. and R. M. Shackelford. The Inheritance of Blood Factors D, E and G in the Domestic Mink. (In press.)

Rapacz, J. and R. M. Shackelford. "Normal" Antibodies against Red Blood Cells in the Domestic Mink. (In press.)

Rapacz, J. and R. M. Shackelford, Erythrocyte Antigen Mosaicism in the Domestic Mink. (In press.)

Saison, R. (1962). J. Immunol. 89 : 881—885.

Stormont, C. W. C. Weir, and L. L. Lane (1959). Acta Genet. 9 : 47—53.

DISCUSSION

K. Hála: I would like to ask Dr. McDermid, what is his view on the effect of the B system on the economic characters. In your report I noticed that the homozygous animals of the B system were better than the heterozygous animals.

E. M. McDermid: I answer the second part of your question first. There were no homozygotes in test. Where apparent homozygotes appeared in the slides, this was in fact an organisation of the data to demonstrate the presence of maternal effects. The presence of the antigen of the B system on the red cell is only one demonstration of the physiological effects of the B system gene. The antigens are also histocompatibility antigens, tissue antigens and are immunologically competent in the splenomegaly reaction. They presumably control biochemical pathways in the cells of the soma. The most likely genetic explanation of the effects of the B system antigens is that of pleiotropy.

P. Millot: It is very interesting to me that the technique of papainized cells detects an immune-agglutinin in the isoimmunized rat and I want to ask Dr. Eyquem whether this agglutinin is also detected by other techniques such as Coombs's test, etc. — and whether papainized cells are more sensitive?

A. Eyquem: In some of the isoimmune sera, incomplete antibodies could be detected by the Coombs test using a rabbit anti-globulin serum; they were also detected in a dextran, albumin and polyvinyl pyrrolidone solution. The technique of papainized red cells is easier to handle and was not used before in the study of rat blood groups.

J. Moustgaard: Mr. Rapacz, have you studied some further polymorphic characters in minks and have found some individual differences?

J. Rapacz: During two years of study of mink blood groups, eight individuals of different colour shades were checked for haemoglobin variation. No variation was found.

Bo Gahne: Mr. Chairman asked for any genetic variation in mink serum. I have preliminarily examined about 300 mink samples. I did not find any variation in transferrin. In the post-albumin area, there was a variation, which I suppose is determined by three alleles.

A. Eyquem: I wonder if the very interesting results concerning the two populations of red cells can be studied by mixed agglutination especially of the indirect type if some incomplete antibodies are present in the sera.

J. Rapacz: Direct saline agglutination test was used only, to test mink red cells, since all reagents contained complete agglutinins.

BLOOD GROUPS AND SERUM PROTEIN
POLYMORPHISM IN HORSES

APPLICATION OF BLOOD TYPING AND PROTEIN TESTS IN HORSES [1]

C. STORMONT, Y. SUZUKI and J. RENDEL[2]

Serology Laboratory, School of Veterinary Medicine, University of California, Davis, California

Studies of blood groups in horses were begun in this laboratory in 1958. They were initiated, in part, with the purpose of developing tests which might be used in the solution of problems of questionable parentage arising in the registration and breeding of horses.

Sixteen specifically different blood-typing reagents were prepared from equine isoimmune antisera and heteroimmune antisera produced in rabbits (Stormont. Suzuki and Rhode, 1964). Some of the antibodies acted exclusively as agglutinins, whereas others acted exclusively as lysins. However, the antibodies in most of the reagents reacted both as agglutinins and lysins. With respect to reagents of the latter category, the choice of technique (agglutination or lysis) depended upon the clarity of the two kinds of reactions. Details of the techniques of equine blood typing used in this laboratory may be found in the report just cited.

The notations used in designating the 16 blood factors are A_1, A_2, A', C, D, H, J, K, P_1, P_2, P', Q, R, S, T and U. In these notations, antibodies for the blood factors A_1, C, D, H, J and K are known to be equivalent, respectively, to the reference reagents for blood factors A, C, D, H, J and K selected by Podliachouk and Hesselholt (1962). Antibodies for the remaining ten blood factors have no equivalents amongst the remaining European reference reagents, anti-E, anti-F, anti-G, anti-H and anti-I. As indicated in the symbolism, blood factors A_1, A_2 and A' are related as subtypes and the same holds for P_1, P_2 and P'.

The inheritance of the 16 blood factors in 639 offspring representing 103 stallion families was studied by Stormont and Suzuki (1964). The 16 blood factors fell into eight genetic systems named A, C, D, K, P, Q, T and U. Four of the eight systems, namely A, D, P and Q, involve multiple alleles, and the genotype-phenotype relationships in those systems are shown in Tables 1 through 4. Each of the remaining four systems (C. K, T and U) presently

[1] Supported by grants from The Jockey Club, The California Thoroughbred Breeders Association and The American Shetland Pony Club.

[2] Present address, Department of Animal Breeding, Agricultural College, Uppsala. Sweden.

involves but a single pair of alleles. The eight loci are autosomal and there was no evidence for linkage.

The genetic theory of multiple alleles for the A, D, P and Q systems was examined by means of Chi-square tests for goodness of fit (Stormont and Suzuki, 1964). With only one of the Chi-square tests approaching significance at the one percent level of probability, namely, that for the Q system in Shetland Ponies, the observed phenotypic frequencies were in good agreement with the expected frequencies, thereby confirming the genetic theory.

Table 1

Genotype-phenotype relationships in the A system of equine blood groups

Phenotypes	Reactions with				Genotype
	A_1	A_2	A'	H	
—	0	0	0	0	aa
A_1	+	+	0	0	$a^{A_1}a^{A_1}$, $a^{A_1}a$
A'	0	+	+	0	$a^{A'}a^{A'}$, $a^{A'}a$
H	0	0	0	+	a^Ha^H, a^Ha
A_1A'	+	+	+	0	$a^{A_1}a^{A'}$
A_1H	+	+	0	+	$a^{A_1}a^H$
$A'H$	0	+	+	+	$a^{A'}Ha^{A'H}$, $a^{A'H}a^{A'}$, $a^{A'H}a^H$, $a^{A'H}a$, $a^{A'}a^H$
$A_1A'H$	+	+	+	+	$a^{A_1}a^{A'H}$

Table 2

Genotype-phenotype relationships in the D system of equine blood groups

Phenotypes	Reactions with		Genotype
	D	J	
—	0	0	dd
D	+	0	d^Dd^D, d^Dd
J	0	+	d^Jd^J, d^Jd
DJ	+	+	d^Dd^J

Table 3

Genotype-phenotype relationships in the P system of equine blood groups

Phenotypes	Reactions with			Genotypes
	P_1	P_2	P'	
—	0	0	0	pp
P_1	+	+	0	$p^{P_1}p^{P_1}$, $p^{P_1}p$
P'	0	+	+	$p^{P'}p^{P'}$, $p^{P'}p$
P_1P'	+	+	+	$p^{P_1}p^{P'}$

The frequencies of alleles at each of the eight loci are given in Table 5. As may be noted, the two breeds of horses differ considerably with respect to the frequencies of most of the alleles.

During the later phases of this program, studies were begun on serum proteins using the techniques of starch-gel electrophoresis. The first polymorphisms explored were those involving the iron-binding proteins (transferrins or siderophilins), as reported by Braend and Stormont at the VIII European Conference of Animal Blood Groups (Ljubliana, 1962) and eventually

Table 4

Genotype-phenotype relationships in the Q system of equine blood groups

Phenotypes	Reactions with			Genotypes
	Q	R*)	S*)	
—	0	0	0	qq
Q	+	0	0	$q^Q q^Q$, $q^Q q$
R	0	+	0	$q^R q^R$, $q^R q$
S	0	0	+	$q^S q^S$, $q^S q$
QR	+	+	0	$q^{QR} q^{QR}$, $q^{QR} q^Q$, $q^{QR} q^R$, $q^{QR} q$, $q^Q q^R$
QS	+	0	+	$q^Q q^S$
RS	0	+	+	$q^{RS} q^{RS}$, $q^{RS} q^R$, $q^{RS} q^S$, $q^{RS} q$, $q^R q^S$
QRS	+	+	+	$q^Q q^{RS}$, $q^S q^{QR}$, $q^{QR} q^{RS}$

*) Some of the reactions with R and S reagents are cryptic. Therefore, it is necessary to perform absorptions with those bloods not lysed by either or both R and S reagents (Stormont, Suzuki and Rhode, 1964).

published elsewhere (Braend and Stormont, 1964). In all, six codominant transferrin alleles, Tf^D, Tf^F Tf^H, Tf^M, Tf^O and Tf^R were indicated but only 16 of the 21 phenotypes, namely, DD, DF, DH, DM, DR, FF, FH, FM, FO, FR, HH, HO, MR, OO, OR and RR, were encountered. All of the remaining phenotypes, namely, DO, HM, HR, MM and MO, have since been encountered in this laboratory. Hitherto unpublished data on the frequencies of the six transferrin alleles in Shetland Ponies and Thoroughbreds are presented in Table 6. With the exception of alleles Tf^H and Tf^O, the two breeds differ considerably on this basis alone.

Using the starch-gel method described by Kristjansson (1963) in his studies of pre-albumin phenotypes in pigs, Stormont and Suzuki (1963) observed phenotypic differences in equine serum proteins which migrated in the region of the albumins. Three phenotypes A, AB and B were observed and these three phenotypes are inherited as if controlled by a pair of codominant autosomal alleles. Estimates of the frequencies of the two alleles are given in Table 7. Those estimates are in close agreement with the earlier estimates (Stormont and Suzuki, 1963) based on smaller samples.

Table 5

Estimates of the frequencies of alleles at the eight blood group loci in Shetland Ponies and Thoroughbreds†)

Loci	Alleles	Frequencies in	
		Shetlands	Thoroughbreds
A	a^{A_1}	0·3107	0·7050
	$a^{A'}$	0·2852	0·0290
	a^H	0·0358	0·0036
	$a^{A'H}$	0·0601	0·0000
	a	0·3082	0·2624
D	d^D	0·1392	0·0000
	d^J	0·1215	0·1503
	d	0·7394	0·8497
P	p^{P_1}	0·3415	0·2058
	$p^{P'}$	0·0483	0·0910
	p	0·6102	0·7031
Q	q^Q	0·1519	0·5082
	q^R	0·3869	0·0000
	q^S	0·0103	0·1038
	q^{QR}	0·1306	0·0756
	q^{RS}	0·1893	0·3125
	q	0·1310	0·0000
C*)	$_cC$	0·6521	0·7317
K*)	k^K	0·1796	0·0635
T*)	t^T	0·4505	0·6594
U*)	u^U	0·3174	0·1485

*) The frequency of the alternative allele(s) at these loci is simply 1 minus the figures shown.

†) From Stormont and Suzuki (1964).

Table 6

Frequencies of the six transferrin alleles in two breeds of horses in United States*)

Alleles	Frequencies in	
	Shetlands	Thoroughbreds
Tf^D	0·1722	0·2667
Tf^F	0·4597	0·5633
Tf^H	0·0256	0·0268
Tf^M	0·0311	0·0000
Tf^O	0·1081	0·0900
Tf^R	0·2033	0·0533

*) Data on 273 Shetland Ponies and 150 Thoroughbreds.

Table 7

Frequencies of the two albumin alleles in two breeds of horses in United States*).

Alleles	Frequencies in	
	Shetlands	Thoroughbreds
Alb^A	0·3865	0·2103
Alb^B	0·6135	0·7897

*) Data on 251 Shetland Ponies and 145 Thoroughbreds (cf. Stormont and Suzuki, 1963).

On the basis of the data on the eight red cell systems and the two serum protein systems, the combined number of possible blood types in horses is 1.032,192. That number will increase considerably when the blood factors E, F, G, H and I studied by Podliachouk (1957, 1958), and other blood factors yet to be discovered, are taken into consideration.

Estimates of the efficacy of any genetic system in solving problems of questionable parentage are dependent upon the number of alleles, their frequencies and the genotype-phenotype relationships. The authors have obtained such estimates in a manner similar to the estimates derived by Rendel and Gahne (1961) in studies of blood groups and transferrin types in Swedish cattle. The estimates are given in Table 8 for each of two breeds of horses in United States. Because of the cryptic nature of many of the reactions for blood factors R and S in the Q system, we have considered only blood factor Q in deriving the estimates for that system. In deriving the estimates for the transferrin system we used information only on the four most frequent alleles Tf^D, Tf^F, Tf^O and Tf^R in order to reduce the number of calculations. Therefore the probabilities of excluding false paternity by use of the transferrin system are slightly underestimated and, accordingly, so are those for the transferrins and albumins combined, and for all ten systems combined.

From the data in Table 8 it is clear that the transferrin system is presently the most effective of all systems in solving problems of questionable parentage in horses. It is also clear from those data, and as already indicated by the gene-frequency data in Table 5, that the percentage of exclusions, and thereby

Table 8

The expected incidence of exclusions, by genetic systems, for equine paternity cases involving two stallions (the true sire and one other stallion)

Genetic systems	Percentage of exclusions expected in	
	Shetland Ponies	Thoroughbreds
A	21·96	3·83
C	0·96	0·38
D	16·98	7·83
K	8·14	4·88
P	11·85	16·52
Q	7·86	2·97
T	4·11	0·89
U	6·89	7·80
Tf*)	36·60	29·06
Alb	18·09	13·85
Red cell systems alone	57·25	37·83
Tf and Alb systems alone	48·07	38·88
All systems combined	77·80	62·01

*) The percentages shown for the Tf system are underestimates. Accordingly, those shown for Tf and Alb alone and for all systems combined are also underestimates.

successful solutions, is expected to be higher in Shetland Ponies than in Thoroughbreds. It may also be seen that the combination of the transferrin and albumin tests is essentially as effective as the combination of the tests for the eight systems of red cell antigens.

In Table 9 are summarized the pertinent data relative to 18 equine paternity cases which have now been processed in this laboratory. As indicated in the table, 15 of the 18 cases were solved.

Table 9

Results of 18 equine paternity cases processed in the Serology Laboratory

Case No.	Exclusions (+) based on the systems									
	A	C	D	K	P	Q	T	U	Tf	Alb
1	0	0	0	0	0	0	0	0	0	+
2	0	0	0	0	0	0	0	0	+	0
3	0	0	0	0	0	0	0	0	0	+
4	0	0	0	0	+	0	0	0	0	0
5	0	0	0	0	0	0	0	0	0	+
6	0	0	0	0	+	+	0	0	+	+
7	0	0	0	0	0	0	0	0	0	0
8	0	0	0	0	0	0	0	0	0	+
9	0	0	0	0	0	0	0	0	+	0
10	0	0	0	0	0	0	0	0	0	0
11	0	0	0	0	0	0	0	0	0	0
12	0	0	0	0	0	0	0	+	+	0
13	0	0	0	0	0	0	0	0	+	0
14	0	0	0	0	0	0	0	0	+	0
15	0	0	0	0	0	0	0	0	+	
16	0	0	0	0	+	0	0	0	+	
17	0	0	0	0	0	0	0	0	+	0
18	0	0	0	0	+	0	0	0	+	0
Totals	0	0	0	0	4	1	0	1	10	5

Cases 1 through 14 in Table 9 involved Thoroughbreds and the first 12 of those 14 cases were concerned with double-registered horses (i. e. horses that are registered to two different stallions). In case number 1, one of the two stallions was excluded on the basis of genetic incompatibility involving albumin (Alb) alleles. In case number 2, one of the two stallions was excluded on the basis of genetic incompatibility involving transferrin (Tf) alleles, and so on.

Cases 13 and 14 in Table 9 each involved white foals and the question was raised whether a Palomino stallion, used as a teaser, might have sired the two foals. In each of the two cases, the Palomino stallion was excluded on the basis of genetic incompability at the transferrin locus.

Case 15 involved Morgan Horses and cases 16 through 18 involved Quarter Horses. In each of those four cases, the question asked was which of the two stallions used in breeding the mare qualified as the true sire. And in each case it was possible to render a verdict. (The animals in cases 15 and 16 were not tested for albumin phenotypes.)

In order to obtain a more satisfactory measure of the observed efficacy of these tests in solving paternity cases and also to show that the tests do not make the error of excluding the true sire, The Jockey Club, The California Thoroughbred Breeders Association and The American Shetland Pony Club were asked to set up a number of simulated paternity cases, each involving the true sire and one other stallion selected at random. A total of 59 simulated cases involving Thoroughbreds was submitted by The Jockey Club and The California Thoroughbred Breeders Association, and a total of 55 simulated cases involving Shetland Ponies was submitted by The American Shetland Pony Club. All blood samples were coded so as not to reveal the identity of the true sire.

The percentage of the simulated cases solved, by breeds, is shown in Table 10 for the blood-typing tests alone, the electrophoretic tests (albumin and transferrin phenotypes) alone, and both methods combined. The observed percentages of solutions shown in Table 10 agree rather closely with the expected percentages in Table 8. But perhaps of most importance from the standpoint of the owners of the horses and the breed organizations, in no

Table 10

Summary of the results of simulated paternity tests which were set up to determine the accuracy and efficacy of blood typing and electrophoresis tests in resolving problems of questionable parentage*.

Breeds	Percentage of cases solved by		
	Blood typing alone	Electrophoresis alone	Both methods combined
Shetlands	63·64	60·00	81·82
Thoroughbreds	32·20	44·07	67·80

*) Data based on 59 Thoroughbred cases and 55 Shetland Pony cases.

instance did the tests exclude the true sire. Thus, it would appear that these equine blood-typing tests used in conjunction with the electrophoretic tests for albumin and transferrin phenotypes are reliable and effective in solving problems of questionable parentage in horses.

As a result of the studies reported here, The Jockey Club, which is the registry organization for Thoroughbred horses in United States, has recently authorized changing the registration papers of double-registered animals whenever these tests show conclusively that only one of the two stallions qualifies as the true sire.

References

Braend, M. and C. Stormont (1964). Nord. Vet.-Med. 16 : 31—37.

Kristjansson, F. K. (1963). Genetics 48 : 1059—1063.

Podliachouk, L. (1957). Les antigenes de groupes sanguins des equides et leur transmission hereditaire. Univ. Paris. Thesis.

Podliachouk, L. (1958). Ann. Inst. Pasteur 95 : 7—22.

Podliachouk, L. and M. Hesselholt (1962). Immunogenet. Letter 2 : 69—71.

Rendel, J. and B. Gahne (1961). Animal Prod. 3 : 307—314.

Stormont, C. and Y. Suzuki (1963). Proc. Soc. Exp. Biol. Med. 114 : 673—675.

Stormont, C. and Y. Suzuki (1964). Genetics (in press).

Stormont, C., Y. Suzuki and E. A. Rhode (1964). Cornell Vet. 54 : 439—452.

THE BLOOD GROUPS OF EQUIDAE

L. PODLIACHOUK

Pasteur Institute, Paris

Red cell antigens: The numerous investigations on horse blood groups performed in the years 1910 to 1955 have now but a historical interest.

In 1950, Eyquem and myself undertook the study of blood groups of equidae (horse, donkey and mule). We designated the eleven red cell antigens identified by us, by capital letters in the chronological order of their discovery.

Each of the species, horse and donkey, possesses its own specific antigens, but all of them can be observed in the mule and hinny. Ten of them were found in the horse, while only the antigen B has so far been observed in the donkey (Podliachouk 1957).

In about 1958, the study of blood groups in horses was undertaken by Sirbu and Popovici in Rumania (Sirbu 1959, Sirbu, Popovici 1958), by Adams (1958) and Franks (1959) in England, By Wojciechowska (1960), Kownacki (1960) and Wroblewski (1959) in Poland.

All these investigators isolated between two and seven red cell antigens. In 1960 they sent their reference sera to the Pasteur Institute to be compared with our sera (Podliachouk 1963).

In 1961, Hesselholt in Denmark prepared a certain number of isoimmune sera. During his stay in Paris, we unified the nomenclatures, choosing as a basis the nomenclature of the Pasteur Institute (Podliachouk 1962a).

Since then we established a collaboration with other investigators:
Queval (Podliachouk 1961) of the Republic of Tchad, East Africa
Schmid (Podliachouk 1963) of the Institute of Research and Animal Breeding in Munich (Germany)
Kaczmarek and Zwolinski (Podliachouk 1962) of the Agronomic Institute of Poznan (Poland)
Wadowski of the Agricultural High School of Olsztyn (Poland)
Salerno of the Agricultural Faculty of Portici-Naples (Italy)
Dabczewski of the Agronomic Institute in Dublin (Ireland).

The objects of our collaboration with these investigators were: identification, exchange of reference sera, study of blood groups in various races of horses and donkeys, and search for natural isoagglutinins.

In Japan, Yamaguchi and Matsumoto (1962) found 2 agglutinating and haemolyzing factors.

In the United States, Gilman (1960) in 1960, studied the blood groups of thoroughbred horses.

A very important study was made by Stormont (1963) and his team in Davis (California). They prepared sixteen antibodies corresponding to the blood group factors which they designated by the symbols A_1, A_2, B, C, ... N, O and X. A_1, A_2, B and C belong to the A system of this nomenclature, D, M and X to the D system. The antigens F and G, respectively, to the F and G systems, etc. ...

In 1962, Stormont asked Hesselholt and myself to let him have our reference sera to be compared with his own. The results of this comparison were published in Immunogenetics Letter (Suzuki 1962). The nomenclature adopted by Stormont is based upon that of the Pasteur Institute.

At present, we are in possession of fourteen reference sera: eleven are specific for the horse and three for the donkey.

The anti-G, I and K sera have been discovered by Hesselholt, anti-J_2 by Schmid (1964).

Table 1

Bood groups of Equidae
Equine antigens (A, F, I), (J_1, J_2), C, D, E, G, H, K
Asinine antigens B, M, N

Among the equine factors, those designated A, F and I belong to one system; J_1 and J_2 are subgroups. C, E, H and K are independent factors. The asinine factors, B, M and N are independent factors as well.

The hereditary transmission of the antigens was studied on thoroughbred horses, draught horses of Southern Germany (Podliachouk, Schmid 1963) and of the Boulogne race, on families of mule producing mares with their offspring and on donkey families (Podliachouk 1957, 1958).

Reference sera: All reference sera were studied using the agglutination test and haemolysis. In the latter case, the complement was provided by the guinea pig or rabbit serum, suitably absorbed and diluted when necessary.

Certain sera only agglutinated the erythrocytes, others both agglutinated and lysed them. So far we did not observe sera that would be haemolytic only.

In our routine work (determination of blood groups, study of antibodies). the agglutination test in tubes at room temperature (Gilman et al. 1960) is mainly used.

The purpose of our present study was to establish whether the agglutinating or haemolyzing power of the globulins responsible for specific activity of the sera depended upon their origin or upon the method of their preparation, and to check whether a relationship existed between their serological character and molecular weight (7S or 19S).

All reference sera, whatever their origin, were treated by β-mercaptoethanol (Deutsch, Morton 1957).

Their specific anti-red cell activity after treatment was compared with a sample of untreated serum which underwent the same manipulation as the treated one (such as dialysis, etc.).

Certain of these sera lost completely their specific activity after the β-mercaptoethanol action; certain others retained the same antibody titre as the untreated samples.

In the first case we can admit that the antibody activity was associated with the 19S globulins; in the second case with the 7S globulins. In one serum, we noticed a decrease in activity, which means that both types of globulins, 7S and 19S, are responsible for the antibody activity. In most cases (8 out of 9), we observed that the 19S antibodies agglutinated erythrocytes, while the 7S both agglutinated and lysed them.

We used as reference sera iso- and heteroantibodies, either natural or immune. We are therefore in a position to classify the reference sera, according to their origin, in 4 different types.

1. **The natural antibodies are agglutinins.** They are rare and of a low titre (not higher than 1/32) in the horse and donkey and more frequent and of a high titre (up to 1/128) in the mule. Isoagglutinins anti-A and anti-C are most frequent in the sera of a large number of horses and mules A-negative and C-negative.

The other isoagglutinins are rare: certains, such as anti-B, were found only in one out of 3,000 animals (6). Generally, these natural antibodies are 19S globulins.

2. **The natural heteroantibodies are 19S agglutinins.** In normal human serum, the heteroagglutinins against equine A and C factors reach the titre of 1/2,000 (8).

During a study of 13 animal species, Hesselholt and myself (1962) noticed the presence of agglutinins specific for antigens A, C, D and J titrating up to 1/128 in the serum of some of these animals. One cattle serum contains a heteroantibody (anti-C) of the type 7S plus 19S.

3. **The isoimmune sera** were obtained by intravenous immunization: small amounts of blood (20—25 ml.) were injected once a weak. The sera ob-

Table 2

Blood groups of Equidae						
Reference sera						
Antibody	Ref.	Origin	Aggl.	Haemol.	Titre	Type
AF	781	Horse anti-meningo	+	−	500	19 S
A	937	Mule	+	−	64	19 S
A	147	Horse isoimmune	+	+	2000	7 S
C	14	Human	+	−	500	19 S
C	682	Bovine	+	−	100	19 S + 7S
C	761	Rabbit anti-horse	+	+	200	7 S
D	528	Horse anti-tetanic	+	−	64	19 S
E	52	Horse anti-tetanic	+	−	32	7 S
G	A1	Horse isoimmune	+	−	64	19 S
H	107	Horse isoimmune	+	+	4000	7 S
H	765	Rabbit anti-horse	+	+	200	7 S
I	A36	Horse isoimmune	+	−	64	19 S
J	1446	Horse isoimmune	+	−	16	19 S
K	976	Horse isoimmune	+	+	500	7 S
K	A28	Horse isoimmune	+	−	16	19S

tained after 3—4-week-immunization agglutinated the erythrocytes and reached a titre up to 1/128. They were 19S globulins. Prolonged immunization (7—8 weeks) led to the formation of 7S globulins.

During the titration by agglutination a marked prozone was observed, which disappeared after heating for 30 minutes at 56°C. These sera, of a very high titre (up to 1/2,000) were haemolytic in the presence of complement.

4. **The heteroimmune sera** were prepared in rabbits immunized for 3—4 weeks by the intravenous route. One ml. of a 50% red cell suspension (washed 4 times) was injected twice a week.

Heteroimmunization led to a simultaneous formation of specific immune antibodies against equine blood factors and specific species-antibodies (rabbit anti-horse) having a titre of more than 1/2,000.

In order to remove the latter, the immune serum was absorbed first 2 or 3 times at room temperature for 1 or 2 hours, and then one hour at 4°C. It is indispensable to heat the serum which is to be absorbed, for 30 minutes at 56°C and it is recommended to dilute it 2 to 5 times.

In the absorbed fraction the specific agglutinins against equine factors having a titre up to 1,000 are left.

Table 3

Blood groups of Equidae			
Reference sera			
Antibodies	Agglutinating	Haemolysing	Type
Natural isoant.	+	−	19 S
Natural heteroant.	+	−	19 S
Isoimmune ant. 3−4 weeks immun. 8 weeks immun.	+ +	− +	19 S 7 S
Heteroimmune ant. 3−4 weeks immun. (rabbit)	+	+	7S

We noted that the immune response was satisfactory when the donor horse did not possess more than one red cell factor. When the horse possessed 2 or 3 factors, the antibodies produced were weak or null. As horses with one factor are rare, we used mule donors, which have a low frequency of equine factors. Specific heteroimmune sera against horse red cell factors prepared by 3−4-week immunization of rabbits, both agglutinated and lysed the erythrocytes. They are 7S globulins.

Conclusion

Natural iso- and heteroantibodies and isoimmune sera obtained by short-term immunization (3−4 weeks) are agglutinins and generally belong to the 1S9 globulin group.

Isoimmune sera obtained after a prolonged immunization (8 weeks) and heteroimmune sera prepared in rabbits immunized for 3—4 weeks are both agglutinating and haemolytic antibodies and belong to the 7S globulin group.

Therefore, the agglutination test is suitable for all reference sera, whatever their origin and method of preparation; the haemolytic technique can be used with only a few of them.

Summary

At the present time, fourteen red cell antigens were identified in the Equidae by means of either natural or immune iso- and heteroantibodies.

The antigens A, C, D, E, F, G, H, I, K, J_1 and J_2 are specific for the horse; B, M and N for the donkey. All of them may coexist in the mule and in the hinny.

The antigens A, F and I belong to one system; J_1 and J_2 to another. The factors C, E, G, H and K are independent of each other. The same holds true for the asinine antigens. Hereditary transmission of these antigens in various races was investigated.

The agglutinating or haemolytic power of a reference serum depends upon its origin and the method of preparation.

The nature of 7S and 19S globulins responsible for specific activity of the serum was checked by β-mercaptoethanol treatment.

Natural iso- and heteroantibodies and isoimmune sera obtained by short-term immunization (3—4 weeks) are agglutinins and generally belong to the 19S globulin group.

Isoimmune sera obtained by a prolonged immunization (8 weeks) and heteroimmune sera prepared in rabbits immunized for 3—4 weeks are both agglutinating and haemolytic antibodies and belong to the 7S globulin group.

References

Adams, D. J. (1958), J. Comp. Path. 68 : 242—252.
Deutsch, H. F. and Morton, J. I. (1957). Science 125 : 600.
Franks, D. (1959). J. Comp. Path. 69 : 353—366.
Gilman, M. A., Schwarz, A. and Wallerstein, H. (1960). Amer. J. Veter. Res. 21 : 393—396.
Kownacki, M. V. and Szeniawska, D. (1960). Roczniki Nauk Rolniczych 76 : 25—39.
Podliachouk, L. (1957). Thèse doct. ès-sci., Paris.
Podliachouk, L. (1958). Ann. Inst. Pasteur 95 : 7—22.

Podliachouk, L. (1960). Ann. Inst. Pasteur 99 : 883—890.

Podliachouk, L. and Hesselholt, M. (1962a). Immunogenetics Letter 2 : 69—71.

Podliachouk, L. and Hesselholt, M. (1962b). Ann. Inst. Pasteur 102 : 742—748.

Podliachouk, L., Kaczmarek, A. and Zwolinski, J. (1962). Ann. Inst. Pasteur 103 : 949 to 749.

Podliachouk, L. and Schmid, D. O. (1963). Ann. Inst. Pasteur 104 : 427—431.

Podliachouk, L. Sirbu, Z., Kownacki, M. and Szeniawska, D. (1960). Ann. Inst. Pasteur 98 : 861—867.

Podliachouk, L. and Queval, R. (1961). Rev. El. Méd. Vét. Pays Trop. 4 : 445—447.

Schmid, D. O. (1964). (In press.)

Sirbu, Z. (1959). Inst. Path. Hyg. Anim. 9 : 189—211.

Sirbu, Z. and Popovici, V. (1958). Inst. Path. Hyg. Anim. 8 : 237—247.

Stormont, C., Suzuki, Y. and Rhode, E. A. (1963). XI. Intern. Congress of Genetics (Holland).

Suzuki, Y. and Stormont, C. (1962). Immunogenetics Letter 2 : 138—140.

Woyciechowska, S. and Lille-Szyckowicz, I. (1960). Roczn. Nauk. Roln., Warsaw, 69-E-4 : 457—472.

Wroblewski, A. et al. (1959). (Personal communication.)

Yamaguchi and Matsumoto. Quoted by Hosoda, T. (1962). The 8th Animal Blood Group Conference in Europa (Ljubljana, Yugoslavia).

BLOOD GROUP STUDIES IN HORSES

D. O. SCHMID

Institute for Animal Blood Group Research — Livestock Breeding Research Organization Munich

In our investigations of the antigen structure of horse red blood cells, since two years we are looking for isoimmune antibodies and normal antibodies in a greater number of isoimmune sera, which we found when immunizing the horses with blood of other horses and in a greater number of sera of not immunized horses (Schmid and Erhard 1962). In three immunization periods 45 horses at the Bavarian horse-breeding station Schwaiganger were immunized. 63 first-immunizations, 16 reimmunizations and 3 repeated reimmunizations in 32 horses of Southern-German "Kaltblut", 11 Haflinger-horses and 2 Rottaler-horses were made.

Immunizations were made once a week by intramuscular injections of horse red blood cells washed in saline throughout three to four weeks. These red blood cells contained the following blood group antigens A, C, D, E, F, G, H, I, J and K according to the nomenclature of Podliachouk and Hesselholt (1961). In contrast to the recommended intravenous immunization with 300—500 ml. blood (Franks 1959), horses were immunized intramuscularly with small amounts (10—15 ml.) of blood, in agreement with our good results in obtaining cattle reagents. No pathological reactions or shock symptoms were observed. In the last immunization period, unwashed red blood cells were used without any pathological symptom.

After isoimmunization with horse blood, both agglutinins and haemolysins, directed against blood group antigens, are detectable in the horse serum. In some horse immune sera only agglutinins were found. In general, agglutinins alone can be proved first, and later haemolysins are also detectable, We think that this is due to the change from 19S to 7S globulins during isoimmunization. Fenton and coworkers (1964) demonstrated that 19S globulins cause only agglutination and 7 S globulins cause agglutination and lysis. During reimmunization, six months after the first immunization, one horse formed only haemolysins. In the first immunization we found both agglutinins and haemolysins.

In the 63 first immunizations we obtained 40 isoimmune sera, i. e. 63·4% and 19 reimmunizations resulted in 16 isoimmune sera, i. e. 84·2%. The A-factor was the strongest antigen. Before immunization, normal anti-

Table 1

Donor	Recipient	Normal antibody	First Immunization				First Reimmunization				Second Reimmunization			
			1	2	3	4	1	2	3	4	1	2	3	4
Daisy ACDFHI	Dahlie CD	$A^{A}_{1/1}$	*January 1962* $A^{A}_{1/1}$	$A^{A}_{1/1}$	$A^{A}_{1/4}$ $H^{A}_{1/1}$	$A^{A}_{1/4}$ $H^{A}_{1/1}$	*June 1962* $A^{A}_{1/4}$	$A^{A}_{1/4}$ $A^{H}_{1/512}$ $H^{A}_{1/4}$ $H^{H}_{1/32}$	$A^{A}_{1/4}$ $A^{H}_{1/512}$ $H^{A}_{1/4}$ $H^{H}_{1/32}$	$A^{A-}_{H1/256}$ $H^{A-}_{H1/32}$	*February 1964* $A^{A}_{1/8}$ $A^{H}_{1/128}$ $H^{A-}_{1/2}$ $H^{H}_{1/64}$	$A^{A}_{1/128}$ $A^{H}_{1/128}$ $H^{A}_{1/32}$ $H^{H}_{1/32}$	$A^{A}_{1/128}$ $A^{H}_{1/256}$ $H^{A}_{1/32}$ $H^{H}_{1/32}$	$A^{A}_{1/128}$ $A^{H}_{1/256}$ $H^{A}_{1/32}$ $H^{H}_{1/32}$

Donor	Recipient	Normal Antibody	First Immunization				First Reimmunization				Second Reimmunization			
			1	2	3	4	1	2	3	4	1	2	3	4
Daisy ACDFHI	Dina DH	—	*January 1962* —	—	$A^{A}_{1/2}$	$A^{A}_{1/2}$ A^{H}_{-} $C^{H-}_{1/32}$	*June 1962* $A^{A}_{1/2}$	$A^{A}_{1/4}$ $A^{H}_{1/256}$ $C^{A}_{1/2}$ $C^{H}_{1/64}$		A^{A}_{H} $C^{H-}_{1/32}$	*February 1962* $A^{A}_{1/16}$ $A^{H}_{1/256}$ $C^{A}_{1/2}$ C^{-}	$A^{A}_{1/64}$ $A^{H}_{1/256}$ $C^{A}_{1/4}$ $C^{H}_{1/32}$	$A^{A}_{1/256}$ $A^{H}_{1/512}$ $C^{A}_{1/4}$ $C^{H}_{1/16}$	$A^{A}_{1/512}$ $A^{H}_{1/512}$ $C^{A}_{1/4}$ $C^{H}_{1/64}$

Donor	Recipient	Normal antibody	First Immunization (January 1962)				First Reimmunization (June 1962)				Second Reimmunization (February 1964)			
			1	2	3	4	1	2	3	4	1	2	3	4
Daisy ACDFHI	Rani C	—	—	—	A H	A H	A_A1/1 A_H—	A_A1/2 A_H1/16 H_A1/16			A_A1/4 A_H1/4 H_A1/4 H_H—	A_A1/64 A_H1/64 H_A1/32 H_H1/32 D_A1/4 D_H—	A_A1/128 A_H1/128 H_A1/128 H_H1/32 D_A1/8 D_H—	A_A1/64 A_H1/64 H_A1/128 H_H1/32 D_A1/4 D_H—

Donor	Recipient	Normal Antibody	First Immunization (July 1962)				First Reimmunization (November 1962)			
			1	2	3	4	1	2	3	4
Hofrat-Elkar CDHJ	Dorothea C	A_A1/2 A_H—		A_A1/2 A_H1/16 D_A1/64 H_A1/2 H_H1/32	A_A1/32 A_H1/64 D_A1/64 D_H1/64 H_A1/64 H_H1/64	A_A1/63 A_H1/512 D_A1/64 D_H1/128 H_A1/128 H_H1/128	A_A1/4 A_H1/1 D_A1/1 D_H— H_A1/1 H_H—	A_A1/4 A_H1/2 D_A1/2 D_H— H H_A1/1 H_H—	A_A1/8 A_H1/4 D_A1/128 D_H1/128 H_A1/32 H_H1/32	A_A1/4 A_H1/4 D_A1/32 D_H1/32 H_A1/64 H_H1/32

bodies with very weak titre were found in 12 horses by agglutination, 11 anti-A and 1 anti-H. After many absorptions with polyvalent isoimmune sera we found 39 anti-A, 8 anti-C, 11 anti-D, 9 anti-H and 4 anti-K agglutinating and lysing. Now it was possible to isolate monovalent reagents for blood typing of horses. These reagents were used for parentage control of valuable animals and in forensic cases. In the last two years, we blood typed more than 500 horses.

Table 2

Blood group normal antibodies in horse sera (1962—1964).

	n	Normal antibodies		anti-A	anti-C	anti-D	anti-H	Unknown or incomplete
January	71	34	47·88%	21	4		5	4
February	24	5	20·83%	2	1	1		1
March	78	12	13·84%	3	3	1	1	4
April	45	12	26·66%		8	1		3
May	37	13	35·13%	1	5		3	4
July	102	11	10·78%	2	5	2		2
August	111	12	10·81 %	2	6		2	2
September	44	2	4·54%	2				
October	216	13	6·01%	4	4		1	4
November	32	8	25·00%	2	1			5
December	47	12	27·65%	5	3			4
Total	807	134 =	16·60%	44 5·45%	40 4·95%	5 0·61%	12 1·48%	33 4·08%

To prove the blood group antigens A, C, H and K in horses we prefer the lysis technique. The haemolysin titre of the isoimmune sera is much higher than the agglutinin titre in the same sera. For the lysis we incubate the tubes at 26—27°C after repeated shaking. The first reading is made after 30 minutes, the second after 60 minutes. In the tubes one drop of a 2% red cell suspension in saline is added to two drops of the monovalent reagent. After 15 minutes one drop of complement must be added. In the lysis it is also possible to use a lipaemic serum.

Pseudoagglutinations and cold agglutinations, which mostly disturb the reaction, are not present in the lysis.

It was striking that normal antibodies in the lysis are not detectable. In normal sera only agglutinins occur, which are directed against single red blood cell antigens and seldom against two antigens. The blood group antigen D, E, J and L are detectable only by agglutination.

When testing sera for specific antibodies against horse blood group antigens, we found some sera with unexpected reactions. In two of these sera, antibodies could be isolated for the detection of two till now unknown horse blood group antigens. One factor may be classified into the J-system, which thus contains two factors J_1 and J_2 as subgroups, while the second isolated blood group factor occurs independently of the previously known blood group systems and is called L.

For preservation of horse red blood cells we recommend the ACD-stabilizer with streptomycine. Horse red blood cells, which are preserved in the Rous Turner solution with formaline, are not suitable for the lysis. Rabbit serum is the complement, which is absorbed with horse red blood cells twice for 10 minutes at 4°C. After preservation in ACD-stabilizer, the horse red blood cells are washed in saline and then the sediment is added to the same volume of the rabbit serum and repeatedly shaken. To avoid lysis it is necessary to use a refrigerator-temperature and fresh red blood cells for absorption.

Using non-absorbed rabbit serum, all horse red blood cells will be lysed by all isoimmune and normal antibodies. Only with anti-A, the lysis was inhibited by the corresponding horse blood group antigen A. If other antibodies were present simultaneously, for example anti-C and anti-H, there was no reduction of the reaction with the corresponding antigens.

If more antibodies are present in the same immune serum, the antibody titres are, as a rule, different. Among the immunized horses there were several pregnant mares. Two of these mares did not form serum and colostrum antibodies. With a serum agglutinin titre anti-A 1/32, only one mare showed a colostrum agglutination titre anti-A 1/512 and a colostrum lysis titre anti-A 1/16,384. While after our observations the physiologic neonatal icterus in foals does not appear on the mucous membranes and the sclerae in one foal were clear yellow after suckling the colostrum with a high antibody titre 1/512 and 1/16,384. The foal was healthy, haemolytic icterus indicated an antigen-antibody reaction. Some days later, the icterus on the conjunctiva disappeared. This observation shows that a haemolytic icterus in foals can be benign also in the case of a high colostrum titre after isoimmunization of pregnant mares. When immunizing a horse with the blood type CD, using blood of another horse with blood type A, C, E, F, I, K, we found a cattle-specific heterohaemolysin anti-Z with the titre 1/4 in addition to a strong horse-specific anti-A, which was isolated by absorption. This horse immune serum was inactivated and unabsorbed. Unabsorbed rabbit serum was the source of complement. The specificity was proved by absorption with Z-positive cattle red blood cells. In the colostrum of this mare, heterohaemolysin was not detectable. After absorption of immune serum with horse blood containing the horse blood group antigens A, C, D, E, F, H, J, and K, it was proved that no antigenic relationship existed between these blood group factors and the blood group factor, which reacted with the described heterohaemolysin.

We found heteroagglutinins against blood group antigens of horses in normal sera of ducks (n = 90, 47 anti-A + C + D, 43 anti-A + C) and turkeys (n = 33, 33 anti-C). These results are proved by absorptions.

When testing pig red blood cells with 9 different monovalent horse reagents anti-A, all pig red cells were agglutinated by 5 of these isoimmune sera. Four anti-A sera and all anti-C, anti-E and anti-H sera showed negative results.

In the investigation of normal antibodies in sera of 807 German, Yugoslavian and Hungarian horses we found normal antibodies in 16·6% of cases. Gilman, Schwarz and Wallerstein (1960) could demonstrate only 6·1% of normal antibodies in 341 horse sera. Woyciechowska and Lille-Szyszkovicz (1960) tested 175 horse sera with the result of 20·4%. Podliachouk, Kaczmarek and Zwolinski (1962) found normal antibodies in 28·2% of 188 horse sera.

Unlike in cattle and humans, no significant seasonal variation in the occurrence of normal antibodies was found in horses. We observed the highest frequency in January and May and the lowest frequency in September, followed by a continuous increase in frequency up to January.

Levine and coworkers (1955) demonstrated a N-similar blood group antigen in horse red blood cells with Vicia graminea extracts. According to the rule of Landsteiner, anti-M was also found in the serum of N-positive horses (Levine, Celano, Lange and Berliner 1957). In cooperation with Uhlenbruck, who isolated a mucoid from the stroma of the horse red blood cells, which reacted with the N-blood-group specific extract from Vicia graminea (Uhlenbruck and Krüpe 1963) we studied in the last weeks whether this n-blood group active mucoid was active with the known horse blood group reagents in the inhibition test. The investigations were performed with the specific reagents anti-A, C and H in the lysis and anti-A, C, D, E, H, J_1, J_2, L and Mü-4 in the agglutination. Contrary to the mucoids from cattle red blood cells, which all were blood group active (Uhlenbruck and Schmid 1962) we found no blood group activity in the mucoid of horse red blood cells. The horse mucoid was obtained by phenol-sodium chloride extraction in the heat and in a following alcohol fractionation in the same way as the mucoid from cattle red blood cells. It is striking that the horse mucoid had no blood group activity. The problem of the position of horse red blood cell antigens remains therefore unsolved. The horse mucoid only seems to contain the heterophillic human antigen N_{Vg}. Uhlenbruck (1964) assumes that all red blood cells of horse and chimpanzee contain the N_{Vg}-receptor, which, in part genetically determined, is, as a cryptantigen, detectable only after RDE-treatment.

At the same time and independent of the transferrin-studies of Braend and Stormont (1962–1964), we have also initiated the study of serum transferrin polymorphism in horses (Schmid 1962). Comparison of the results shows that our type AA is identical with the type DD, the type BB with FF and the type CC with HH. In addition to the homozygous Tf-types, we observed the heterozygous combination types AB (DF), AC (DH) and BC (FH). The Tf-types CC, AC and BC also had one band in addition to the slow moving one. Until now, we did not know about this fact.

According to the fact that horse adult haemoglobin consists of two components we have found in every case, also in the newborn foals, two haemoglobin-bands in starch gel electrophoresis.

References

Braend, M. and C. Stormont (1962). VIII. European Conference of Animal Blood Groups Ljubljana.

Braend, M. and C. Stormont (1964). Nord Vet. Med. 16 : 31.

Fenton, J. W., Duggleby, C. R. and D. Kracht (1964). Immunogenetics Letter 3 : 116.

Franks, D. (1962). Ann. N. Y. Acad. Sciences 97 : 235.

Gilman, M. A., Schwarz, A. and H. Wallerstein (1960). Amer. J. Vet. Res. 82 : 393.

Levine, P., Celano, M. J., Lange, S. and Berliner V. (1957). Vox Sang. 2 : 433.

Levine, P., Ottensooser, F., Celano, M. J. and W. Pollitzer (1955). Amer. J. Physic. Anthropol. 13 : 29.

Podliachouk, L. and M. Hesselholt (1962). Immunogenetics Letter 2 : 69.

Podliachouk, L., Kaczmarek, A. and J. Zwolinski (1962). VIII. European Conference of Animal Blood Groups Ljubljana.

Podliachouk, L. and D. O. Schmid (1963). Ann. Inst. Pasteur 104 : 427.

Schmid, D. O. (1964). Z. Immun. forsch. 126 : 408.

Schmid, D. O. and L. Erhard (1962). Z. Immun. forsch. 124 : 149.

Uhlenbruck, G. and M. Krüpe (1963). Z. Immun. forsch. 124 : 342.

Uhlenbruck, G. and D. O. Schmid (1962). Z. Immun. forsch. 123 : 466.

Woyciechowska, S. and I. Lille-Szyszkovicz (1960). Roczn. Nauk Roln., Warszawa, 69 E-4 : 457.

References

SERUM PROTEINS IN EQUIDAE: SPECIES, RACE AND INDIVIDUAL DIFFERENCES

M. KAMINSKI

(With the technical assistance of N. Brunet)
Laboratory of Histophysiology, University of France, Paris

Previous studies of erythrocyte antigens and serum proteins of two Equidae species: horse and donkey, and of their hybrids: mule and hinny showed that

1. with respect to blood groups, horse and donkey erythrocytes bear different antigens, all of which can be detected on mule or hinny erythrocytes (Podliachouk 1957 et 1964). Races or breeds of horses can be differentiated by studying frequencies of a given antigen.

2. immunochemical comparison of several serum proteins failed to reveal antigenic differences between the two parental species, and all four systems appeared as immunologically identical (Podliachouk 1964).

3. comparative examination of serum esterases by starch gel electrophoresis permitted, on the contrary, a clear differentiation between horse and donkey sera; the latter is lacking the faster-migrating esterase (Kaminski 1964). Both hybrids have the same zymogram as the horse, which probably means that the presence of the fast esterase is a dominant character.

The confrontation of these results, provided by three different kinds of methods, both immunological and biochemical, shows the inheritance of specific parental characters by hybrids. No factor specific for the hybrid alone has been detected so far.

Continuing this investigation, the starch gel proteinograms were compared. Serum protein polymorphisms, mainly in the β-globulin i. e. transferrin region, were already described in horses (Ashton 1958, Schmid 1962, Braend 1964). The purpose of this investigation was to see whether the intraspecific variations were always smaller than the interspecies differences.

Three series of sera from horses of different races (samples provided by Mme Podliachouk), two series of donkey sera (from geographically distant populations in Ireland, sent by Dr. Dabczewski) and a series of mule sera from Pasteur Institute, provided by Mme Podliachouk, were examined.

Several buffer systems were tried and the best results were obtained with Tris-citrate buffer containing lithium hydroxide for the gel and borate-lithium buffer in electrode vessels. When 300 volts were applied to a 22 cm. long plate, the albumin spot migrates about 11 cm. in 6 hours. Heating was avoided by placing ice-containing bags above and under the plate. The gel was cut into three slices, one stained with Amido-Black, two others treated with different reagents, in order to localize esterases, haptoglobin, lipids etc.

The proteinograms of horses, donkeys and mules were directly compared, as 10 samples were run on the same plate.

Starch-gel proteinograms of horses

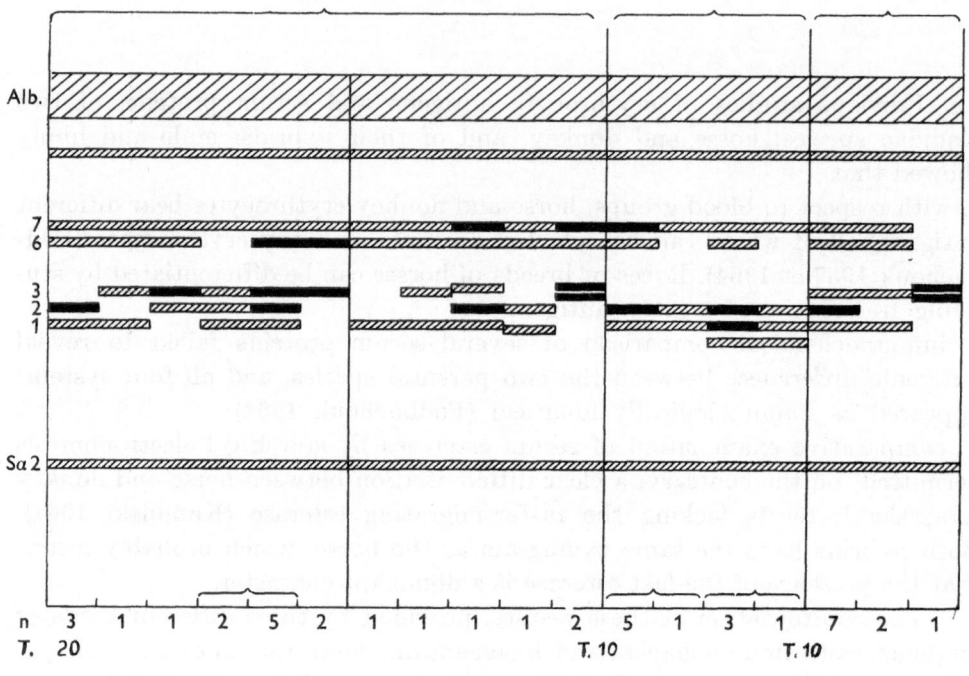

Fig. 1.

The diagram I and tables 1 and 2 show the results obtained with horse sera. Two bands, called 6 and 7, are generally present together; band 6 is absent from 4 sera; both vary in intensity. The main variable system of bands is designated 1 to 3. Bands 1 and 2 are found in all three groups of horses, but in some sera they are somewhat faster, and are then designated 1a and 2a. In the first group, two types of pattern are observed: one is composed of bands 1, 2 and 3; the other of 1 or 1a, 2a and 3. No correlation seems to exist between these bands, they are present or absent independently of each other; there are also variations in intensity. Bands 1 and 1a are always the less intensely stained.

The second group looks more homogeneous; its characteristic feature is the absence of band 3 from all analyzed sera. All sera, on the other hand, have bands 1 and 2. Some sera have also a slower band.

Table 1

Frequencies of individual bands

Species	Band	Total N° of sera		%
Horse	1	present in 34 sera out of 40		85
	2	30		75
	3	25 (absent from 2nd group)		
	"3a"	10 (3d group only)		
	6	36		90
	7	40		100
Donkey	3	38	45	85
	4	42		93
	5	26		58
	6	43		95
	8	18		44
	9	45		100
Mule	1	7	31	22
	2	20		64
	3	22		71
	4	23		74
	5	4		13
	6	26		84
	7	30		99
	8	4		13
	9	29		96

Table 2

Frequencies of main patterns

Species	Pattern				Total N° of sera		%
Horse	1, 2, 3		6, 7	present in 8 out of 20 (1st group)			
	1, 2, 3 or 3a		6, 7	7	10		
	1, 2		6, 7	9	10 (2nd group)		
	1, 2		6, 7	are common to 25	39		64
Donkey		3, 4, 5, 6	9		22	45	
		3, 4, 6	9		15		
		3, 4, 6	9		37	45	82
Mule		3, 4, 6, 7	9		9	31	
	1, 2, 3, 4,	6, 7	9		4		
	2, 3, 4,	6, 7	9		5		
		3, 4, 6, 7	9		18		58
		3, 4, 6,	9		19		61
	1, 2,	6, 7			6		19

Starch-gel proteinograms of donkeys

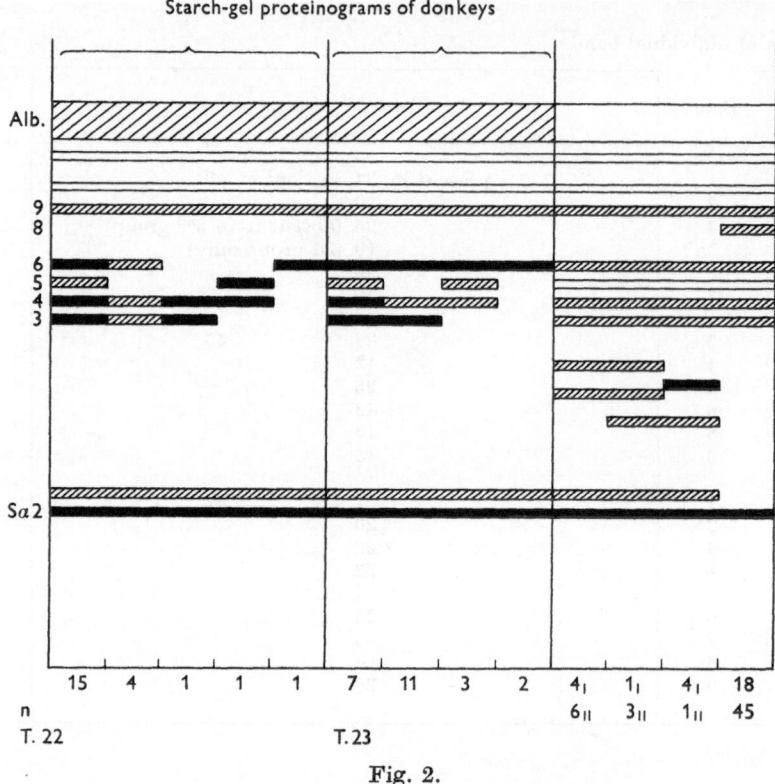

Fig. 2.

In the third group of sera, the band 3 is replaced by a large, fuzzy spot, looking sometimes as two adjacent bands. In addition, slower migrating bands are observed. The majority of sera have both bands 1 and 2.

Frequencies of individual bands and the most frequent patterns are listed in tables 1 and 2. Similar differences appear to exist between proteinograms of the three examined groups of horses. Nevertheless, bands 1, 2, 6 and 7 are present together in 64% of sera; these are thus considered as characteristic for horse sera.

Table 3

Bands				I group (22 sera)	II group (23 sera)
3	4	5	6	15	7
	4	5		1	0
3	4			1	0
			6	1	2
3	4		6	4	11
	4	5	6	0	3

Diagram 2 shows the results obtained with donkey sera. The bands are not the same; all sera contain band 9; the main variable system consists of bands 3, 4, 5, 6; in some sera, slower bands are present. The most frequently observed are bands 3, 4, 6 and 9, but in 8 haemolyzed sera bands 3 and 4 were

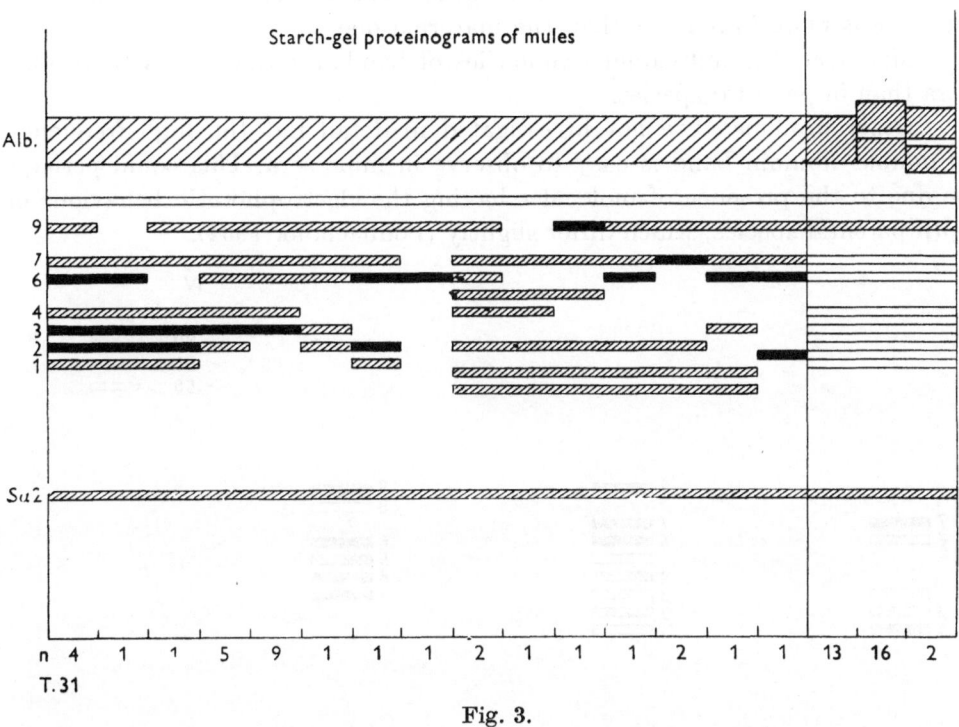

Starch-gel proteinograms of mules

n 4 1 1 5 9 1 1 1 2 1 1 1 2 1 1 13 16 2
T.31

Fig. 3.

indistinguishable from a large spot of haemoglobin; the pattern 3, 4, 6, 9 is a common base of 82% of sera. Thus, donkey sera seem to be characterized by the presence of bands 4, 9 and sometimes 5 and by the absence of bands 1, 2 and 7.

The two groups of donkey have the same bands, but they can be differentiated by their frequencies:

When compared with horses, the donkeys appear more homogeneous, but other races should be examined.

As shown on diagram 3, mule sera contain bands characteristic for horses together with those characteristic for donkeys (see diagram 4). Bands 1a and 2a of some horse sera were not, however, observed in mule sera.

The mule patterns are the most heterogeneous in comparison with horses and donkeys: 9 patterns out of 14 observed were each represented by only one serum. The most frequently observed pattern contains bands 3, 4, 6, 9 and is present in 58% of sera; the complete pattern is present in only 13%

of sera, while in horses 50% have bands 1, 2, 3, 6, 7 and in donkeys the bands 3, 4, 5, 6, 9 are also present in 50% of sera.

Bands 3, 4, 6, 9 simultaneously present in 82% of donkey sera are found in 61% of mule sera; bands 1, 2, 6, 7 present at the same time in 64% of horse sera, are found in only 19% of mule sera. It seems therefore that the paternal heritage is more important than the maternal one.

Moreover, the individual frequencies of bands are always lower in mule sera than in parental species.

Other proteins were not investigated in detail but the high frequency of double albumin band is easy to observe in mule sera. This could perhaps be due to the presence of molecules having the electrophoretic behaviour of both parental species, which differ slightly (Podliachouk 1964).

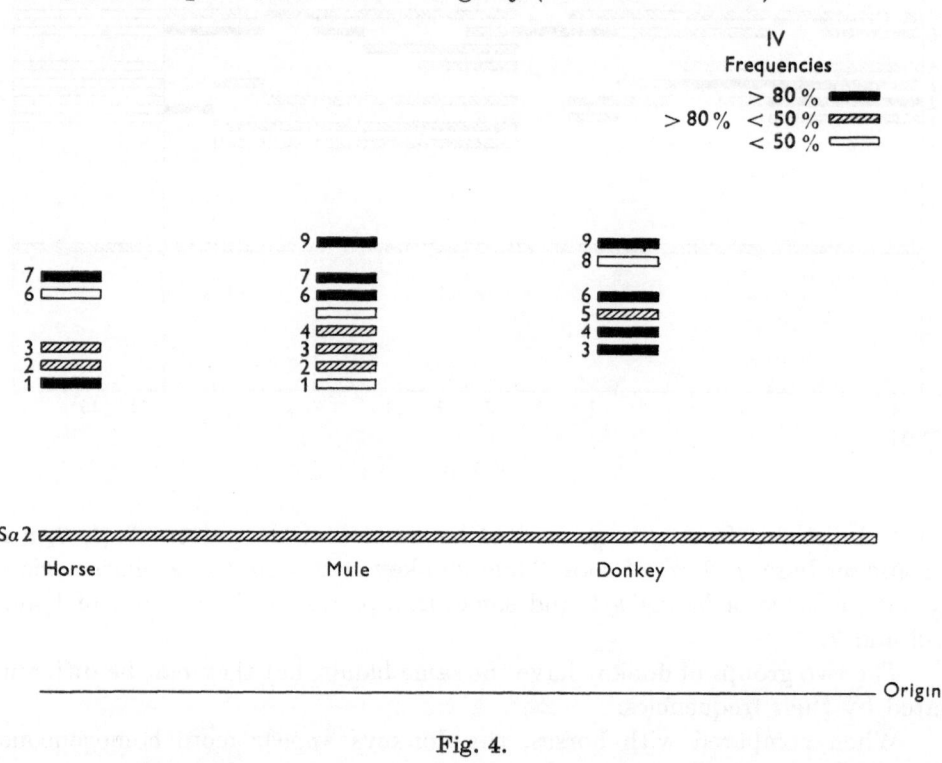

Fig. 4.

To summarize the reported results, horses and donkeys give different patterns of protein bands, mainly in the β-globulin region; all bands from parental species can be observed in the hybrid sera, but in lower frequencies. No component specific for the hybrid was detected.

In horse and donkey sera, differences between races or populations were observed; these differences, however, are less important than those between the two species.

References

Podliachouk, L. (1957). Thèse doct. ès-sci. Paris.

Podliachouk, L. et Kaminski, M. (1964). Etudes immunologiques sur les Equidés. Ann. Inst. Pasteur 106 : 497.

Kaminski, M. and Gajos, E. (1964). Comparative examination of Carboxylic Esterases in Sera of Horse, Donkey and their Hybrids, Nature, G. B., 201 : 716.

Ashton, G. C. (1958). Serum Protein Variations in Horses, Nature 182 : 1029.

Schmid, D. O. (1962). Über den Serumtransferrin-polymorphismus bei Pferden, Zt. Immun. exper. Therap. 124 : 219.

Braend, M. and Stormont, C. (1964). Studies on Hemoglobin and Transferrin Types of Horses (8th European Conference for Animal Blood Group Research Ljubljana 1962) Nord. Vet. — Mad. 16 : 31.

HAEMOGLOBINS, HAPTOGLOBINS AND ALBUMINS OF HORSES*

M. BRAEND and G. EFREMOV**

Department of Medicine, The Veterinary College of Norway, Oslo

Horse haemoglobins have been studied by Cabannes and Serain (1955), Bangham and Lehmann (1958), Efremov (1963) and Braend and Stormont (1964). Cabannes and Serain reported the findings of two phenotypes; one composed of two components was observed in 45 horses and the other composed of one component was observed in three horses. All the other investigators reported one phenotype only, namely that having two components.

Haptoglobin of horses was studied by Braend and Stormont (1964) who found 30 horses all to have the same haptoglobin (Hp) type.

Polymorphism of albumin was found independently by Stormont and Suzuki (1963) and by Braend (1964). Three phenotypes were observed which could be explained by two alleles.

The present report deals with further studies on haemoglobins, haptoglobin and albumins of horses. The primary intentions were to search for variation in haemoglobin and haptoglobin besides genetic studies of albumins.

Materials and Methods

Haemoglobins were studied in fresh blood samples from 147 adult horses and two newborn foals. Of these, 78 plus the two foals belonged to the Döle breed. Thirty two were Fjordings as 37 were thoroughbreds. Haemoglobin was prepared by mixing distilled water and red blood cells washed three times in saline. The haemolysates contained 1 to 2 g. Hb per 100 ml.

A total of 114 horses were investigated as to haptoglobin type. Of these 53 were Döle horses, 41 were Fjordings and 20 were of foreign origin (thoroughbreds, ponies and Belgians). Haptoglobin solution was prepared by mixing one drop of Hb solution with 4 drops of serum or plasma. The Hb solution was

*) These investigations were supported in part by a research grant from The Agricultural Research Council of Norway.

**) On leave from Department of Physiology, University of Skoplje, Yugoslavia.

made from one part washed and packed erythrocytes and 9 parts distilled water.

Samples from a total of 290 horses were investigated for albumin variation. Our two native breeds Döle and Fjording were represented by 130 and 88 samples, respectively. In addition, 12 horses of foreign origin, mainly thoroughbreds and Belgians were investigated.

The technique was horizontal starch gel electrophoresis. Starch prepared by hydrolysis of commercial Norwegian potato starch was used. The trays for gels were made from plastic plates and plastic lists. The inside dimensions were 20 cm $\times$ 13 cm. $\times$ 0·4 cm., for the albumins. For haemoglobins varying sizes were used. Insertions were made on filterpaper Whatman No 3 and 4 cm. from the bridge. The buffer solutions used for haemoglobins were based upon the continuous system described by Gahne et al. (1960). For haptoglobin and albumins the buffers of Poulik's (1957) discontinuous system were used. Stock solutions were prepared as described by Kristjansson (1963), A having 10·5 citric acid per litre and B with 23 g. Sigma 7—9 per litre. The bridge buffer was also as reported by Kristjansson. Gels for haptoglobin were made from 29 g. starch, 10 ml. A, 50 ml. B and 190 ml. distilled water, pH. 8·9. Voltage output of power supply was 220 volts. The electrophoresis was stopped when the front edge of free Hb had travelled 4 cm. Staining of haemoglobin and haptoglobin was done with benzidine. For albumins the gels contained 20 g. starch, 15 ml. A, 10 ml. B and 175 ml. dist. water, pH 5·9. The starting voltage was 200 volts. Increase to 300 was made when the boundary line was 2 cm. before start line. Two cm. after start line voltage was increased to 400. Electrophoresis was ended when boundary line had moved 9 cm. in front of start line. Staining was done with amido-black. Destaining took place in Smithies (1955) acetic acid-methanol-water solution.

Results

A photograph showing variations in horse haemoglobin is presented in Fig. 1. The one band phenotype was found in 13 horses, but only in Döle. We have called this component A_1. The two band phenotype shows different ratios of the two components. Usually the proportions were approximately 70% A_1 and 30% A_2. In the two horses, whose Hb types are given in the photograph, the percentages A_1 of total Hb are approximately 80 and 60, respectively. We did not, however, attempt to divide the two component phenotypes in subclasses. Table 1 shows the distributions of the one and two component phenotypes in three breeds. The two newborn foals both showed the same phenotype, a type indistinguishable from the common two band Hb phenotype.

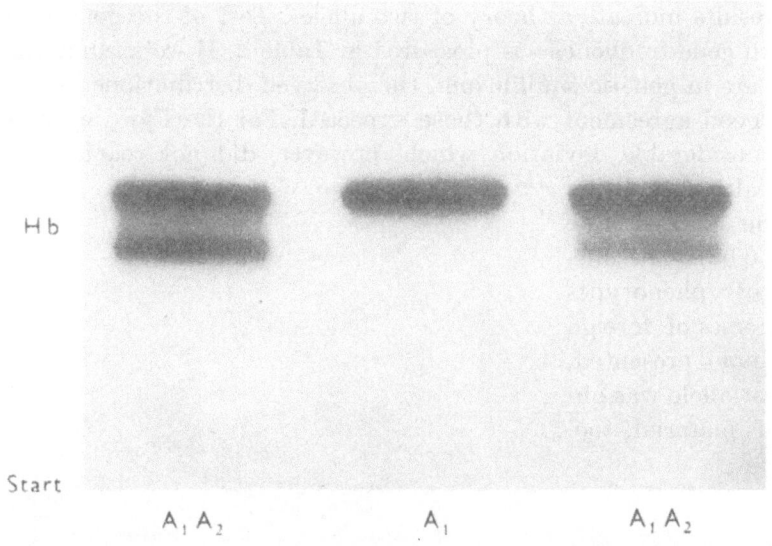

Fig. 1. Photograph of a gel stained with benzidine where there different Hb phenotypes are shown.

The haptoglobin studies revealed one type of Hp only. Usually this appeared as one band but in some gels a very faint band was seen between the major Hp band and free Hb. The positions of Hp bands relative to the free Hb components may be seen in Fig. 2.

Three albumin phenotypes were observed. A photograph showing these is presented in Fig. 3. These phenotypes are designated FF, FS and SS. The FF type has two major bands with a faint one in front. Of the two major bands, the fastest is usually the strongest. The FS type has four bands of about the same thickness and intensity of staining. A faint fifth band is indicated in the front. The migration rates of the two fast bands in the FS type are the same as those for the two major bands of the FF type. The bands of the SS type show a similar appearance as the FF bands, the only difference being rates of migration. The SS bands match the two slowest bands of the FS type.

Table 1

Distribution of haemoglobin phenotypes

Breed	Hb components	
	A_1	A_1A_2
Döle	13	65
Fjording	—	32
Thoroughbred	—	37

These results indicate a theory of two alleles. Test of the genetic theory together with gene frequencies is presented in Table 2. If we assume that the populations are in genetic equilibrium, the observed distributions in the Döle breed show good agreement with those expected. For the Fjording material, there was considerable deviation which, however, did not reach statistical significance although being near to the border line (0·1 > P > 0·05). The distributions of phenotypes among the horses of foreign origin are not presented, but the rarest allele was observed in this material, too.

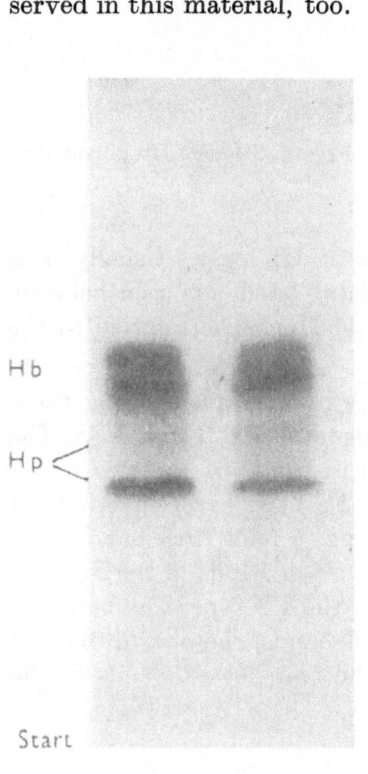

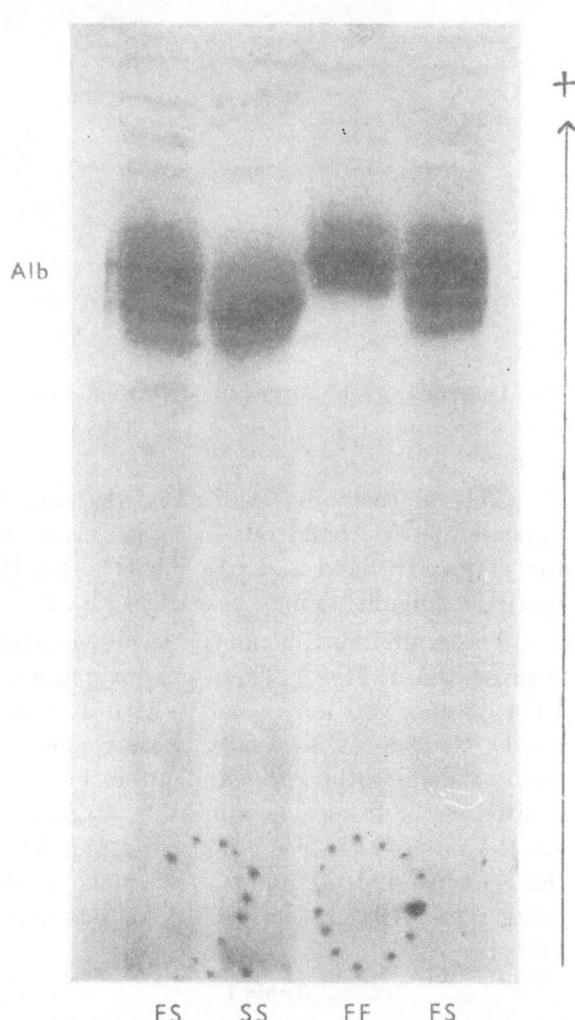

Fig. 2. Photograph of a stained gel showing position of Hp bands in two samples.

Fig. 3. Photograph showing the appearance of horse albumin phenotypes on stained starch gels.

Family data also confirmed the genetic theory. Table 3 shows the albumin phenotypes in the family of Bloc de Lootenhulle who was heterozygous in

256

Table 2

Gene frequencies and observed and expected distributions of albumin phenotypes in Norwegian horses

Breed		Phenotypes			Total	Gene frequency
		FF	FS	SS		
Döle	Obs	46	60	24	130	$Al^F = 0.58$
	Exp	44	64	22		$Al^S = 0.42$
Fjording	Obs	15	32	41	88	$Al^F = 0.35$
	Exp	11	40	37		$Al^S = 0.65$

Table 3

Albumin and transferrin phenotypes on the family of Bloc de Lootenhulle

Sire: Al type FS, Tf type FR

	Al type	Tf type		Al type	Tf type
Dam 1	SS	FH	Dam 7	FF	FF
Offs 1	SS	FH	Offs 7	FS	FR
Dam 2	FS	HR	Dam 8	SS	FH
Offs 2	FF	FR	Offs 8	SS	FF
Dam 3	SS	HR	Dam 9	SS	FH
Offs 3	FS	HR	Offs 9	SS	FR
Dam 4	FF	FO	Dam 10	SS	FH
Offs 4	FS	OR	Offs 10	FS	FR
Dam 5	FF	FO	Dam 11	FF	HH
Offs 5	FS	FR	Offs 11	FF	HR
Dam 6	FF	FO			
Offs 6	FS	FR			

both the albumin and transferrin systems. The albumin phenotypes of the offspring were all in agreement with those of the respective parents.

In this same family, test for independence of albumin and transferrin genes could be performed. The albumin genes segregated independently of transferrin genes. Only one mating resulted in independent segregation of Al^S and Ff^F alleles as five matings showed independence of Al^F and Tf^R alleles. It must, however, be realized that the family is small.

Discussion

Bangham and Lehmann (1958) proposed a theory of two closely linked genes for horse haemoglobins. This theory may still be correct. The differences in expression or lack of A_2 bands may partly be caused by environmental factors and a possible explanation is to assume the existence of regulator or inhibitor genes.

Whether any variation is going to be found for horse haptoglobin is too early to predict. However, realizing that we have selected horses belonging to many quite different breeds we are inclined to believe eventual new Hp genes to be rare.

The material used for studying albumin variation was obtained from many breeding districts and represented elite horses as well as common horses. In our opinion, it should be fairly representative.

The units of inheritance in the albumin system were named Al^F and Al^S where F stands for fast and S for slow. They were designated with the letters A and B in previous reports (Stormont and Suzuki 1963, Braend 1964). We prefer, however, a nomenclature where eventual new alleles can get logical symbols which might not be possible if they were named A and B.

When testing the genetic theory, the deviation from expected numbers in the Fjording material was rather large although not reaching statistical significance. We consider this deviation to be caused by a too small sample although superiority of the heterozygote might be a possible explanation.

Summary

Horse haemoglobin, haptoglobin and albumin were studied by means of starch gel electrophoresis. Variation of horse haemoglobin was found for Döle horses where 13 of 78 showed one Hb component only. Fjordings and thoroughbreds all had two Hb components.

No variation was found for haptoglobin, although 114 horses representing native Norwegian breeds as well as horses of foreign origin, were investigated. Haptoglobin appeared as one band with a very faint one in front, both travelling slower than free Hb on starch gel.

Three albumin phenotypes were observed. Genetic studies confirmed the existence of two codominant autosomal alleles which were called Al^F and Al^S.

Each allele controls two major and a weak third band on starch gels. The heterozygous type shows four major bands, and a fifth faint one. Gene frequencies were Al^F = 0·58 in the Döle breed and Al^F = 0·35 in the Fjording breed. In the family of a heterozygous stallion the albumin genes segregated independently of the transferrin genes.

References

Bangham, A. D. and Lehman, H. (1958) Nature 181 : 267.

Braend, M. (1964). Nord. Vet-Med. 16 : 363.

Braend, M. and Stormont, C. (1964). 16 : 31.

Cabannes, R. and Serain, C. R. (1955). Soc. Biol., Paris, 149 : 1193.

Efremov, G. (1963). Thesis. University of Beograd.

Gahne, B., Rendel, J. and Venge, O. (1960). Nature 186 : 907.

Kristjansson, F. K. (1963). Genetics 48 : 1059.

Poulik, M. D. (1957). Nature 180 : 1477.

Smithies, O. (1955). Biochem. J. 61 : 629.

Stormont, C. and Suzuki, Y. (1963). Proc. Soc. Exp. Biol. Med. 114 : 673.

DISCUSSION

J. G. Hall: I think Dr. Schmid says his standard haemolytic test lasts 60 minutes. May I ask if this is experimentally determined as the optimum time. I would have thought that leaving the red cells, antibody and complement together for several hours would give a more sensitive test. Because this is true for haemolytic systems in general.

D. Schmid: We have found 50—60 minutes to be the optimum incubation time for horse red blood cells in the immune haemolytic test. Later you may get unspecific reactions.

C. Stormont: It is true, as Dr. Schmid has said, that it is necessary to complete the readings within a comparatively short period of time. The reason for this, as discussed in a recent paper (Stomont and Suzuki, 1964. Serology of horse blood groups. Cornell Vet. 54 : 439—452), is that rabbit serum contains one unidentified lytic component which is not readily absorbed at 0° to 4°C by horse red cells. This unidentified lytic component seems to act on all horse red cells about equally and the lysis begins to make its appearence around 60 minutes. Therefore it is necessary to complete the recording of the specific lytic reactions usually by 60 minutes. I felt somewhat obliged to assist Dr. Schmid in commenting on this question since he learned from us about the method of using absorbed rabbit serum as complement in horse blood-typing tests. But in that connection I should also make it clear that we do not use rabbit complement with all of our horse blood-typing reagents. Some are used as straight saline agglutinins and some are used as haemolysins employing guinea pig rather than rabbit complement. The tests are set up in three sections.

M. Hesselholt: I only want to call the attention to the haemolytic disease of the newborn foal. Apparently it is not any problem in Denmark. We have not diagnosed this disease yet. I should like to know the incidence of this disease in the U. S. A., France and Germany. Further I should like to know which antigenic factors are involved in the intravascular haemolysis?

L. Podliachouk: In France 2 cases of haemolytic disease of foals were observed. Among 6 sera, sent by Bruner — U. S. A., however, 4 were anti-A and showed an immune character.

D. Schmid: Haemolytic disease in foals represents no problem in Germany.

C. Stormont: On theoretical grounds one might expect a higher incidence of haemolytic disease in Shetland Ponies than in Thoroughbreds because of the much greater heterogeneity in blood types of Shetland Ponies. Yet, we are not aware of a single case

261

in that breed and have seen no indirect evidence (occurrence of haemolysins in Shetland Pony mares) for it in tests on the bloods of many Shetland Ponies. On the other hand, we have seen indirect evidence for this disease in typing Thoroughbreds and Arabians. Since the Thoroughbred horse is derived largely from Arabian stock, it would appear that Arabian horses are the progenitors of the disease.

J. Bouw: Answer to Hesselholt about Schmid's paper. In the Netherlands we observed a number of cases of haemolytic disease. They all occurred in Racing horses and in these cases the disease was due to matings with imported American stallions.

C. Stormont: I merely remind everyone that there were no horses in America when America was discovered.

H. Buschmann: Referring to Dr. Schmid's lecture, where he told about testing pig blood samples with horse sera, I want to report of my experience in testing pig blood samples with cattle reagents. Thereby it appeared that several internationally controlled cattle reagents of different specificity gave reactions with A-positive pig bloods. How can this fact be explained?

S. N. Naik: I have two questions to Dr Braend.
1. Whether the slow moving fraction is different in mobility from that of the foetal haemoglobin in horse?
2. May I know whether the heterozygous and homozygous haemoglobins in horse behave identically when subjected to alkali-denaturation technique?

M. Braend: We have investigated haemoglobin in two newborn foals. No difference was found between foetal and adult haemoglobin.
We have not done any alkali denaturation of the haemoglobin of the two foals. However, this has previously been done by Efremov for horses as well as many other domestic animals.

G. Efremov: Investigations which were done by myself some years ago did not show any differences between haemoglobin of adult horse and haemoglobin of newborn foal. Electrophoretic migration rate of alkali denaturation and amino acid composition were the same in the haemoglobin of newborn foal and adult horse. The same was found in other species as pig and rabbit where electrophoretic and other investigations did not show any differences between haemoglobin of newborn and adult animals. It is most probable that in these species there is only one type of haemoglobin which is synthetized during the whole life.

M. Hesselholt: Dr. Braend, are the albumins in the horse described by you and those described by Stormont identical?

M. Braend: We use a different technique. In that connection I would like to mention that the product of the albumin allele with our technique is as follows: a strong one in the middle, a weaker slower and a faint more or less indistinct in front. The slowest of these three might correspond to that appearing as a postalbumin with Stormont technique.

M. Nikolajczuk: May I ask Dr. Eyquem to inform us about his horse haemoglobin studies.

A. Eyquem: Studies of the horse haemoglobin, made by the paper electrophoresis techniques in the years 1960—1962 in the Pasteur Institute in Paris, made it possible to prove, in the majority of horses, a haemoglobin, comprising two components, of which the more rapid one is dominant. In three rare cases we found one unique rapid fraction.

In the donkeys we have detected only one haemoglobin, the migration of which is similar to the migration of the slow horse fraction, 2 fractions exist in the mules of which the slower one dominated.

By means of a passage on the amberlite column I RC 50 XE 97, we obtain 4 or 5 bands in mules and 3 in horses, which possess 2 electrophoretic fractions.

The occurrence of the genetic relationships among certain components of the haemoglobins in horses and mules justifies a deeper study by means of a more sensitive technique.

We didn't observe electrophoretic differences between the foetal haemoglobin of foals and haemoglobin of adult horses.

SERUM PROTEIN POLYMORPHISM IN MAN, CATTLE, SHEEP, GOATS, PIGS AND CANIDAE

SERUM PROTEIN POLYMORPHISM IN MAN AND OTHER PRIMATES

M. HARBOE

University of Oslo, Institute for Experimental Medicial Research, Ullevaal Hospital, Oslo

The topic of serum protein polymorphism in man and other primates is so vast that any review has to be confined to certain aspects of it. The present review is only concerned with protein polymorphism occurring in healthy individuals, and it will be restricted to some of the main serum proteins that are easily demonstrated by physicochemical and immunological techniques. Enzyme polymorphism will not be included. Two main aspects will be considered: a) proteins where studies of genetically determined variations have given valuable information on general questions concerning genetic control of protein synthesis, and b) polymorphisms that have been reasonably well characterized in man, whereas considerably less is known in the various animal species. These systems appear to be valuable objects for further studies in animals.

The haptoglobins

In 1955, Smithies reported that electrophoresis of serum in starch gel separates the protein components much more effectively than other electrophoretic methods because the "sieving" effect of the gel is added to differentiation due to electrical charge. This quickly led him to the discovery of the inherited haptoglobin types (Smithies and Walker 1956). Three different phenotypes were observed which were called Hp 1—1, Hp 2—1 and Hp 2—2, and a simple hypothesis involving two co-dominant autosomal genes Hp^1 and Hp^2 was suggested. In general, this hypothesis has been well substantiated by extensive family studies.

However, in rare instances aberrant phenotypes were observed, as Hp 2—1 modified, Hp Johnson, and Hp Carlsberg. It was also evident that mixture of a Hp 1—1 serum and a Hp 2—2 serum in vitro produced an electrophoretic pattern which differed from that of a Hp 2—1 serum. The genetic basis of the haptoglobin types thus appeared more complex than originally presumed.

By reduction of purified haptoglobin with mercaptoethanol in the presence of urea, two types of polypeptide chains were obtained which could be separated by electrophoretic procedures Smithies, Connell and Dixon (1962,) Connell, Dixon and Smithies (1962). The so-called beta chain was apparently unaffected by the genes at the Hp locus, whereas the alpha chain showed distinct electrophoretic properties in different individuals. Haptoglobins separated from individual persons of type Hp 1—1 and subjected to reductive cleavage in the presence of urea, gave distinct patterns upon starch gel electrophoresis. Some gave a fast moving, single band called hp 1F; some gave a slow moving band called hp 1S, and some gave two bands with slightly different mobility. It was thus evident that the gene Hp¹ was not a unique entity, but could have one of two closely similar forms called Hp¹F and Hp¹S. After reduction in the presence of urea, a mixture of Hp 1F—1F and Hp 2—2 proteins gave an electrophoretic pattern identical with that of Hp 2—1F protein treated in the same way.

Studies of purified haptoglobins from individuals with various phenotypes by this technique have indicated that there are at least five alternative alleles at one locus: Hp¹F, Hp¹S, Hp², Hp²J and Hp²M. Further subdivisions of Hp² have been described by Nance and Smithies (1963). Studies on Hp Carlsberg by this technique have apparently not yet been published. Phenotype patterns of the various isolated haptoglobins in intact and reduced form are given by Giblett and Brooks (1963).

In a classic paper, Smithies, Connell and Dixon (1962) reported their biochemical investigations on isolated alpha chains from haptoglobins of various types.

hp1F alpha and hp1S alpha chains were isolated, digested with alpha-chymotrypsin, and subjected to fingerprinting. The resulting peptides in the two substances were indistinguishable, except for the replacement of one peptide (F) in hpF alpha by another peptide (S) in phS alpha. Amino acid analyses suggested the replacement of a single lysine residue in hp1F alpha by an acidic amino acid (or its amide) in hpS alpha. The difference between these two polypeptide chains thus appear to be caused by a "point" mutation, and the difference in the corresponding DNA chain may be a single base change.

The molecular weight of the hp1F and hp1S alpha chains is identical (8,860 ± 400), whereas that of the hp2 alpha chain is distinctly higher (17,300 ± 1,400). All peptides in hp1F alpha + hp1S alpha are present in hp2 alpha but with small differences in the fingerprint pattern wich are easily explained if almost the whole of the amino acid sequence of the two hp1 alpha polypeptides occurs in a single molecule of ph2 alpha. These biochemical findings strongly indicate that gene Hp² resulted from non-homologous crossing-over in a heterozygous individual of type Hp¹F/Hp¹S.

Several consequences of the intragenic duplication probably involved in gene Hp² can be predicted by considering other systems involving duplications. The "Bar" allele in *Drosophila* is associated with a cytologically demonstrable duplication. Homozygotes for the "Bar" allele are relatively frequently involved in unequal but homologous crossing-over which give rise, either to the formation of "double-bar" associated with cytological triplication, or to a reversion to "wild type" which lacks the original "Bar" duplication. Homozygous individuals of genotype Hp²/Hp² are in many ways analogous to homozygotes for the "Bar" allele. Unequal crossing-over would therefore be expected to occur occasionally in such individuals, either with the production of a triplicated haptoglobin gene, or a reversion to the "wild type" genes Hp¹F or Hp¹S. Biochemical studies of alpha polypeptides obtained from isolated Hp-Johnson indicate that this polypeptide is considerably larger than ph2 alpha. It was therefore suggested that gene Hp²J could be the triplication formed by unequal but homologous crossing-over in a homozygous Hp²/Hp² individual during displaced synapsis. This view gets some support from the finding of Hp Johnson in widely separated ethnic groups. It seems improbable that they have a common ancestor. Yet they appear to be identical as judged from the phenotypes. The cause of their identity could be a common mechanism of formation rather than descent from a common ancestor.

The haptoglobin serum group system thus provides examples of the effect on the structure of a single polypeptide chain of both a "point" mutation and chromosomal rearrangements. Their study is of great importance for the understanding of protein polymorphisms in general, and the mechanisms involved are probably of great importance in the evolution of protein structure.

The transferrins

Variation in electrophoretic mobility of beta-globulins in human sera is also demonstrable by starch gel electrophoresis. Identification of these proteins as transferrins may be obtained by addition of ^{59}Fe prior to electrophoresis. One half of the gel is stained with a protein stain, and compared with the autoradiograph of the other half of the gel (Giblett, Hickman and Smithies 1959).

Genetically determined variations in human transferrins have recently been reviewed by Giblett (1962). Fourteen different molecular species of human transferrins have been identified. Of these, transferrin C is frequently occurring, and most individuals are transferrin C homozygous. The other types are rare. In the families studied, the majority of the pedigrees contain only two transferrins; one being a variant, and the other being the common transferrin C. The fact that no human serum has been conclusively shown to contain three transferrins provides some indirect evidence that the 14 transferrins represent the products of a series of 14 alternative dominant genes at a single locus.

Transferrin polymorphism has been observed in man and several other primates and in a wide variety of animal species (see Giblett 1962). In many instances, the transferrin alleles are associated with more than one protein band. In each homozygous type of cattle there are four detectable zones; the two with the slowest mobility being most heavily concentrated, and the fastest being very faint. These multiple bands might be due to polymer formation, a situation which would be analogous to the haptoglobin polymers found in man, but other possibilities exist. It appears that careful biochemical comparison between human transferrin, where one gene corresponds to one protein band, and transferrin from homozygous individuals of other species that give multiple bands might give important information about the chemical basis of these differences. Such information might be of great interest in relation to general questions concerning genetic control of protein synthesis.

The group specific component

The group specific component was discovered by Hirschfeld in 1958 by means of a modified immunoelectrophoretic technique later described in detail (Hirschfeld 1959). The distinction between the different types within this system, the so-called Gc types, is based on the appearance in the immuno-electrophoretic slides of an alpha$_2$-globulin. By Hirschfeld's technique, Gc type 1—1 appears as a single fast moving component, and type 2—2 as a single slow mowing component. Gc type 2—1 appears as a long, flattened or double-peaked arc which is identical with the picture given by a mixture of equal amounts of sera of the two other types.

Hirschfeld, Jonsson and Rasmuson (1960) showed that the Gc types were genetically determined. Their family analysis indicated that the two genes, Gc^1 and Gc^2, were co-dominant autosomal alleles. The fast and slow moving components represent the two homozygous Gc types, and the 2—1 type is found in heterozygous individuals. Extensive genetic studies from a number of laboratories have verified this hypothesis. At least 1,013 families with 2,345 children and 4,479 mother-child combinations have been studied, and no exceptions from the postulated rules of inheritance have been demonstrated (see Reinskou 1964a).

Of 2,549 Norwegians, 1,371 were of type Gc 1—1, 1,005 of type Gc 2—1 and 173 of type Gc 2—2 giving the gene frequencies 0,735 for gene Gc^1 and

0·265 for gene Gc² (Reinskou 1964a). Considerable variation has been observed with regard to gene frequencies in different populations (Cleve and Bearn 1962, Hirschfeld 1962).

Technically, the reproducibility of Gc typing is excellent provided that good antisera are available. Antisera well suited for demonstration of the Gc types can be obtained without too much difficulty by immunization of rabbits with pooled human serum. For optimal results, antisera should be selected that give a strong Gc line and few other lines with proteins in the alpha region. They ought, however, to contain antibodies against albumin and alpha₂-macroglobulin which are good "landmarks" in the evaluation of the slides.

More detailed studies of the group specific component would be facilitated by introduction of other techniques. Schultze, Biel, Haupt and Heide (1962a) used a vertical starch gel electrophoretic procedure and succeeded in resolving the postalbumins of normal human sera into three clearly separated bands. Their interpretation of these findings was that the product of the Gc¹ gene corresponded to the second, and that of the Gc² gene to the third band. Cleve and Bearn (1962) showed by a similar technique that isolated Gc proteins of types 1−1 and 2−2 gave sharp bands with clearly distinct mobility within the postalbumin region. Other proteins do, however, give bands in the post-albumin region and may contribute to the individual variations originally observed by Smithies (1959). At the present time, Gc typing by immuno-electrophoresis is far superior to starch gel electrophoresis. The two methods may, however, supplement each other in the study of variants and isolated proteins.

An important finding was made by Reinskou (1963a) using a further modification of the immunoelectrophoretic technique. By prolonged separation time, he could clearly demonstrate a heterogeneity of the fast moving Gc band. With the conventional technique, Gc 1−1 sera show a single, fast moving arc. By the modified technique, sera of this type produce a double peaked precipitate. Sera of type Gc 2−1 show a triple peak precipitate, whereas the slow moving Gc 2−2 component appeared homogeneous by this technique. The heterogeneity of the Gc 1 precipitate was present in all of 35 sera of types Gc 1−1 and Gc 2−1 tested. According to Reinskou, the Gc 1 component should be considered to consist of two different proteins with slightly different electrophoretic mobility, that both are determined by the Gc¹ gene.

Rare variants in the Gc system have been described. Their elucidation has been facilitated by the use of Reinskou's modified technique (Reinskou 1963a, Reinskou 1964b).

In Norway, a mother-child combination was studied with aberrant Gc patterns. The mother had only the slow moving component of the Gc1 protein. Her child was of type Gc 1−1. This serum showed a double peaked Gc 1−1 precipitate, but had more of the slow moving component than usually observed (Reinskou 1964b).

Sera described as GcY by Hirschfeld (1962) and Gc Chippewa (Cleve and Bearn 1962, Cleve, Kirk, Parker, Bearn, Schacht, Kleinman and Horsfall 1963) were shown to contain only the fast moving part of the Gc 1 protein (Reinskou 1964b). The Gc Aborigine component (Cleve, Kirk, Parker, Bearn, Schacht, Kleinman and Horsfall 1963) was also found to be heterogeneous, both components moving slightly faster than the corresponding Gc 1 components. Finally, the Gc X protein (Hirschfeld 1962) was found to be homogeneous moving slightly slower than the slowest of the Gc 1 components.

The Gc protein has been purified independently by three groups of investigators, Cleve and Bearn (1962), Schultze, Biel, Haupt and Heide (1962b) and Reinskou (1963b). A method for quantitation of the protein has recently been described by Reinskou (1964c). This method appears to provide a good basis for the search for variation of Gc protein concentration in disease. This might give some clues concerning its function which is entirely unknown at the present time. The Gc protein is produced in the liver (Prunier, Bearn and Cleve 1964, van Furth 1964).

Variations in relative electrophoretic mobilities of alpha$_2$-globulins have been demonstrated by immunoelectrophoretic analyses of sera from other primates. In *Macaca irus*, a system of variants was observed, which appeared to be strikingly similar to the Gc system in man (Hirschfeld 1962). Experiments using various antisera indicated a close immunological relationship between the Gc protein of human serum and this protein in *Macaca irus* serum.

Brummerstedt-Hansen and Hirschfeld (1961) have observed similar alpha$_2$-globulin variations in swine. Variations in alpha$_2$-globulins, which may be an equivalent to the Gc protein in humans, have also been observed in rabbits (Hirschfeld 1959). It appears therefore that the Gc protein might be a promising object for comparative studies in various animal sera.

The beta-lipoproteins

In 1961, Allison and Blumberg described a serum (C de B) from a patient with refractory anaemia of unknown cause who had been given multiple transfusions. This serum gave a precipitate with some, but not all human sera by precipitin tests in gel. The active substance in serum C de B was identified as a gamma$_{ss}$-globulin and was considered to be an antibody which had been formed after stimulation with foreign serum proteins introduced by the transfusions. Sera which gave a precipitate were called Ag(a+), where

Ag is an abbreviation of "antigenic", and those which did not give a precipitate Ag(a—). Family studies indicated that the gene responsible for formation of positively reacting protein expressed itself in single and double dose. The antigen was presumed to be alpha$_2$-macroglobulin. It was later identified as low density beta-lipoprotein (Blumberg, Dray and Robinson 1962). More recently, Blumberg, Alter, Riddell and Erlandson (1964) studied eleven sera from polytransfused individuals that gave precipitatin lines with human beta-lipoproteins. They indicated that these antisera demonstrate "minimum five and possibly seven" different beta-lipoprotein antigens. The relation between these different antigens was only vaguely indicated, and Hirschfeld (1963) does not agree with their conclusion.

Hirschfeld (1963) found a serum from a polytransfused patient (L L) that gave a pricipitin line with some, but not all human sera. Its reactions were compared with those of serum C de B, the specificities in the two antisera being referred to as anti-Ag(a) in serum C de B, and anti-Ag(x) in serum L L. Three phenotypes were observed: Ag(a + x +), Ag(a + x —) and Ag(a—x—), whereas the fourth possible combination Ag(a—x+) was not observed. This is analogous to the findings in the Gm system. Family data showed that the property Ag(x) was genetically determined and that the responsible gene expressed itself in single and double dose. From these findings, it would be reasonable to assume that genes Ag^a and Ag^x were closely linked. In a series of experiments, Hirschfeld (1964) was able to show that serum C de B did not contain a single antibody but three distinct antibodies reacting with beta-lipoproteins. These were provisionally designated anti-Ag(a$_1$), anti-Ag(x) and anti-Ag(z). Serum L L appeared to contain only anti-Ag(x). Genetic analyses indicated that the Ag^{a1} and Ag^x genes are present at two distinct loci, and that the apparent relationship in phenotypes is a serological artefact because of multiple specificities of the antibodies present in serum C de B.

Berg (1963) applied another principle to the study of human beta-lipoproteins. He immunized rabbits with isolated beta-lipoprotein from single individuals hoping that antibodies might be formed against a) antigenic determinants present on all human beta-lipoproteins and b) antigenic determinants present only on some individuals' beta-lipoproteins. The antisera were subsequently absorbed with beta-lipoprotein from various individuals in an effort to remove antibodies of type a) and to leave antibodies of type b). By this technique, antisera were obtained that gave a good precipitin line with some human sera, but not all. Family studies showed that the beta-lipoprotein type which manifests itself by giving a precipitate with his antisera, is genetically determined. Positive reactors were called Lp(a+), where Lp is an abbreviation of lipoprotein. Negative reactors were called Lp(a—). The genetic studies showed that the responsible gene, Lp^a, was able to express itself in single and double dose (Berg 1963). The property Lp(a) is clearly distinct from Ag(a), and they are present on different beta-lipoprotein molecules in a given serum (Berg 1964).

Beta-lipoproteins thus show genetically determined, individual variations in humans. Similar findings have been made in monkeys. This protein is apparently a good antigen, if administered in the form of blood-or plasma transfusions between individuals of the same species. Studies of polymorphisms in this protein appear highly promising in animal species where good and moderately good precipitating antibodies are formed. Berg's principle of heteroimmunization followed by absorption has proved valuable for detection of human beta-lipoprotein polymorphism and may be useful in other species, too.

The immunoglobulins

The immunoglobulins of human serum consist of three main proteins that possess antibody activity, namely gamma$_{ss}$-globulin, gamma$_1$-macroglobulin and gamma$_{1A}$-globulin. A related protein, Bence Jones protein, is excreted in large amounts in the urine of some patients with malignant proliferation of plasma cells in the disease multiple myeloma. Small amounts of similar protein are present in normal serum and urine. Recent investigations have clarified the structural relationship between these proteins: By reduction of gamma$_{ss}$-globulin with mercaptoethanol in the presence of urea (Edelman and Poulik 1961), or by reduction with mercaptoethanol at low pH (Fleischman, Pain and Porter 1962), two types of polypeptide chains can be isolated from this protein. These chains have been called L (light) and H (heavy) chains because of difference in molecular weight. Identical L chains are found in all four immunoglobulins. This similarity in chemical composition of L chains is the basis of the well known immunological similarity, or "cross-reaction", between these four proteins. The H chain of gamma$_{ss}$-globulin is characteristic of this molecule and gives the molecule its distinctive properties. Gamma$_1$-macroglobulin contains its characteristic H chain, and the third major type of H chain is found in gamma$_{1A}$-globulin. Bence Jones proteins consist of L chains only. (For references, see Fahey 1962.)

Our knowledge of genetic factors in immune globulins has been rapidly increasing since the discovery of the Gm(a) factor in 1956 (Grubb 1956, Grubb and Laurell 1956). Studies of these factors have contributed a great deal to our knowledge of gamma-globulin structure and may resolve many questions concerning the remarkable heterogeneity of these proteins. The topic is very complex at the present time, and only some informative aspects will be referred to. Recent reviews (Grubb 1961, Steinberg 1962, Harboe and

Osterland 1963) contain numerous references to various aspects of these factors.

The test system is somewhat odd and consists of human group O Rh positive red cells that are sensitized by selected incomplete anti-Rh antibodies. Such coated cells are agglutinated by selected sera from patients with rheumatoid arthritis or selected sera from other individuals. Some normal sera will inhibit this type of agglutination, and these are referred to as Gm(+). Other sera do not inhibit the agglutination and are called Gm(−). The specificity of the test system is determined by the combination of agglutinator and the anti-Rh antibody used to sensitize the red cells. By trial and error sets of reagents have been found, each consisting of an agglutinating serum and an incomplete anti-Rh antibody, that determine a number of factors whose genetic basis has been clarified through the study of large family materials.

The factors Gm(a) and Gm(b) are determined by genes which behave as classical co-dominant alleles. Gene Gm^x appears to be closely linked with Gm^a, and Gm^e (Ropartz, Rivat and Rousseau 1962) appears to be closely linked with Gm^b. The structure of the Gm(b) factor shows small variations in different populations which are only partly understood and highly complex. The Gm(c) factor, previously known as Gm-like, occurs almost exclusively in Negroes. It has been definitely demonstrated that the Gm(a), Gm(x) and Gm(b) determining sites are located on the H chain of $gamma_{ss}$-globulin, and these factors are only found on $gamma_{ss}$-globulin (Harboe, Osterland and Kunkel 1962a, Franklin, Fudenberg, Meltzer and Stanworth 1962, Cohen 1963).

The factors Inv(a) and Inv(b) appear to be determined by genes which also behave as classical alternative alleles. Population and family studies have indicated that these genes are present at another locus than the Gm genes previously referred to. It has been demonstrated that Inv factors are present, not only on $gamma_{ss}$-globulin, but on all four classes of immunoglobulins, and further that the determining sites are located on the L chains of $gamma_{ss}$-globulin which is known from other studies to be common to all these proteins (Harboe, Osterland and Kunkel 1962a, Franklin, Fudenberg, Meltzer and Stanworth 1962, Cohen 1963). The concept, one gene — one polypeptide chain, appears thus to be applicable to the structure of immune globulins.

Some data are however difficult to reconcile with this concept and present concepts of immunoglobulin structure. The Gm(p) factor (Waller, Hughes, Townsend, Franklin and Fudenberg 1963) appears to be caused by a gene which may be located at a locus distinct from the classical Gm locus. Still, its determining site has been demonstrated on the H chain of $gamma_{ss}$-globulin. Data from the study of isolated myeloma proteins and isolated myeloma proteins and isolated antibodies strongly indicate that the $gamma_{ss}$-

globulin molecules in a given individual are heterogeneous with respect to genetic factors. For instance, in a Gm(a−b+) individual which from family studies apparently is homozygous for gene Gm^b, only a small proportion of the gamma$_{ss}$-globulin molecules appear to contain the Gm(b) factor (Martensson 1961, Harboe, Osterland and Kunkel 1962b, Allen, Kunkel and Kabat 1964). If there is only one type of H chains in gamma$_{ss}$-globulin, and the concept one gene-one polypeptide chain is correct for this protein, one would certainly expect all molecules to contain the Gm(b) character in such individuals. Finally, recent data on the Gm(f) character (Grubb 1964) are very difficult to understand in the light of present concepts of gamma$_{ss}$-globulin structure. I am convinced that further studies of these genetic factors will provide most valuable information for a change of current concepts of gamma-globulin structure.

Gamma-globulin polymorphism has been demonstrated in various other species including monkeys, mice, guinea pigs and rabbits. In the latter species, at least six different factors are well characterized.

Reagents for determination of gamma-globulin types in rabbits are obtained by classical immunization procedures, e. g. by immunization of rabbits by isolated gamma-globulin from individual animals incorporated in Freund's adjuvant. In humans, the agglutinators found in non-rheumatoid subjects appear to be classical iso-antibodies being produced as a result of stimulation by foreign gamma-globulin introduced by blood transfusions (Allen and Kunkel 1963) or transplacental passage (Steinberg and Wilson 1963). Immunization procedures might therefore give good reagents for demonstration of gamma-globulin types in various other animal species. This might be of considerable genetic interest and might contribute to current studies of genetic control of synthesis of gamma-globulins and specific antibodies.

References

Allen, J. C. and Kunkel, H. G. (1963). Science 139 : 418—419.
Allen, J. C., Kunkel, H. G. and Kabat, E. A. (1964). J. exp. Med. 119 : 453—465.
Allison, A. C. and Blumberg, B. S. (1961). Lancet I : 634—637.
Berg, K. (1963). Acta path. microbiol. scand. 59 : 369—382.
Berg, K. (1964). Proc. Xth Congr. Int. Soc. Blood Transf., Stockholm.
Blumberg, B. S., Dray, S. and Robinson, J. C. (1962). Nature, Lond. 194 : 656—658.
Blumberg, B. S., Alter, H. J., Riddell, N. and Erlandson M. (1964). Vox Sang. 9 : 128 — 145.
Brummerstedt-Hansen, E. and Hirschfeld, J. (1961). Acta veter. scand. 2 : 317—322.
Cleve, H. and Bearn, A. G. (1962). Progr. Medical Genetics 2 : 64—82.

Cleve, H., Kirk, R. L., Parker, W. C., Bearn, A. G., Schacht, L. E., Kleinman, H. and Horsfall, W. R. (1963). Amer. J. hum. Genet. 15 : 368—379.

Cohen, S. (1963). Biochem. J. 89 : 334—341.

Connell, G. E., Dixon, G. H. and Smithies, O. (1962). Nature, Lond. 193: 505—506.

Edelman, G. M. and Poulik, M. D. (1961). J. exp. Med. 113 : 861—884.

Fahey, J. L. (1962). Advanc. Immunology 2 : 41—109.

Fleischman, J. B., Pain, R. H. and Porter, R. R. (1962). Arch. Biochem. Biophys. Suppl. 1 : 174—180.

Franklin, E. C., Fudenberg, H., Meltzer, M. and Stanworth, D. R. (1962). Proc. nat. Acad. Sci., Wash. 48 : 914—922.

Furth, R. van (1964). The formation of immunoglobulins by human tissues in vitro. Thesis. Leiden, The Netherlands, p 75.

Giblett, E. R. (1962). Progr. Medical Genetics 2 : 34—63.

Giblett, E. R., Hickman, C. G. and Smithies, O. (1959). Nature, Lond. 183 : 1589—1590.

Giblett, E. R. and Brooks, L. E. (1963). Nature, Lond. 197 : 576—578.

Grubb, R. (1956). Acta path. microbiol. scand. 39, 195—197.

Grubb, R. (1961). Arthr. Rheumatism 4 : 195—202.

Grubb, R. (1964). Ann. N. Y. Acad. Sci. In press.

Grubb, R. and Laurell, A.-B. (1956). Acta path. microbiol. scand. 39 : 390—398.

Harboe, M., Osterland, C. K. and Kunkel, H. G. (1962a). Science 136 : 979—980.

Harboe, M., Osterland, C. K. and Kunkel, H. G. (1962b). J. exp. Med. 116 : 719—738.

Harboe, M. and Osterland, C. K. (1963). IIIrd Int. Symp. Immunopathology, La Jolla, Cal. Schwabe, Basel, Switzerland, pp. 13—24.

Hirschfeld, J. (1960). Lecture, 1958 referred to in Nord. Med. 63 : 246.

Hirschfeld, J. (1959). Acta path. microbiol. scand. 46 : 224—238.

Hirschfeld, J. (1962). Progr. Allergy 6 : 155—186.

Hirschfeld, J. (1963). Science Tools, Stockholm 10 : 45—54.

Hirschfeld, J. (1964). Proc. 2nd Scand. Congr. Forensic Medicine, Oslo.

Hirschfeld, J., Jonsson, B. and Rasmuson, M. (1960). Nature, Lond. 185 : 931—932.

Martensson, L. (1961). Acta med. scand. 170 (suppl. 367): 87—93,

Nance, W. E. and Smithies, O. (1963). Nature, Lond. 198 : 869—870.
    1005—1007.

Prumier, J. H., Bearn, A. G. and Cleve, H. (1964) Proc. Soc. exp. Biol., N. Y. 115 : 1005—1007.

Reinskou, T. (1963a). Acta path. microbiol. scand. 59 : 526—532.

Reinskou, T. (1963b). IX Congr. Europ. Soc. Haemat., Lisboa 1963. Sangre, Barcel. in press.

Reinskou, T. (1964a). Proc. 2nd Scand. Congr. Forensic Medicine, Oslo.

Reinskou, T. (1964b). Proc. Xth Congr. Int. Soc. Blood Transf., Stockholm.

Reinskou, T. (1964c). Scand. J. Haemat., in press.

Ropartz, C., Rivat, L. et Rousseau, P.-Y. (1964). Proc. IXth Congr. Int. Soc. Blood Transf., Mexico City 1962. Karger, Basel/New York.

Schultze, H. E., Biel, H., Haupt, H. und Heide, K. (1962a). Naturwissenschaften 49: 16—17.

Schultze, H. E., Biel, H., Haupt, H. und Heide, K. (1962b). Naturwissenschaften 49: 108.

Smithies, O. (1955). Biochem. J.: 61 : 629.

Smithies, O. (1959). Biochem. J. 71 : 585—587.

Smithies, O. and Walker, N. F. (1956). Nature, Lond. 178 : 694.

Smithies, O., Connell, G. E. and Dixon, G. H. (1962). Amer. J. hum. Genet. 14 : 14—21.

Steinberg, A. G. (1962). Progr. Medical Genetics 2 : 1—33.
Steinberg, A. G. and Wilson, J. A. (1963). Science 140 : 303—304.
Waller, M., Hughes, R. D., Townsend, J. I., Franklin, E. C. and Fudenberg, H. (1963).
    Science 142 : 1321—1322.

STUDIES ON PROTEIN POLYMORPHISM IN PIGS, HORSES AND CATTLE

M. A. GRAETZER, M. HESSELHOLT, J. MOUSTGAARD and
M. THYMANN

Department of Physiology, Endocrinology and Blood Grouping,
The Royal Veterinary and Agricultural College, Copenhagen

A. Pigs

1. Haematin-binding proteins

By means of starch gel electrophoresis and benzidine stain of pig sera, to which haemoglobin was added prior to electrophoresis, 10 protein components were observed. According to convention the components were designated Hp 1—Hp 10. Grouped variations involving Hp 1, Hp 2 and Hp 3 in the individual sera were observed. Family investigations supported the hypothesis that the polymorphism exhibited was under simple genetic control and determined by 3 allelic genes (Kristjansson, 1961). During later Danish investigations a new component Hp 0 was found. Family investigations comprising pigs of Danish Landrace have supported the hypothesis that the synthesis of Hp 0, Hp 1, Hp 2 and Hp 3 is determined by 4 alleles: Hp^0, Hp^1, Hp^2, and Hp^3 with codominant mode of inheritance (Brummerstedt-Hansen et al. 1962 a, b; Hesselholt 1963a, b).

Until recently, the above mentioned proteins have been supposed to be haemoglobin-binding proteins, haptoglobins. Recent in vitro and in vivo experiments have shown that pH 5 was able to react with newly prepared haemoglobin. Hp 0, Hp 1, Hp 2 and Hp 3 did not combine with newly prepared haemoglobin, but reacted with haemoglobin, which has been stored, alkaline haematin and methaemoglobin (King, 1963; Imlah, 1963; Hesselholt, 1963b). Thus Hp 5 is the suilline haptoglobin. Until further characteristic are given, the other proteins will be called haematin-binding proteins.

The purpose of the present investigation was to determine the heredity of the haematin-binding proteins in pigs of Danish Landrace and further to elucidate the genetic structure of this breed with respect to the genes, which control the synthesis of these proteins.

Technique. Preparation of haemoglobin, addition of haemoglobin to the serum sample prior to electrophoresis, starch gel electrophoresis and staining of the gel was carried out as described in detail elsewhere (Hesselbolt, 1963a, b).

Table 1

Distribution of haematin-binding protein phenotypes from various mating classes in pigs of Danish Landrace

Mating classes	No. of matings	No. of animals in class	Hp 0-0 obs.	exp.	Hp 1-1 obs.	exp.	Hp 2-2 obs.	exp.	Hp 3-3 obs.	exp.	Hp 2-1 obs.	exp.	Hp 3-2 obs.	exp.	Hp 3-1 obs.	exp.	Hp 0-1 obs.	exp.	Hp 0-2 obs.	exp.	Hp 0-3 obs.	exp.	X²
3-3 × 3-3	10	47							47	47													
3-3 × 3-1	20	88							45	44					43	44							0·02
3-3 × 3-2	5	20							9	10			11	10									0·20
3-3 × 2-1	1	4											2	2	2	2							
3-3 × 2-2	1	4											4	4									
3-3 × 1-1	7	47													47	47							
3-3 × 0-1	1	4													3	2					1	2	
3-3 × 0-2	2	8											3	4							5	4	
3-3 × 0-3	4	16							9	8											7	8	
3-1 × 3-1	20	86			17	21·5			24	21·5					45	43							0·25
3-1 × 3-2	8	31							8	7·75	5	7·75	8	7·75	10	7·75							1·33
3-1 × 2-2	4	16									7	8	9	8									1·65
3-1 × 2-1	10	45			11	11·25					11	11·25	12	11·25	11	11·25							0·25
3-1 × 0-1	3	12			3	3									4	3	4	3			1	3	
3-1 × 0-2	1	4									0	1	2	1			1	1			1	1	
3-1 × 0-3	5	23							3	5·75					5	5·75	5	5·75			10	5·75	0·07
3-1 × 1-1	7	37			18	18·50									19	18·50							0·03
3-2 × 3-2	2	4					1	1	1	1			3	2									
3-2 × 2-1	2	8					2	2			2	2	2	2	2	2							
3-2 × 0-1	1	8																	3	1			
3-2 × 1-1	1	4									2	1			1	1							
2-2 × 2-1	1	4					2	2			2	2											
2-2 × 1-1	3	4									4	4											
2-1 × 2-1	3	12			2	1	4	3·25			6	6											
2-1 × 0-1	3	13			6	3·25					6	6·50					4	2	2	1			
2-1 × 0-3	1	4									2		2	1	0	1	1	1	1	1			
1-1 × 1-1	1	4			4	4																	
1-1 × 0-2	1	4															2	1					
0-1 × 0-3	1	4	1												1	1	1	1			3	1	
Total	127	569																					

Results. By means of starch gel electrophoresis of sera from 900 adult pigs, selected at random, and from a family material comprising 127 matings with 569 piglets, ten general types of pattern with respect to Hp 0, Hp 1, Hp 2 and Hp 3 were obtained. The ten phenotypes are shown in figure 1. According to the above four allele-hypothesis, the phenotypes shown in figure 1 thus represent animals with the genotypes shown in the figure.

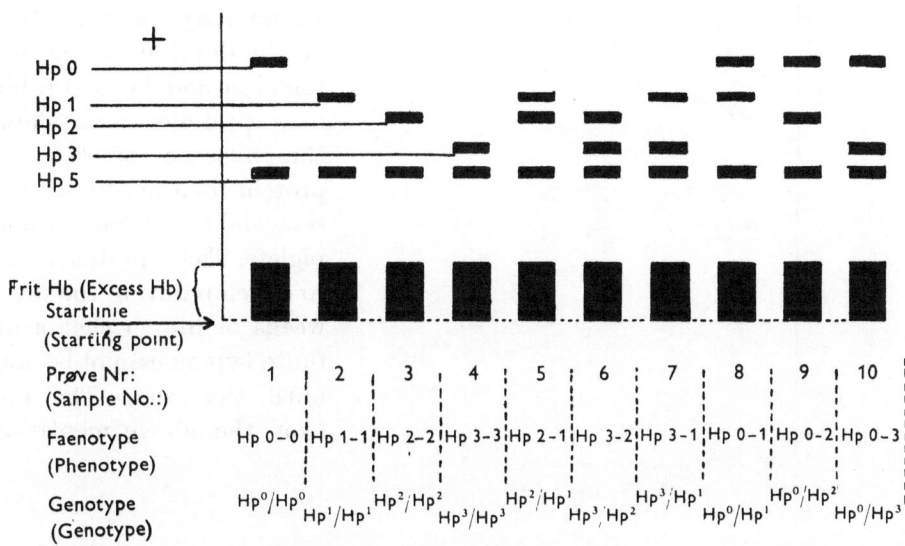

Fig. 1. Diagram of the haematin-binding protein phenotypes observed in pigs of Danish Landrace.

To support the hypothesis, the family material was typed. The samples were at the same time typed for blood-group antigens. The result is seen in table 1. Table 1 indicates that the progeny phenotypes observed in every mating class are only those to be expected from the hypothesis. Further good agreement was found between the expected distribution of the progeny calculated on the basis of the above hypothesis and the distribution observed.

Table 2

The observed and on basis of gene frequencies expected distribution of haematin-binding protein types among 900 pigs og Danish Landrace (Hesselholt, 1963b)

Hp 0–0 obs. exp	Hp 0–1 obs. exp.	Hp 0–2 obs. exp.	Hp 0–3 obs. exp.	Hp 1–1 obs. exp.	Hp 2–1 obs. exp.	Hp 3–1 obs. exp.	Hp 2–2 obs. exp.	Hp 3–2 obs. exp.	Hp 3–3 obs. exp.
3 1·5	29 24·9	5 8·4	33 36·7	99 105	76 70·7	312 309·2	14 11·7	98 104	231 227·1

$X^2 = 3·01.$

The following gene frequencies were found among 900 pigs selected at random: $q_{Hp}0 : 0.041$, $q_{Hp}1 : 0.342$, $q_{Hp}2 : 0.115$, $q_{Hp} : 0.503$.

As indicated in table 2, good agreement between the observed and on the basis of gene frequencies expected distribution of haematin-binding protein types among 900 pigs was obtained.

Ahaptoglobinaemia. Preliminary investigations on the development of haptoglobin and haematin-binding proteins have shown the absence of the two protein fractions from foetal sera and sera from newborn piglets. These proteins seem to develop during the first 3 weeks of life so that a definite typing cannot be done until this age. The sera from the above mentioned

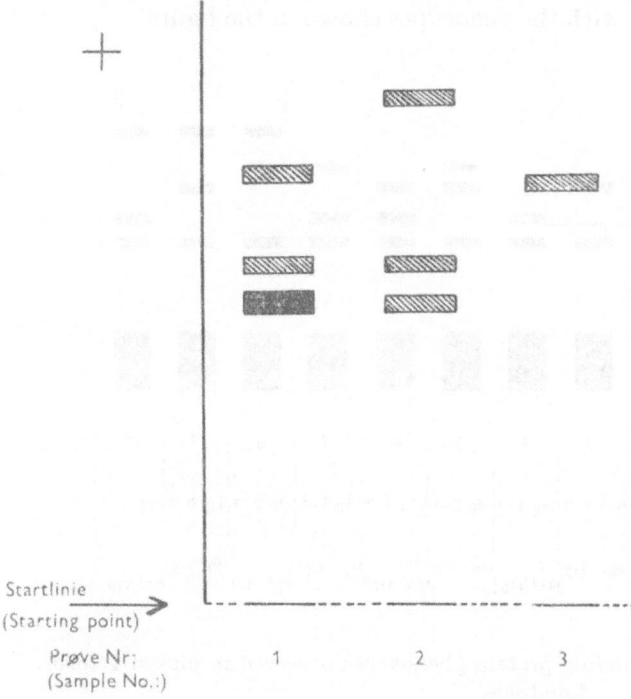

Startlinie
(Starting point)

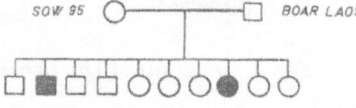

Prøve Nr:
(Sample No.:)   1   2   3

Fig. 2. Diagram of starch gel electrophoresis of sera with different haptoglobin content. 1: Normal content (Hp 3—1). 2: Low content (Hp 0—3). 3: Absence of haptoglobin (Hp 2—2).

Fig. 3. Pedigree of the mating where two of progeny piglets are ahaptoglobinaemic. Shaded symbols indicate ahaptoglobin- aemic individuals.

genetic investigations were all from pigs older than 8 weeks. In no case was a lack of haematin-binding proteins observed nor were there any apparent quantitative variations in these proteins. The suilline haptoglobin showed individual quantitative variations ranging from very heavily stained bands to very faintly stained bands (see figure 2). In sera from three 4-months-old pigs, complete absence of haptoglobin was observed. Two of the ahaptoglobin- aemic pigs were from the same litter. Serum from the sow and the boar showed an apparent normal haptoglobin content. The third ahaptoglobinaemic pig came from a litter sired by the same boar. A blood sample from the sow could not be obtained (figure 3). The genetic and physiological basis for ahaptoglobinaemia in pigs will be studied in the future.

2. Ceruloplasmin

The copper-binding protein, an α_2-globulin called ceruloplasmin (Holmberg and Laurell, 1948), can be located following zone electrophoresis because it possesses oxidase activity (Holmberg and Laurell, 1951). It will oxidize para-phenylenediamine (PPD) to give an intensively coloured compound called Bandrowsky's base (Uriel, 1958). This reaction has been utilized in looking for variations in human ceruloplasmin after separation by chromatographic or electrophoretic methods, or a combination of these. Heterogeneity of human ceruloplasmin has been reported by various groups (Broman, 1958; McAllister et al. 1961; Morell and Scheinberg, 1960; Poulik and Bearn, 1962; Sass-Kortsak et al., 1960); Uriel (1963) has presented evidence arguing against such heterogeneity.

Imlah has found grouped qualitative variations in pig ceruloplasmin following starch gel electrophoresis of serum. Individual sera exhibited one of three patterns of bands after PPD staining. This suggested that a genetically controlled heterogeneity might exist in ceruloplasmin in pigs (Imlah, 1963.) In the following preliminary investigations, we have surveyed sera from pigs of Danish Landrace to see if similar variations in ceruloplasmin could be demonstrated in them.

Materials and methods. Sera from 69 matings with 258 progeny were available. These were all from pigs of the Danish Landrace and all were older than 8 weeks of age.

The buffer system used was the same as that described for horse transferrin. A starch concentration of 16% was employed. A voltage of 10 v/cm. was applied for the first hour of electrophoresis; this was raised to 17 v/cm. for the remainder of the run. Electrophoresis was stopped when the buffer boundary was 11 cm beyond the insertion point. The gels were sliced horizontally, and one half was incubated with PPD for 1/2—1 hour at 37°C. The PPD was dissolved in acetate buffer pH 7·5 which was warmed to 37°C, and used immediately (Uriel, 1958).

Results. Three patterns of PPD oxidase activity were observed in the pig sera surveyed. They correspond to those observed by Imlah and are shown in figure 4. Two of the three consist of 3 bands, the middle band being the most heavily stained in both cases. The third pattern appears to be a combination of these two but, because of some overlapping of the bands, it has 5 instead of 6 bands. The fast ceruloplasmin fraction was designated as fraction 1 and the slow fraction as fraction 2. The hypothesis advanced was that the shown polymorphism was genetically controlled and determined by 2 alleles with a codominant mode of inheritance. Therefore, the phenotypes shown in figure 4 would represent the two homozygotes and a heterozygote with respect to the two alleles governing the synthesis of ceruloplasmin in pigs.

The great majority of our family material sera exhibited the slow ceruloplasmin fraction, as can be seen in table 3. The fast fraction has been observed in only 5 families in which samples from both parents were available for analysis. We have observed a few sera with the fast fraction only, but again these did not happen to occur in our family material. Even though few families have shown the fast component as yet, the data presented in table 3 tend to support the hypothesis that synthesis of ceruloplasmin in pigs may be governed by 2 alleles, Cp^1 and Cp^2.

Based on a sampling of 130 parents, the gene frequencies are: fast component $q_{Cp}1 : 0.0154$; slow component $q_{Cp}2 : 0.9846$. All sera used in this study were typed for haematin-binding proteins. Furthermore, all pigs were bloodtyped. No attempt has been made as yet to correlate the ceruloplasmin type with the other known blood and serum types.

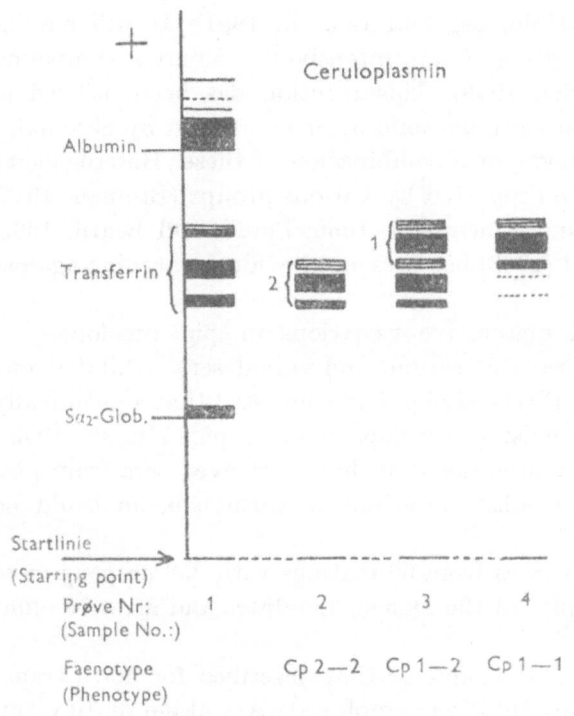

Fig. 4. Diagram of the ceruloplasmin phenotypes observed in pigs of Danish Landrace. Sample No 2, 3 and 4 are stained with PPD. Sample No 1 is stained with amido black.

Table 3

Distribution of ceruloplasmin types of progeny from various mating classes in pigs of Danish Landrace

Mating-classes	No. of matings	No. of progeny	Progeny phenotypes						X^2	d. f.
			2−2 obs.	exp.	2−1 obs.	exp.	1−1 obs.	exp.		
2−2 × 2−2	64	241	241	241					0	
2−2 × 2−1	5	17	9	8·5	9	8·5			0·6	1
Total	69	258	250	249·5	8	8·5				

3. A new serum protein polymorphism in pigs

During a preliminary survey of ceruloplasmin type in pig sera, a pattern of clear bands appeared in starch gels which had been placed at 4°C overnight following electrophoresis and staining for ceruloplasmin, as described above.)* That these clear bands might constitute a genetically controlled polymorphic system was suggested by several observations: 1. every pig serum tested displayed at least one band, 2. no more than two bands were ever observed in a single serum, and 3. the bands could be located at 3 positions. Using the buffer system described for ceruloplasmin, the bands were well-separated and lay in the region between transferrin and the slow α_2-globulin bands (figure 5).

The position of the clear bands was noted in a large number of sera from pigs of the Danish Landrace, including those of the family material studied for ceruloplasmin variations. The fastest, or most anodic band, has been designated X 1, the middle area X 2, and the slowest band X 3. The five

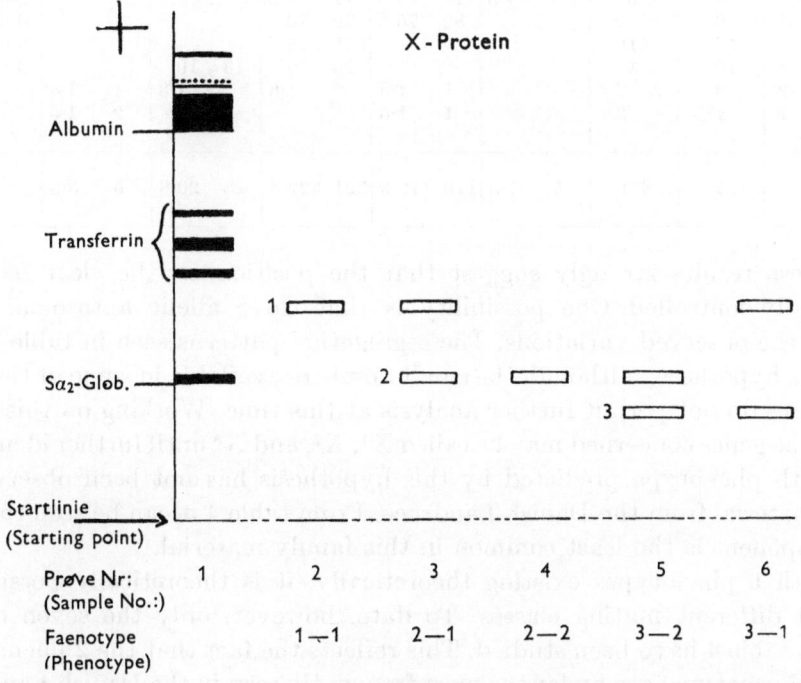

Fig. 5. Diagram of the 5 X-protein phenotypes observed in pigs of Danish Landrace. Sample No. 1 is stained with amido black. Samples No. 2—6 were treated as described in the text.

---

*) To our knowledge the same pattern was observed by Dr. J. King and collaborators, Edinburgh, Scotland.

phenotypes that have been observed to date, are shown in figure 5, and they are, according to our designation, 1, 2—1, 2, 3—2, and 3—1. The typings are reproducible. Some of the sera have been fresh but most have been stored at —20°C for periods up to 2 years. The family material was previously used for blood typing and typing of haematin-binding proteins.

The distribution of the phenotypes among the progeny of the matings studied is shown in table 4. It is obvious that this distribution is non-random and that, in fact, no band is present in a progeny serum that is not also present in serum of one of the parents.

Table 4

Distribution of X phenotypes of progeny from various mating classes in pigs of Danish Landrace

Mating-classes	No. of classes	No. of progeny	1—1 obs.	1—1 exp.	2—1 obs.	2—1 exp.	2—2 obs.	2—2 exp.	3—2 obs.	3—2 exp.	3—1 obs.	3—1 exp.	X^2	d. f.
1–1 × 2-2	5	18			18	18							0·00	0
2–1 × 2-1	8	30	5	7·5	14	15	11	7·5					2·53	2
2–2 × 2-1	40	152			82	76	70	76					0·95	1
2–2 × 2-2	57	218					218	218					0·00	0
2–2 × 3-2	10	38					20	19	18	19			0·11	1
2–1 × 3-2	2	7			1	1·8	2	1·8		1·8	4	1·8		
1–1 × 3-2	1	3			1	1·5					2	1·5		
Total	123	466	5	7·5	116	112·3	321	322·8	18	20·8	6	3·3		

These results strongly suggest that the position on the clear bands is genetically controlled. One possibility is that three allelic autosomal genes control the observed variations. The segregation patterns seen in table 4 support this hypothesis, although the small numbers available in some of the mating classes do not permit further analysis at this time. Working on this hypothesis, the genes concerned may be called X^1, X^2, and X^3 until further identified. The sixth phenotype predicted by this hypothesis has not been observed as yet in material from the Danish Landrace. From table 4 it can be seen that the X3 component is the least common in this family material.

With 6 phenotypes existing theoretically, it is theoretically possible to have 21 different mating classes. To date, however, only the seven classes listed in table 4 have been studied. This reflects the fact that the 2 phenotypes (or X^2X^2 genotype) are by far the most frequently seen in the Danish Landrace. Based on a sampling of 175 parents, the gene frequencies (where frequency of gene 1,

$$q_1 = \frac{2(X^1X^1) + (X^2X^1) + (X^3X^1)}{2N} \text{ etc.)}$$

are $q_1 : 0·13$, $q_2 : 0·84$, and $q_3 : 0·03$. By expanding the expression $(q_1 + q_2 + q_3)^2$, we can arrive at an estimate of the expected frequencies of the various phenotypes. These were calculated and applied to the sampling of 175 parents, as shown in table 5. There is good agreement between the numbers expected and those observed in the various categories. Assuming that our sampling was effectively random, this agreement can then be taken as a further indication that the genetic hypothesis suggested is plausible.

Table 5

The observed and on basis of gene frequencies expected distribution of X phenotypes among 175 pigs of Danish Landrace

1 — 1		2 — 1		2 — 2		3 — 1		3 — 2		3 — 3	
obs.	exp.	obs.	exp.	obs.	exp.	obs.	exp.	obs.	exp.	obs.	exp.
1	3	44	38·6	120	123·5	0	1·3	10	8·4	0	0·1

Concerning the proposed X 3 phenotype, it can be calculated from its expected frequency that it would occur once in 1,250 randomly-selected sera.

A sampling of 85 pig sera from a slaughterhouse in Iceland also revealed the presence of this system although no 3 band was observed. The gene frequencies were $q_X1 : 0·14$, and $q_X2 : 0·86$. Preliminary investigations have revealed at least one similar band in cattle, horse and dog sera.

Efforts at the present time are concentrated on identifying this system. The fact that it appears as clear bands in a starch gel suggested the presence of a starch splitting enzyme such as amylase, which, along with a variety of other enzymes, is found in serum (Fishman 1960.) In our gels, however, there is a soft, whitish area, around the insertion point of the serum-soaked filter paper, which appears to be the result of amylase digestion. The bands in the system do not show such a softening even after prolonged storage. The appearance of the bands, furthermore, is not inhibited by adding EDTA to the incubation medium, as would be expected if an amylase was involved (White et al., 1954). After 1 hour incubation at 37°C in acetate buffer (pH 5·7) the system appears as white areas in the gel; if incubation is continued at room temperature or 37°C overnight, the white areas remain. If the incubation is at 4°C, clear areas appear in their place.

The system is precipitated by TCA (equal vols. serum and 10% TCA) and is completely precipitated by 60% $(NH_4)_2SO_4$; most is precipitated by 50% saturated $(NH_4)_2SO_4$. Two-dimensional starch gel electrophoresis (Poulik et al., 1958) has shown that this component migrates in the albumin region in both agar and paper electrophoresis. After the two-dimensional run, agar to starch gel, the component could be stained with amido black, which further indicates its protein nature. Apparently under these conditions it is concentrated sufficiently to make possible visible staining for protein.

As has been pointed out by Uriel, a single enzyme may be carried or complexed to other proteins or protein fragments (Uriel, 1963). In the present case, since the system appears to be due to what may be enzymic activity, we have to consider that perhaps only one enzyme form is involved but that it may be complexed to another protein which itself exists in different genetically controlled forms. It is hoped that further fractionation and purification will lead to the early identification of this system.

B. Horses

Transferrin types of the Icelandic horse

Individual serum protein variations in horses were first observed by Ashton in a starch gel electrophoretic study of 19 horse sera. Although some of the differences seemed to be grouped, no attempt was made to determine the genetic basis of the observed variations (Ashton, 1958). 6 different transferrin phenotypes in horse sera were later demonstrated and a hypothesis of three alleles as basis for the observed polymorphism was advanced (Schmid, 1962). Starch gel electrophoretic studies on American and Norwegian horse sera revealed 16 transferrin phenotypes and family data supported the theory that the observed differences were attributed to the action of six codominant alleles designated Tf^D, Tf^F, Tf^H, Tf^M, Tf^O, and Tf^R (Braend and Stormont, 1964; Braend, 1964).

The object of the following investigation was to determine the occurrence of transferrin phenotypes in the Icelandic horse breed and to elucidate the heredity of these proteins with the purpose of utilizing transferrin typing in conjunction with blood grouping in the parentage control.

Material and methods. Plasma samples from 10 stallions, 92 mares and from 109 of their progeny, all of the Icelandic horse breed, were subjected to horizontal starch gel electrophoresis in a modified discontinuous buffer system of Poulik. Gel buffer: 1·73 g/l Sigma 7—9 (tris) and 0·85 g/l citric acid, pH 7·6. Vessel buffer: 0·3 M boric acid and 0·1 M sodium hydroxide, pH 8·6. Voltage gradient: 15—20 v/cm. Time: 1·5—2 hours. Temperature: 4—5°C. Staining methods: Gels were kept either 2 minutes in a saturated solution of amido black 10 B or 15 minutes in a 0·05% solution of nigrosin. As solvent and washing solution methanol-water-glacial acetic acid (5 : 5 : 1) asw used. In some cases it was necessary to concentrate the plasma samples by means of dialysis against Carbowax 1500.

Results. After starch gel electrophoresis of the above material 21 different types of transferrin pattern were observed (figure 6). As can be seen from the figure the types of pattern differ from each other both by the number of bands and by the position of the bands.

The first 6 samples represent individuals with the smallest number of transferrin components. They possess a relatively thick band followed by a thin band. The samples differ from each other by the mobility of the bands.

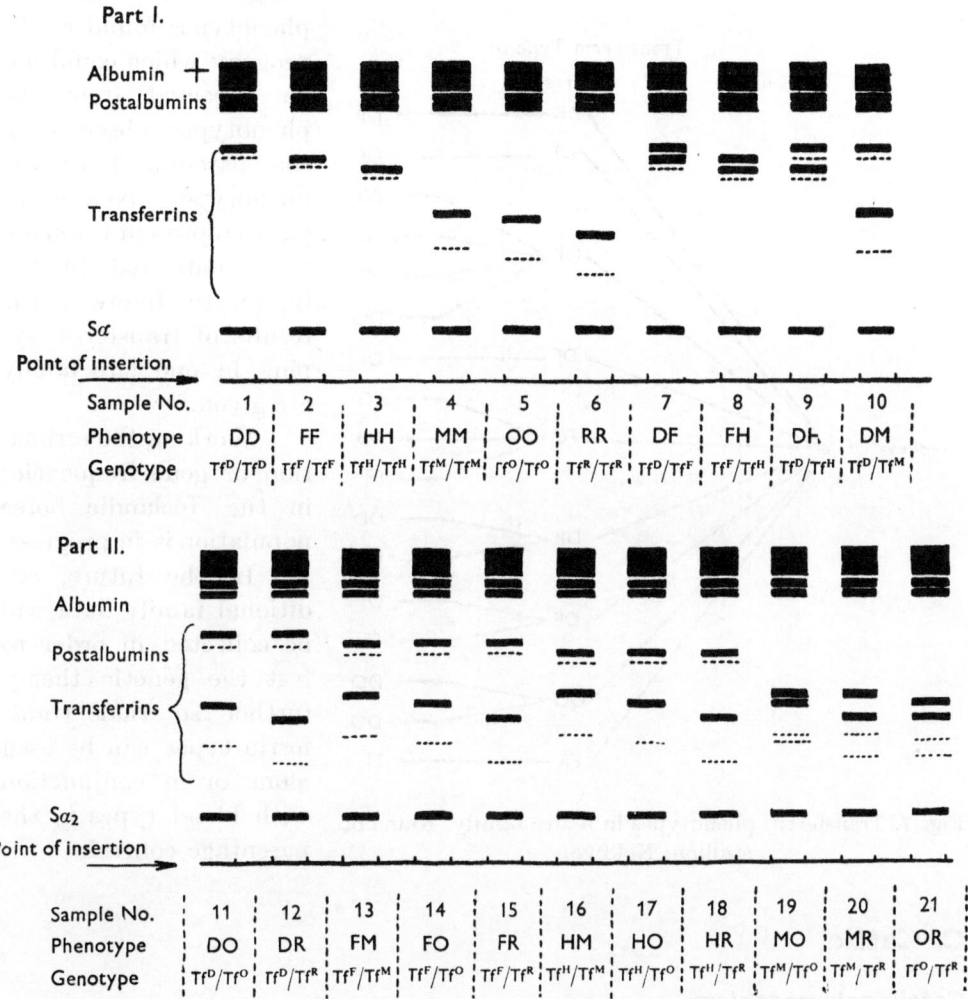

Fig. 6. Diagram of the transferrin phenotypes observed in the Icelandic horse by means of starch gel electrophoresis.

Samples No. 7—21 represent the different combinations between the first 6 types. No sample was devoid of transferrin components.

According to the above findings, the theory was advanced that the synthesis of the transferrin in horse sera is under genetic control and determined by 6 codominant alleles. As our observations are in good agreement with those of Braend and Stormont, the alleles will be designated in accordance with the nomenclature proposed by these authors.

In order to test the genetic hypothesis, the family data were analyzed. The individuals were at the same time tested for the known blood-group antigens. In no case were phenotypes found in the progeny which could not be expected from the phenotypes observed in the parents. Thus the phenotypes given in figure 6 represent the genotypes indicated in the figure. In figure 7 the results of transferrin typing in one sire family are given.

Work on the estimation of gene frequencies in the Icelandic horse population is in progress.

In the future, additional family data will be collected in order to test the genetic theory further so that transferrin types can be used alone or in conjunction with blood types in the parentage control.

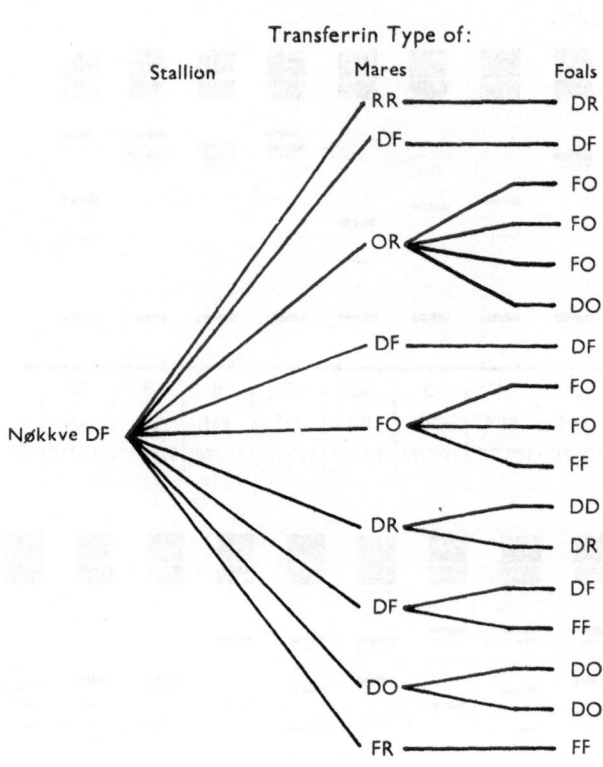

Fig. 7. Transferrin phenotypes in a sire family. Name of stallion: Nøkkve.

C. Cattle

Casein polymorphism

By means of paper electrophoresis, individual genetically determined differences in β-casein have been demonstrated by Aschaffenburg (1963). It was shown that β-casein could be separated into three electrophoretically different components A, B, and C in the order of decreasing mobility. The bands always appeared singly or in pairs, in accord with the hypothesis that three allelic genes without dominance control the synthesis of β-casein. The six possible phenotypes expected in accordance with this concept have all been found.

Using starch gel electrophoresis, Thompson (1962) discovered polymorphism in α_s-casein. Individual samples show six different types, and the hypothesis for heredity is the same as for the β-casein synthesis. Family data support this concept (Kiddy et al., 1964). Figure 8 shows the different phenotypes for α_s and β-casein. This report gives the results from preliminary investigations of gene frequencies in Danish cattle, and a simple technique for using undiluted whole milk on urea-starch gel is described.

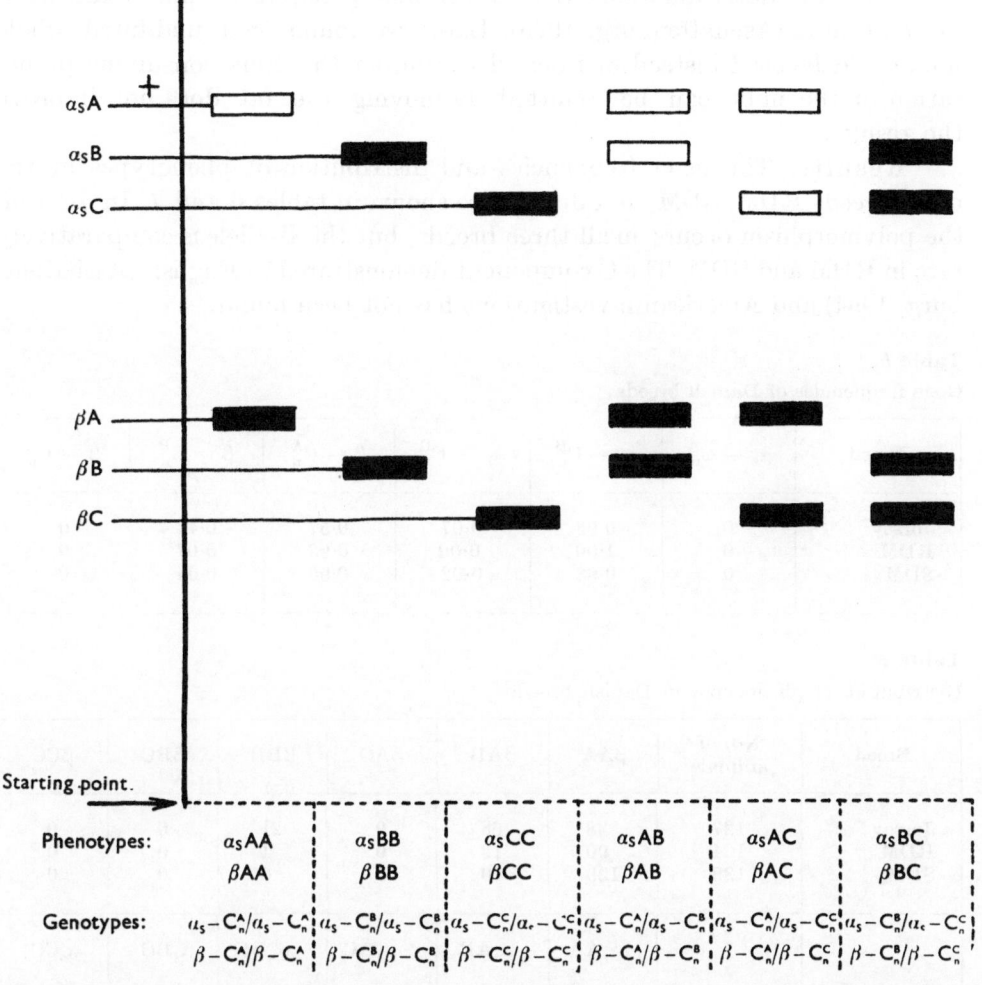

Fig. 8. Starch gel pattern demonstrating the different phenotypes on α_s- and β-casein. The black squares show the types found until now. The α_s-CC and β-AC were not found in Danish breeds, but in Yugoslavian cow milk. The β-BC and β-CC were found in Hungarian cow milk.

Materials and methods. 370 individual samples from 4 different live-stock equally distributed over the three breeds RDM, SDM and Jersey have been investigated. Wake and Baldwin's (1961) technique for horizontal urea-starch gel electrophoresis has been used in a modified form. The following has been changed: Gel buffer: Tris-(hydroxymethyl) aminomethane 15·13 g/l, Na_2EDTA 1·5 g/l, H_3BO_3 1·15 g/l, pH 8·9 (Thompson, 1964). Voltage gradient 15—20 v./cm. Time: $2\frac{1}{2}$—3 hours. Insertion material: cellulose acetate paper. Thickness of gel: 1·8 mm. Slicing of the gel is ommitted and the gel has to be read from the underside.

For these determinations the casein was precipitated and redissolved in 7 M urea (Aschaffenburg, 1963). Later we found that undiluted whole milk could be used instead of isolated casein, so the time consuming preparation of the milk can be omitted. Removing the fat does not improve the results.

Results. The gene frequencies and distribution of phenotypes in the three breeds RDM, SDM, and Jersey are shown in tables 6 and 7. In β-casein the polymorphism occurs in all three breeds, but the B-allele is comparatively rare in RDM and SDM. The C component demonstrated in English (Aschaffenburg, 1964) and American investigations has not been found.

Table 6

Gene frequencies of Danish breeds

Breed	$\alpha_s - C_n^A$	$\alpha_s - C_n^B$	$\alpha_s - C_n^C$	$\beta - C_n^A$	$\beta - C_n^B$	$\beta - C_n^C$
Jersey	0	0·93	0·07	0·57	0·43	0
RDM	0	1·00	0·00	0·93	0·07	0
SDM	0	0·98	0·02	0·96	0·04	0

Table 7

Distribution of phenotypes in Danish breeds

Breed	No. of animals	βAA	βAB	βAC	βBB	βBC	βCC
Jersey	127	38	68	0	21	0	0
RDM	104	90	13	0	1	0	0
SDM	139	129	10	0	0	0	0

Breed	No. of animals	$\alpha_s AA$	$\alpha_s AB$	$\alpha_s AC$	$\alpha_s BB$	$\alpha_s BC$	$\alpha_s CC$
Jersey	127	0	0	0	110	17	0
RDM	104	0	0	0	104	0	0
SDM	139	0	0	0	134	5	0

In α_s-casein the B-allele is most common. All RDM are homozygotes α_s-BB. Type α_s-BC has been found in 3·6% of SDM and in 13·6% of Jersey cattle. The homozygote α_s-CC has not been observed in Danish breeds; the A-allele has not been found either.

References

Aschaffenburg, R. (1963). J. Dairy Res. 30 : 251.

Aschaffenburg, R. (1964). Nature 192 : 431.

Ashton, G. C. (1958). Nature 182 : 1029.

Broman, L. (1958). Nature 182 : 1655.

Brummerstedt-Hansen, E., M. Hesselholt, B. Larsen, J. Moustgaard, I. Moller, P. B. Nielsen & B. Palludan (1962a). The 8th Anim. Blood Group Conf. in Europe, Ljubljana, Yugoslavia.

Brummerstedt-Hansen, E. & M. Hesselholt (1962b). Annual Report, The Royal Veterin. and Agric. Coll., Inst. f. Sterility Research 211.

Braend, M. & C. Stormont (1964). Nord. Vet.-Med. 16 : 31.

Braend, M. (1964). Nord. Vet.-Med. 16 : 363.

Fishman, W. H. (1960). In The Plasma Proteins 2, Chap. 12, Acad. Press, N. Y. & London.

Hesselholt, M. (1963a). Acta Vet. Scand. 4 : 238.

Hesselholt, M. (1963b). Annual Report, The Royal Veterin. and Agric: Coll., Inst. f. Sterility Research 93.

Holmberg, C. C. & Laurell, C. B. (1948). Acta Chem. Scand. 2 : 550.

Holmberg, C. C. & Laurell, C. B. (1951). Acta Chem. Scand. 5 : 476.

Imlah, P. (1963). Personal communication.

Kiddy, C. A., J. O. Johnston & M. P. Thompson (1964). J. Dairy Sci. 47 : 147.

King, J. W. B. (1963). Personal communication.

Kristjansson, F. K. (1961). Genetics 46 : 907.

McAlister, R., G. M. Martin & E. P. Benditt (1961). Nature 190 : 927.

Morell, A. G. & I, H, Scheinberg (1960). Science 131 : 930.

Poulik, M. D. & O. Smithies (1958). Biochem. J. 68 : 636.

Poulik, M. D. & A. G. Bearn (1962). Clin. Chim. Acta 7 : 374.

Schmid, D. O. (1962). Z. Immunitätsforsch. 124 : 219.

Sass-Kortsak, A., S. J. Jackson, & A. F. Charles (1960). Vox Sang. 5 : 87.

Thompson, M. P. (1962). Nature 195 : 1001.

Thompson, M. P. (1964). Personal communication.

Uriel, J. (1958). Nature 181 : 999.

Uriel, J. (1963). Ann. N. Y. Acad. Sc. 103 : 493.

Wake, R. C. & R. L. Baldwin (1961). Biochim. et Biophys. Acta 47 : 225.

White, A., P. Handler, E. Smith & D. Stetten (1954). Principles of Biochemistry, Chap. 11, p. 260, McGraw-Hill Book Co. N. Y.

A NEW HAEMOGLOBIN VARIANT IN ZEBU CATTLE

S. N. NAIK, L. D. SANGHVI

Human Variation Group, Indian Cancer Research Centre, Parel, Bombay

Haemoglobin polymorphism in different animal species is reported by several investigators: Cabannes and Serain 1956, Bangham 1957, Grimes et al. 1957, Salisbury and Shreffler 1957, Harris and Warren 1955, Lehmann 1959, Huisman et al. 1959, Khanolkar et al. 1963 and Naik et al. 1964. Eleven out of 14 British cattle breeds studied by Bangham (1957) showed only Hb—A. Many cattle breeds studied by us in India showed both Hb—A and Hb—B. Haemoglobin study in most of the animal species is limited to a few samples which have revealed the commonly occurring variants. The first new variant in the adult Zebu cattle (*B. indicus*) was discovered in our laboratory in 1959 and was called Hb—X. A similar variant called Hb—C was reported in U. S. A. by Crocket et al. (1963) in imported Brahman cattle and Brahman-Hereford crosses. The discovery of the new variant prompted us to survey a larger number of cattle to establish its frequency in various Indian breeds and also to look for more variants. This presentation describes one more new haemoglobin variant detected in Zebu cattle.

Materials and Methods

Blood samples (about 1 ml. each) were aseptically collected from ear vein, in 3·8% sodium citrate solution, from 1,060 adult healthy animals constituting 5 common breeds (Malvi, Khillari, Kankrej, Dangi and Gir, see Table 1) of Western India.

The blood samples were washed with 0·9% NaCl solution until the RBCs were free from plasma proteins. The packed cells were lysed with an equal volume of distilled water. The haemolysate was further treated with $\frac{1}{4}$ volume of toluene to separate it from the stoma proteins on centrifugation at 3,000 r. p. m. The haemoglobin solution was adjusted to 10% before it was subjected to paper electrophoresis.

The electrophoretic apparatus and the power unit were locally prepared.

The electrophoretic runs were made in 0·05M veronal buffer pH 8·6 for 10—12 hours at 200 volts and 3 mA, using Whatman No. 3 filter paper. The haemoglobin being a coloured protein, the results were read directly.

Results and Discussion

The slow moving haemoglobin called Hb—A and the fast moving is called Hb—B in conformity with the previous workers (see Fig. 1). The present study revealed 321 cases of Hb—A, 498 cases of Hb—AB, 236 cases of Hb—B besides 4 cases of Hb—AX and one sample containing Hb—A along with a new fraction (see Table 1). This new variant is designated tentatively Hb-Khillari because it was found in a Khillari bullock, and an appropriate designation will be assigned to it on the basis of further work. Unlike the Hb—X fraction, which has a mobility intermediate to that of Hb—A and Hb—B, the Hb-Khillari fraction has a slower mobility than that of Hb—A (see Fig. 2). It is interesting to note that this new variant is also associated with Hb—A like all our 7 cases of Hb—X. The new heterozygous phenotype showed an almost equal amount of Hb—A and Hb-Khillari indicating a similar dosage effect of alleles. This study reveals probably the presence of a fourth gene responsible for the control of the haemoglobin variants in the *B. indicus*. The gene that determines the Hb-Khillari seems to be lower in frequency than that which determines Hb—X.

Table 1

Distribution of haemoglobin variants in the five Indian cattle breeds

Breed	Number examined	Haemoglobin variants				
		Hb—AA	Hb—AB	Hb—BB	Hb—AX	Hb—A+Khillari
Malvi	200	62	99	37	2	0
Khillari	210	63	92	54	0	1
Kankrej	230	77	119	33	1	0
Dangi	220	57	103	59	1	0
Gir	200	62	85	53	0	0
Total:	1060	321	498	236	4	1

The common haemoglobin variants are known to be controlled by two allelomorphic genes, which are co-dominant (Bangham 1957). The factors that determine all these haemoglobin types may probably be allelic with each other. Since Hb—C (Hb—X) is found only in association with Hb—A, Crockett et al. (1963) proposed that the other phenotypes of Hb—C, like Hb—CC and Hb—BC might be lethals. The single case of Hb-Khillari is also found in association with Hb—A. Further studies of these two variants

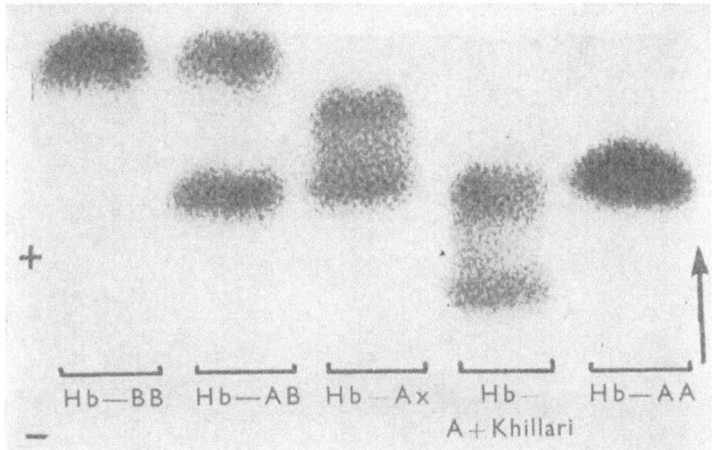

Fig. 1. Paper electrophoretic pattern showing the common haemoglobin variants encountered in Zebu cattle in India.

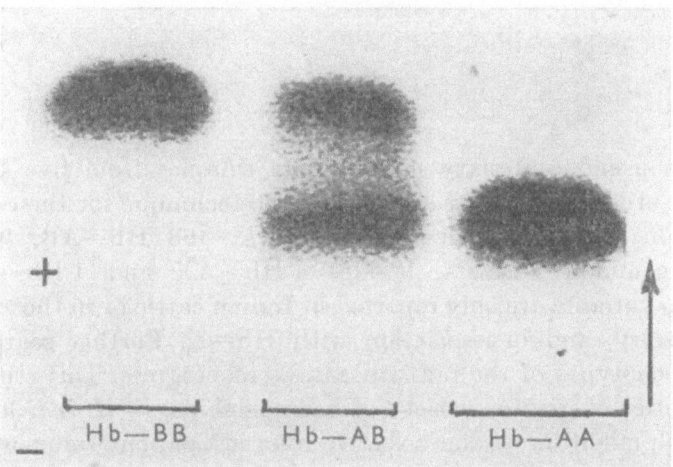

Fig. 2. Paper electrophoretic pattern of the haemoglobin variants so far known in Zebu cattle in India.

are in progress to elucidate more genetic information regarding their other phenotypes. The finding of this fourth haemoglobin in a limited search among the Zebu cattle (*B. indicus*) of the tropical region, indicates the possibility

297

of a greater variety than in *B. taurus* of the temperate region as encountered in man. These polymorphic haemoglobin pigments might provide good opportunities for the study of their selective roles in their environments.

Fig. 3. Photograph of a bullock belonging to the Khillari breed which revealed Hb-Khillari.

Summary

One thousand and sixty haemoglobin samples from five Indian cattle breeds were studied by paper electrophoretic technique for their haemoglobin variants. This study revealed 321 Hb—AA, 498 Hb—AB, 236 Hb—BB which are common variants, besides 4 Hb—AX and 1 Hb—A+Khillari. The last two variants are only reported in Indian cattle or in those with Indian (Zebu) ancestry; and in association with Hb—A. Further search for other probable phenotypes of the rare variants is in progress. This study indicates the possibilities of greater variety of haemoglobins, in *B. indicus* than in *B. taurus*, which might have some selective roles in their environments.

Acknowledgements

Our thanks are due to Dr. S. A. Malandkar for his kind co-operation to collect the blood samples.

References

Bangham, A. D. (1957). Nature 179 : 467.

Cabannes, R. and Serain, C. (1956). C. R. Soc. Biol., Paris, 149 : 1193.

Crockett, J. R., Koger, M., and Chapman, H. L. Jr. (1963). J. Animal Sci. 22 : 173.

Grimes, R. M., Duncan, C. W., and Lassiter, C. A. (1957). J. Dairy Sci. 40 : 1138.

Harris, H. and Warren, F. L. (1955). Biochem. J. 60 : 29.

Huisman, T. H. J., Helm, H. J. van der., Visser, H. K. A. and Van Vliet, G. (1959). A symposium. Blackwell Scientific Publications, Oxford. p. 181.

Khanolkar, V. R., Naik, S. N., Baxi, A. J. and Bhatia, H. M. (1963). Experientia 19 : 472.

Lehmann, H. (1959). MAN No. 91.

Naik, S. N., Sukumaran, P. K. and Sanghvi, L. D. (1964). Blood groups and haemoglobin variants in the bovine animals of Bombay. In press; Quoted by Crockett et al. (1963).

Salisbury, G. W. and Shreffler, D. C. (1957). J. Dairy Sci. 40 : 1198.

TRANSFERRIN TYPES IN SOUTH AFRICAN CATTLE BREEDS

D. R. OSTERHOFF and J. R. H. VAN HEERDEN

Faculty of Veterinary Science, University of Pretoria and Blood Group Laboratory.
Animal Husbandry and Dairy Research Institute, c/o Veterinary Research Institute, Onderstepoort.

Since the initiation of the starch-gel electrophoresis technique by Smithies (1955), genetically controlled polymorphism in sera from normal cattle has been demonstrated by different workers (Hickman and Smithies, 1957; Ashton, 1957, 1958, 1959, 1960, 1961; Ashton and McDougall, 1958; Gahne et al., 1960; Højgaard et al., 1960; Gahne, 1961; Brummerstedt-Hansen et al., 1962; Ashton and Fallon, 1962; Schmid, 1962; Buschmann, 1963; Datta and Stone, 1963).

In the present paper some results of transferrin studies in South African cattle breeds are given.

Procedure

The discontinuous buffer system suggested by Gahne (1962) was used. The authors applied some modifications and the method is briefly described. The electrode buffer consisted of 1·2 g. lithium hydroxide, 11·8 g. boric acid and 1,000 ml. distilled water. As gel buffer one volume of electrode buffer solution was added to nine volumes of the following solution: 6·2 g. tris-hydroxymethylaminomethane, (TRIS), 1·33 g. citric acid, 1,000 ml. distilled water. Connaught hydrolysed starch at a concentration of 14·5 g. per 100 ml. buffer was used throughout. Trays with the internal dimensions 13·5. cm. $\times$ 20 cm. $\times$ 0·6 cm., were made from $\frac{1}{4}''$ section perspex walls with $\frac{1}{16}''$ perspex bottoms.

In a slot made across the gel the serum samples — 24 at a time — were inserted on pieces of Whatman 3 mm. filter paper of the size 0·7 $\times$ 0·7 cm. In addition, three samples of a reference Tf_{AE}-serum stained with bromophenol blue were inserted to indicate the distance migrated in the starch block.

The gels were covered with wet cheese cloth and a fan was played thereon for cooling. A voltage gradient of 7 V/cm was applied and the total running time was four hours.

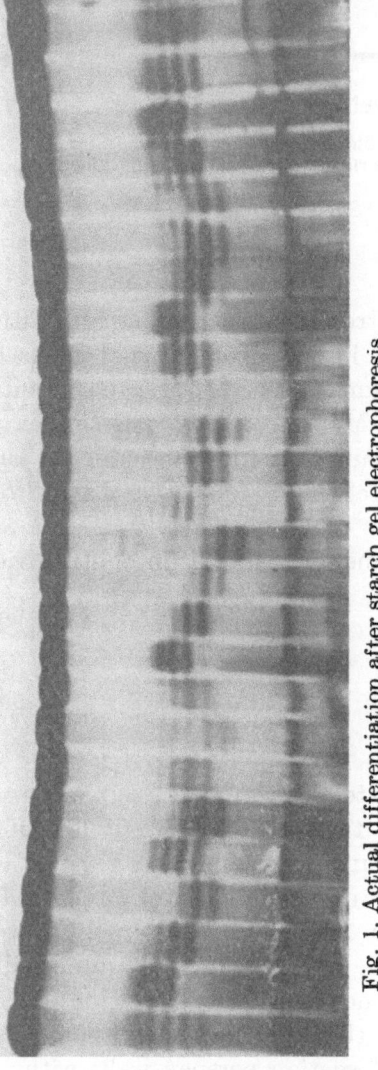

Fig. 1. Actual differentiation after starch gel electrophoresis.

The removal of the filter paper after a certain time, which is recommended by other workers, was abolished and good results have been obtained. After the normal slicing and staining with Amido Black a method for classification and printing suggested by Carr (1963) and published by Johns (1961) has been applied. The gel is soaked in ethanol for several hours, preferably overnight and then placed in a bath of benzyl alcohol and glycerol (2 : 1 v/v). Photographs of gels treated in this way give very good contrasts between the different transferrin bands. Fig. 1 shows a photograph of the protein differentiation after starch gel electrophoresis and ethanol treatment.

Serum samples from 3,146 pure bred animals of 16 different breeds in South Africa taken at random, and from the Boran breed in Kenya, East Africa were collected. The breeds were classified according to their purpose into beef, dairy and dual purpose breeds, and according to their origin into imported, indigenous and crossed breeds. Into the last group the Bonsmara and the Drakensberger were placed, the latter on evidence of blood typing results performed earlier (Osterhoff, unpubl.).

In table 1, the breeds, their classification, the number of samples tested and the number of farms from which the samples originated, are summarized.

Results

A. Dairy Breeds

In the four dairy breeds included in the present study differences could be established. Table 2 gives the actual transferrin types and the calculated frequency of the transferrin alleles Tf^A, Tf^D and Tf^E.

Table 1

Material used for transferrin studies

Breed	Type of breed	Origin of breed	No. of animals tested	Samples collected at
Aberdeen Angus	Beef	Imported	232	3 farms
Afrikaner	Beef	Indigen.	223	5 farms
Ayrshire	Dairy	Imported	244	2 farms
Beef Shorthorn	Beef	Imported	100	1 farm
Bonsmara	Beef	Cross	96	2 farms
Boran	Beef	Indigen.	102	1 farm
Brown Swiss	Dual P.	Imported	271	3 farms
Dairy Shorthorn	Dual P.	Imported	101	2 farms
Drakensberger	Beef	Cross	185	2 farms
Friesian	Dairy	Imported	116	2 farms
Guernsey	Dairy	Imported	99	2 farms
Hereford	Beef	Imported	123	3 farms
Jersey	Dairy	Imported	276	3 farms
Nguni	Dual P.	Indigen.	100	1 farm
Red Poll	Beef	Imported	283	6 farms
South Devon	Dual P.	Imported	324	4 farms
Sussex	Beef	Imported	271	5 farms

Table 2

The transferrin types and gene frequency of the Dairy Breeds

Breeds tested	Transferrin types						Total	Gene frequency		
	AA	DD	EE	AD	AE	DE		A	D	E
Ayrshire	3	160	7	19	1	54	244	.054	.805	.141
Friesian	19	29	1	61	2	4	116	.435	.530	.035
Guernsey	13	49	—	37	—	—	99	.318	.682	—
Jersey	136	39	—	102	—	—	276	.678	.322	—

It can be seen that the Guernsey and Jersey breeds do not possess the transferrin Tf^E and that the Ayrshire has a very high frequency of the transferrin Tf^D.

Table 3 gives a comparison between these frequencies and those obtained in other countries.

The results obtained with South African Friesians are in good agreement with those of Friesians in other countries. The Jersey and Guernsey results compare fairly well with those in other countries, but the Ayrshire in South Africa shows great differences. Ayrshires in Australia, Britain and Canada also show a low frequency of Tf^A and a relatively high frequency of Tf^E, but the most striking difference lies in the very high frequency of Tf^D in the South African Ayrshire. The total number of Ayrshire samples investigated is by far the largest in our material, and no reasons for the discrepancy can be given.

Table 3

South African Dairy Cattle transferrin results compared with those of cattle in other countries

Breed	Country	Gene frequency			Total animals tested
		A	D	E	
Ayrshire	South Africa	.054	.805	.141	244
	Australia (Ashton, 1962)	.241	.583	.176	139
	Britain (Ashton, 1958)	.272	.641	.088	124
	Canada (Smithies, 1958)	.297	.453	.250	42
Friesian	South Africa	.435	.530	.035	116
	Britain (Ashton, 1958)	.463	.474	.062	261
	Canada (Smithies, 1958)	.563	.374	.063	102
	Denmark (Moustgaard, 1960)	.518	.459	.023	238
	England (Ashton, 1962)	.522	.447	.031	179
	Australia (Ashton, 1962)	.362	.628	.010	203
	Sweden (Gahne, 1961)	.485	.500	.015	204
	U. S. A. (Datta, 1963)	.460	.530	.010	473
Guernsey	South Africa	.318	.682	—	99
	Britain (Ashton, 1958)	.494	.507	—	62
Jersey	South Africa	.678	.322	—	276
	Britain (Ashton, 1958)	.737	.264	—	49
	Australia (Ashton, 1962)	.543	.457	—	375
	Denmark (Moustgaard, 1960)	.706	.291	.003	194

B. Beef Breeds

Ashton (1959) investigated Zebu cattle for the first time, and described two new beta-globulin alleles, which he called Tf^B and Tf^F. The two alleles are extremely rare and a number of types have not been found in the material under discussion. The types not found are omitted in table 4.

Table 4

The transferrin types and gene frequency of the Beef Breeds

Breeds Tested	Transferrin Types									Total	Gene Frequency				
	AA	DD	EE	AD	AE	AF	DE	DB	EB		A	D	E	F	B
Aberdeen Angus	94	41	—	92	3	—	2	—	—	232	.610	.379	.011	—	—
Afrikaner	41	32	23	52	41	2	32	—	—	223	.397	.332	.267	.004	—
Beef Shorthorn	52	17	—	31	—	—	—	—	—	100	.675	.325	—	—	—
Bonsmara	28	15	3	20	10	—	20	—	—	96	.448	.359	.193	—	—
Boran	6	38	8	15	9	2	20	3	1	102	.186	.559	.225	.010	.020
Drakensberger	21	48	11	54	18	—	33	—	—	185	.308	.494	.198	—	—
Hereford	6	69	1	32	1	2	12	—	—	123	.192	.739	.061	.008	—
Red Poll	93	13	24	45	81	—	27	—	—	283	.545	.171	.273	—	—
Sussex	34	65	—	170	—	—	2	—	—	271	.439	.557	.004	—	—

The results show great differences between the beef breeds, with Herefords differing the most. The low frequency of Tf^A together with the high Tf^D is strange, but the fact that two animals possessing the F-gene were found was even more strange, because so far, the Tf^E and also the Tf^B were only discovered in Zebu-type cattle. (In the meantime the owner admitted that there was a possibility that the two animals in question were not purebred.)

From the breeds given in Table 4 only three have been investigated by other workers (Ashton, 1958): Aberdeen Angus, Beef Shorthorn and Hereford. In table 5 these results have been compared, and they are apparently in good agreement, except for the Hereford.

Table 5

South African Beef Cattle transferrin results compared with those of cattle in other countries

| Breed | Country | Gene frequency | | | | Total animals tested |
		A	D	E	F	
Aberdeen Angus	South Africa	.610	.379	.011	—	232
	Britain (Ashton, 1958)	.628	.260	.113	—	52
Beef Shorthorn	South Africa	.675	.325	—	—	100
	Britain (Ashton, 1958)	.629	.358	.014	—	141
Hereford	South Africa	.192	.739	.061	.008	123
	Britain (Ashton, 1958)	.387	.606	.008	—	77

Ashton (1959) suggested the relative high frequency of Tf^E in Zebu-type cattle as an indication of the well-known climatic and ecological tolerance of these cattle. This seems to be true for the Afrikaner, Bonsmara, Boran, Nguni and Drakensberger — all the indigenous or crossed breeds listed in table 1 —, but the high frequency of Tf^E in the Red Poll would be difficult to explain along the same lines.

C. Dual Purpose Breeds

The breed differences in this group are very distinct as can be seen in table 6, because these breeds have an entirely different origin.

The results of the South African Brown Swiss are in agreement with the frequencies obtained by Buschmann and Schmid (1961), Tf^A — 0·229, Tf^D — 0·705 and Tf^E — 0·066. A difference between Dairy Shorthorn and Beef Shorthorn is almost non-existent in regard to their transferrin make-up. The Nguni is an indigenous breed of Sanga origin with a high degree of Zebu-

Table 6

The transferrin types and gene frequency of the Dual Purpose Breeds

Breeds tested	Transferrin types									Total	Gene frequency				
	AA	DD	EE	AD	AE	AF	DE	DB	EB		A	D	E	F	B
Brown Swiss	2	181	—	88	—	—	—	—	—	271	.170	.830	—	—	—
Dairy Shorthorn	50	9	—	42	—	—	—	—	—	101	.703	.297	—	—	—
Nguni	14	6	7	18	29	2	22	1	1	100	.385	.265	.330	.001	.001
South Devon	49	121	4	127	8	—	—	—	—	324	.359	.592	.048	—	—

type blood which can be recognized in the high frequency of Tf^E and the appearance of Tf^F and Tf^B.

Summary

Transferrin types of 3,146 serum samples of cattle of 17 different breeds from South and East Africa have been established. The results obtained with several imported breeds agreed with those in other countries. The greatest deviation from earlier results appeared in the South African Ayrshire.

Breeds with a high climatic and ecological tolerance possess a higher frequency of the transferrin allele Tf^E. A few animals with the Tf^B- and Tf^F--alleles were also found and it appears that an original greater polymorphism has been diminished to three alleles Tf^A, Tf^D and Tf^E — through the ages.

References

Ashton, G. C. (1957). Nature 180 : 917.

Ashton, G. C. (1958). Nature 182 : 370.

Ashton, G. C. (1959). Nature 184 : 1135.

Ashton, G. C. (1960). J. Agric. Sci. 54 : 321.

Ashton, G. C. (1961). J. Reprod. & Fert. 2 : 117.

Ashton, G. C. & Fallon, G. R. (1962). J. Reprod. & Fert. 3 : 93.

Ashton, G. C. & McDougall, E. I. (1958). Nature 183 : 945.

Brummerstedt-Hansen, E., Hesselholt, M., Larsen, B., Moustgaard, J., Møller, I, Bräuner Nielsen, P. & Palludan, B. (1962). Proc. 8th Europ. Conf. Anim. Blood Groups, Ljubljana (Mimeogr. report).

Buschmann, H. (1963). Z. Bl. of Vet. Med., Reihe, B, Bd. X : 49.

Buschmann, H. & Schmid, D. O. (1961). Nature 190 : 1209.

Carr, W. R. (1963). Personal communication.

Datta, S. P. & Stone, W. H. (1963). Immun. Letter 3, Part 1 : 26.

Gahne, B. (1961). Anim. Prod. 3, Part 2 : 135.

Gahne, B. (1962). Proc. 8th Europ. Conf. Anim. Blood Groups, Ljubljana. (Mimeogr. report.)

Gahne, B., Rendel, J. & Venge, O. (1960). Nature 186 : 4728.

Hickman, C. I. & Smithies, O. (1957). Proc. Gen. Soc. Canada 2 : 39.

Højgaard, N., Moustgaard, J. & Møller, I. (1960). Instit. for Sterility Res., Annual Report : 99.

Johns, E. W. (1961). J. Chromatog. 5 : 91.

Moustgaard, J., Møller, I. & Havskov Sørensen, P. (1960). Proc. Immunogenetica Edinburgensis: 122.

Schmid, D. O. (1962). Tierärtzl. Umschau 17 : 302.

Smithies, O. (1955). Bio. Chem. J. 61 : 629.

Smithies, O. & Hickman, C. G. (1958). Genetics 43, Part 3 : 374.

NEW HAEMOGLOBIN DIFFERENTIATION IN CATTLE

(RESEARCH NOTE)

D. R. OSTERHOFF, J. A. H. VAN HEERDEN

Faculty of Veterinary Science, University of Pretoria and Blood Group Laboratory,
Animal Husbandry and Dairy Research Institute, c/o Veterinary Research Institute, Onderstepoort

It is generally accepted that besides the foetal haemoglobin three electro-phoretically distinguishable adult bovine haemoglobin phenotypes exist, which are determined by two alleles Hb^A and Hb^B (Bangham 1957, Grimes et al. 1957, 1958, Shreffler et al. 1959, Schmid 1962, 1963).

Using the starch gel method described (Osterhoff et al. 1964) and applying a running time of two and a half hours instead of the normal one hour for haemoglobin determination, a clear differentiation of Type A haemoglobin was possible. It must be mentioned that proper staining with Amido Black and careful destaining is essential.

It should be noted that type A_1/A_1 is identical with type A/A originally described, and it can be seen in Fig. 1 that type A_2/A_2 moves faster than type A_1/A_1 and type A_3/A_3 migrates more slowly than type A_1/A_1, the heterozygotes migrating accordingly.

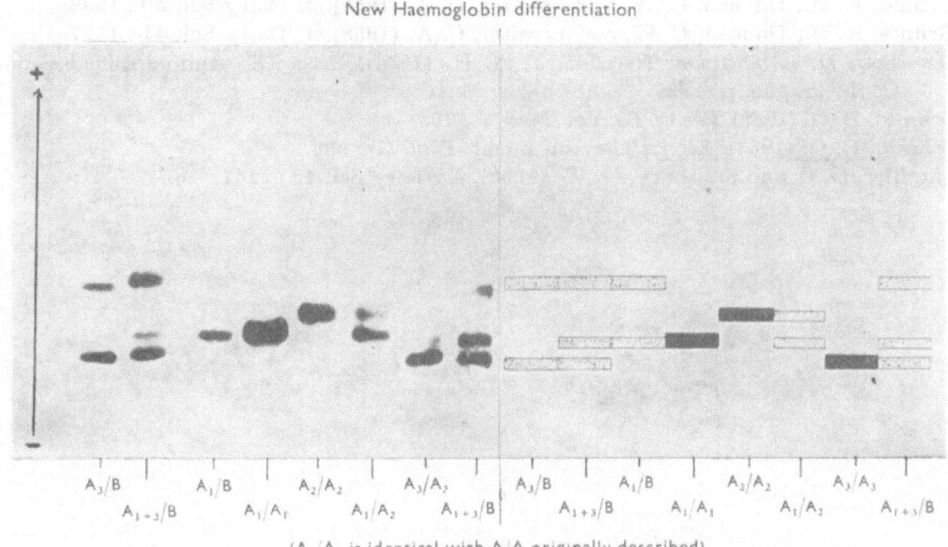

Fig. 1.

Table 1

New haemoglobin differentiation in cattle breeds

Breed	No tested	Haemoglobin phenotype*)								
		A_1/A_1	A_2/A_2	A_3/A_3	A_1/A_2	A_1/A_3	A_{1+3}/B	B/B	A_1/B	A/B
Afrikaner	50	25	2	—	23	—	—	—	—	—
Beef Shorth.	50	32	—	—	—	18	—	—	—	—
Simmentaler (Fleckvieh)	21	—	—	4	—	8	3	2	2	2
Sussex	43	20	—	—	—	23	—	—	—	—
Friesian	100	100	—	—	—	—	—	—	—	—

*) The phenotypes — A_2/B, A_2/A_3, A_{1+2}/B and A_{2+3}/B — have not been found yet.

Until now, 264 haemoglobin samples have been investigated and the following distribution could be obtained.

No explanation for these variations can be given at this stage and investigations on the genetic background of this new differentiation are under way.

References

Bangham, A. D. (1957). Nature 179 : 467.

Grimes, R. M., Duncan, C. W. and Lassiter, C. A. (1957). J. Dairy Sci. 40 : 1338.

Grimes, R. M., Duncan, C. W. and Lassiter, C. A. (1958). J. Dairy Sci. 41 : 1527.

Osterhoff, D. R. and van Heerden, J. A. H. (1964). Proc. IX. Animal Blood Group Conf., Prague. p. 299.

Schmid, D. O. (1962). Zentr. Bl. Vet. Med. 9 : 705.

Schmid, D. O. (1963). Zschr. Tierz. u. Zücht.-Biol. 79 : 286.

Shreffler, D. C. and Salisbury, G. W. (1959). J. Dairy Sci. 42 : 1147.

TF^G- A NEW TRANSFERRIN ALLELE IN CATTLE
(RESEARCH NOTE)

D. R. OSTERHOFF and J. A. H. VAN HEERDEN

Faculty of Veterinary Science, University of Pretoria and Blood Group Laboratory,
Animal Husbandry and Dairy Research Institute, c/o Veterinary Research Institute, Onderstepoort

Polymorphism in bovine serum beta-globulin was described for the first time in 1958 by Smithies and Hickman, and by Ashton. These beta-globulins were proved to be the specific iron-binding proteins of serum (Giblett et al. 1959) and were called transferrins. After Ashton's finding (1959) of two additional alleles in Zebu cattle the identification of the following 15 transferrin phenotypes was possible:

```
AA
        BB
AB              DD
        BD              EE
AD              DE              FF
        BE              DF
AE              DF
        BF
AF
```

Gahne (1961) reported on a new type of transferrin which was found in two animals from Norway. It is somewhat surprising that no other variants were detected during the extensive breed comparison studies by many workers.

During the study of almost 3,200 serum samples of 17 different breeds (Osterhoff et al. 1964) five animals were found with an interesting deviation from the known migration speed.

It can be seen that each of three clearly distinguishable bands of the new homozygous type migrated exactly one step behind the corresponding

Table 1

The distribution of the new phenotypes in different breeds

Breeds	Phenotype			
	GG	AG	DG	EG
Red Poll	1*)	—	1*)	1
Drakensberger	—	—	—	—
Simmentaler (Fleckvieh)	—	—	1	1

*) these two animals originated from the same sire family.

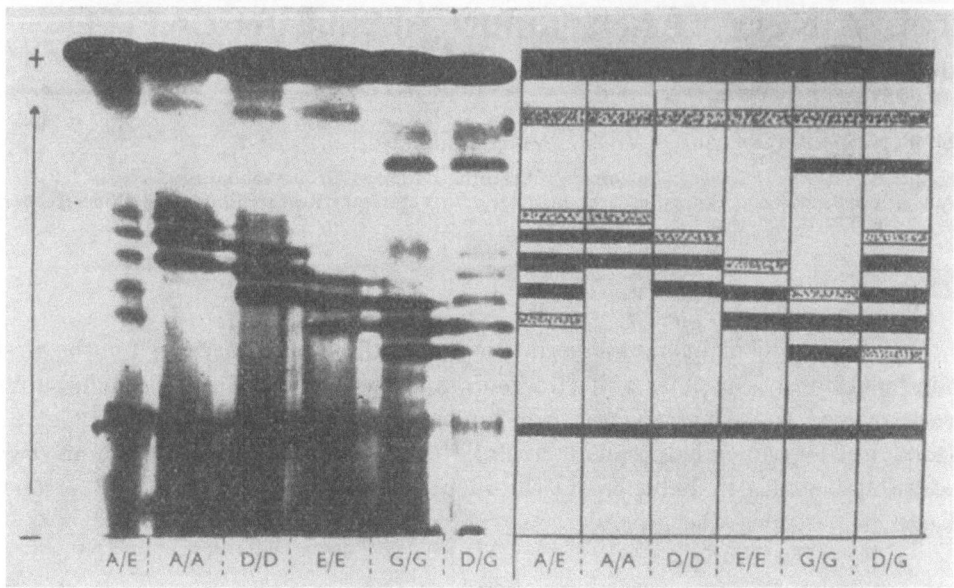

Fig. 1.

bands in the reference serum Tf EE next to it. After the detection of several heterozygous types it was possible to identify an additional transferrin allele, which was called Tf^G.

Due to the rare appearance of this allele, it will be extremely difficult to prove the exact genetic mechanism, but, according to the phenotypic exhibition of three zones in the starch block it can be accepted that we deal with another autosomal allelomorph.

Summarizing these findings it can be said that the new allele, Tf^G, has a very low frequency, about 0·001. It may appear sporadically by mutations or was originally present in different breeds and had been lost by selection through the ages.

References

Ashton, G. C. (1958). Nature 182 : 370.
Ashton, G. C. (1959). Nature 184 : 1135.
Gahne, B. (1961). Anim. Prod. 3 : 135.
Giblett, E. R., Hickman, C. G. and Smithies, O. (1959). Nature 183 : 1589.
Osterhoff, D. R. and van Heerden, J. A. H. (1964). Proc. IX. Animal Blood Group Conf., Prague. p. 299.
Smithies, O. and Hickman, C. G. (1958). Genetics 43 : 374.

HAEMOGLOBINS, TRANSFERRINS AND ALBUMINS OF SHEEP AND GOATS*)

G. EFREMOV** and M. BRAEND

Department of Medicine, The Veterinary College of Norway, Oslo

By means of paper electrophoresis Cabannes and Serain (1955) and Harris and Warren (1955) detected polymorphism of sheep haemoglobin. Three phenotypes explained by two alleles were found. Evans et al. (1956) called the two genes HbA and HbB where HbA controls the fastest moving component. Braend et al. (1964) and Braend and Efremov (1964) reported a new sheep haemoglobin which was named N. This has a slower rate of migration on starch gels than Hb B. The Hb N was found in sheep having the HbA gene, but not in all. It was variously expressed and if present usually found in small quantities. Extensive investigations have been carried out to see if the alleles HbA and HbB have any selective advantages and if they may be correlated with economical important traits (see Ogden 1961).

Transferrins of sheep have been studied by Ashton (1963) who reported the findings of 14 different alleles.

Goats are reported to have two Hb components (Harris and Warren 1955) and two transferrin alleles (Ashton and Mc Dougall 1958).

The present report deals with our studies of genetic polymorphism of the two normal haemoglobins, transferrins and albumins of Norwegian sheep and goats.

Materials and Methods

Samples from adult sheep of the native Norwegian Breeds, Spael, Dala and Steigar were investigated. The Spael sheep is a short tailed rather primitive type which is also called old Norwegian sheep. The two other native breeds are more or less influenced by British sheep. The Spael material consists of two subpopulations, one in North Norway and one in South Norway. From these 153 and 50 animals were examined. A total of 140 Dala and 80 Steigar

*) These investigations were supported in part by a research grant from the Agricultural Research Council of Norway.

**) On leave from Department of Physiology, University of Skoplje, Yugoslavia.

sheep were sampled. From the English breeds Cheviot and Oxforddown 68 and 56 samples were investigated. These two breeds have been kept pure in our country since preceding century. Occasional importations usually of rams have taken place.

A total of 108 goats were examined. They belong to the only native goat breed of this country.

The samples for haemoglobin studies (heparinized) were all fresh at the time of investigation. Plasma and serum samples were freshly frozen and kept at −25°C until used.

The technique was starch gel electrophoresis. Some details were given in another study (Braend and Efremov 1964). For sheep transferrins 25 g. starch, 16 ml. A, 19 ml. B and distilled water 250 ml., pH 8·2 was used in preparation of gels. Voltage was 220 volts for 35 minutes, then increase to 290 volts. When boundary line was 4 cm. beyond start line, a further increase to 380 volts was made. Electrophoresis was ended when boundary line has travelled 13 cm. beyond insertion line. For albumins the gels were made from 24 g. starch, 13 ml. A, 9 ml. B and distilled water ad 250 ml., pH 5·9. Starting voltage output was 220 which was increased to 300 after 15 minutes. When boundary line was 2 cm. beyond insertion line, voltage was increased to 400 volts.

For statistical analyses conventional methods were employed.

Results

Distributions of observed and expected haemoglobin phenotypes together with frequencies of Hb^A are presented in Table 1. There is agreement between observed and expected numbers. Altitudes are also given in Table 1. Spael sheep has a very high frequency of Hb^A although those we investigated live at sea level or at low altitudes. The Dala sheep has a lower frequency of Hb^A

Table 2

Observed and expected transferrin phenotypes in two breeds

Breed		DD	DG	DJ	DM	DP	DS	GG	GJ	GM	GP
Spael	obs.	1	18	11	8			31	27	35	
	exp.	2·49	18·11	7·39	7·65	0·13	0·76	32·95	26·91	27·85	0·47
Dala	obs.	12	19	31	4		1	7	26	7	
	exp.	11·14	18·61	30·18	5·36	0·28	2·25	7·78	25·22	4·48	0·24

Table 1

Observed and expected haemoglobin phenotypes in various sheep breeds. Gene frequencies

Breed		AA	AB	BB	Total	Hb^A	Altitude
Spael South	obs.	151	2	—	153	0·99	150 m
	exp.	150	3	—			
Spael North	obs.	49	1	—	50	0·99	Sea level
	exp.	49	1	—			
Dala	obs.	86	50	4	140	0·79	150 m
	exp.	88	46	6			
Steigar	obs.	45	33	2	80	0·77	Sea level
	exp.	47	29	4			
Cheviot	obs.	15	39	14	68	0·51	750 m
	exp.	18	34	16			
Oxforddown	obs.	—	15	41	56	0·13	100 m
	exp.	1	13	42			

than Spael as has also Steigar which is assumed to have its origin in Sutherland Cheviot. Most interesting may be the Hb^A frequency of Cheviot. The samples were collected from sheep living in three of our main valleys and at rather

Table 2

Observed and expected transferrin phenotypes in two breeds

GS	JJ	JM	JP	JS	MM	MP	MS	PP	PS	SS	Total
	2	9	1	6	4						153
2·78	5·49	11·37	0·19	1·14	5·88	0·20	1·18	0·00	0·02	0·06	153·02
	19	6	1	5			2				140
1·89	20·44	7·26	0·39	3·06	0·65	0·07	0·54	0·00	0·03	0·11	139·98

high altitudes. In the summer months they are pasturing at even higher altitudes.

The transferrin studies revealed 6 different two band patterns and 15 different phenotypes. Fig. 1 shows a photograph of a stained gel where four representative samples were investigated. As it has been reported by Ashton

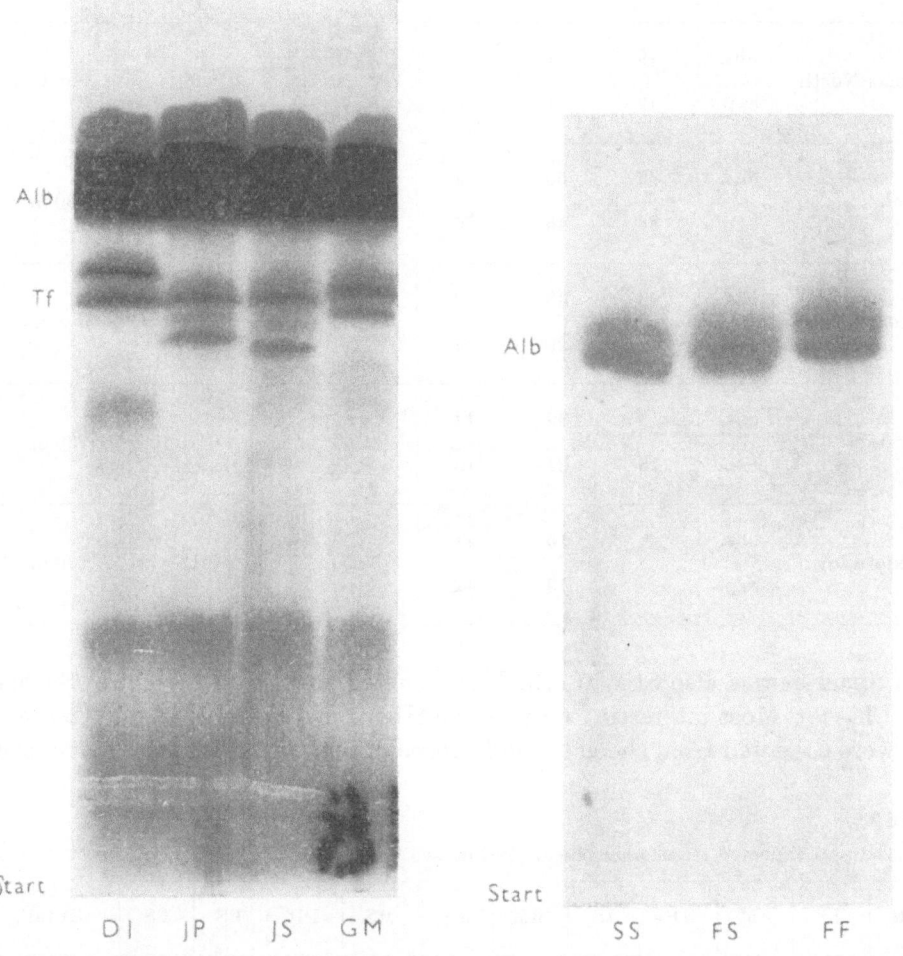

Fig. 1. Photograph of a stained gel showing individual positions of transferrin bands.

Fig. 2. Photographs showing appearance of the three albumin phenotypes on stained starch gels.

(1963), the product of the individual transferrin allele appears as two bands on starch gels, the slowest moving being the strongest one. We have not been able to obtain reference samples and accordingly the transferrins were preliminarily named as we would prefer to have them named. The D bands are the fastest and the S bands the slowest. There is some indication that the J bands are of two types showing very small differences in the rates of migration. Distributions of transferrin phenotypes for Spael and Dala are given in Table 2. There is agreement between observed and expected numbers. For Spael North, Steigar and Oxforddown there was also agreement between observed and expected numbers of Tf phenotypes. For Cheviot there was significant difference from the expected numbers. If we compare the breeds as to Tf frequencies we find significant differences between them all and even between the two subpopulations of Spael sheep (tab. 3).

Table 3

Transferrin gene frequencies in various sheep breeds

Breed	Tf^D	Tf^G	Tf^J	Tf^M	Tf^P	Tf^S	Total
Spael South	0·13	0·46	0·19	0·20	0·003	0·02	153
Spael North	—	0·37	0·30	0.33	—	—	50
Dala	0·28	0·24	0·38	0·07	0·004	0·03	140
Steigar	0·59	0·14	0·18	0·09	—	—	80
Cheviot	0·43	0·13	0·43	—	—	—	68
Oxforddown	0·13	0·40	0·41	0·05	0·02	—	56

Fig. 2 shows a photograph of a stained gel where samples representing the three different albumin phenotypes were investigated. The slowest phenotype (SS) which is the most common appears as two bands, one strong

Table 4

Observed and expected albumin phenotypes in two breeds; Gene frequencies

Breed		FF	FS	SS	Total	Al^F
Spael South.	obs.	2	36	115	153	0·13
	exp.	3	34	116		
Spael North	obs.	—	14	36	50	0·14
	exp.	1	12	37		
Dala	obs.	—	8	132	140	0·03
	exp.	—	8	132		

and one weaker in front. The FF type which was found in two animals only shows a similar appearance, the difference being the rate of migration. The FS type has three bands which match the respective two bands in the FF and SS types. Accordingly, the middle band of the FS type is the strongest.

Distributions of observed and expected albumin phenotypes are presented in Table 4. There is agreement between observed and expected numbers. We have designated the two alleles Al^F and Al^S. The Al^F allele has a low frequency in the Spael breed, but even lower in Dala. It was rarely found in Oxforddown and Cheviot but not in Steigar.

The goats all showed the same Hb type, a type indistinguishable from Hb B of sheep. They all had the same transferrin type which appeared as two bands, one strong slow and a faster weak one. These bands did not match any of the sheep Tf bands, being slightly slower than the J bands. All goats have the same albumin phenotype.

Discussion

Evans et al. (1957) and Efremov (1963) found that lowland breeds were predominantly of Hb type B. In highland breeds the situation was reversed. Huisman et al. (1958) found differences in the oxygen dissociation curves of the two sheep haemoglobins, with the A type having the greater affinity. These results have been explained by selective advantage. The B type should not be able to compete successfully at higher altitudes.

Our studies show that the Hb^B gene is very rare in Spael sheep even though those we examined lived at low altitudes. On the other hand, high frequency of the Hb^B allele was found in Cheviot living at high altitudes. These results seem to contradict the previous mentioned findings. There are, however, many factors, for instance, origin and arificial selection, to be taken into consideration when discussing this problem in detail, which is not possible in this short report.

In our transferrin studies there was general agreement between observed and expected distributions of phenotypes in all breeds except Cheviot. In our opinion, the deviation in the Cheviot sample may be due to simple chance, to too small sample investigated or because the material consisted of small subpopulations in geographically separated areas.

In studies of albumin variations the observed distributions of albumin phenotypes agree with the genetic theory of two codominant alleles. Furthermore, when considering the appearance of the albumin bands on stained starch gels, we are of the opinion that the genetic theory is correct.

Whether the albumin alleles have any adaptive significance is too early to discuss. However, we are most inclined to consider this variation to be another example of genetic polymorphism. The F and S symbols were employed because these letters were used by McIndoe (1962) for chicken and by us for other domestic animals (Braend and Efremov 1964) and man (Efremov and Braend 1964). We also prefer these designations in contrast to, for instance, A and B, if possible new alleles have to be named.

Summary

Gene frequency studies of Hb^A and Hb^B showed the Hb^A gene to have a frequency of 0·99 in the native Norwegian sheep called Spael. The sheep investigated lived at sea level or low altitudes. Other Norwegian sheep as Dala and Steigar had lower frequency of Hb^A. Cheviot sheep even living at high altitudes has a Hb^A frequency of 0·51.

Transferrin studies resulted in findings of 15 different phenotypes explained by the occurrence of 6 codominant alleles. Four of them, preliminarily called Tf^D, Tf^G, Tf^J and J^M were common in most breeds. The alleles Tf^P and Tf^S were of low frequencies and not found in all breeds. There were significant differences between breeds as to Tf gene frequencies.

Polymorphism of albumin is described by the occurrence of three phenotypes. Their appearance on stained starch gels and the observed distributions agree with the genetic theory of two codominant alleles, which were named Al^F and Al^S. The Al^F is rare but most common in Spael sheep where it has a frequency of 0·13.

References

Ashton, G. C. (1963). Genetical Research 4/2 : 240—247.
Ashton, G. C. and McDougall, E. I. (1958). Nature 182 : 945—946.
Braend, M., Efremov, G. and Helle, O. (1964). Nature, in press.
Braend, M. and Efremov, G. (1964). Science, in press.
Braend, M. and Efremov, G. (1964). Prague, 9th European Conference for Animal Blood Group Research.
Cabannes, R. and Serain, Ch. (1955). C. R. Soc. Biol. (Paris) 149 : 1193.
Efremov, G. (1963). Thesis, University of Beograd.
Evans, J. V., King, I. W. B., Cohen, B. L., Harris, H. and Warren, F. L. (1956). Nature 178 : 849.
Evans, I. V., Harris, H. and Warren, F. L. (1957). Biochem. J. 65 : 42 P.

Harris, H. and Warren, F. L. (1955). Biochem. J. 60 : 29 P.
Huisman, T. H. I., Van Vliet, G. and Sebens, T. (1958). Nature 182 : 172.
Ogden, A. L. (1961). Animal Breed. Abstr. 29 : 127.
McIndoe, W. M. (1962). Nature 195 : 353.

GENETIC DETERMINATION OF THE SERUM „THREAD PROTEINS" AND THE SLOW α_2 GLOBULIN POLYMORPHISM IN PIGS

J. SCHRÖFFEL

Laboratory of Physiology and Genetics of Animals, Czechoslovak Academy of Sciences, Liběchov

In all animals, whose sera have been tested so far by means of starch gel electrophoresis, it is possible to discern pronounced fractions between the start and transferrin region. The protein forming this fraction has been designated by Smithies (1959) $S\alpha_2$ globulin (S slow), sometimes it is also called α_2 macroglobulin.

Tests of pig blood serum samples in our laboratory showed that in some animals the $S\alpha_2$ globulins are divided into 2 fractions the relative distance of which is different and displays a certain genetic regularity, the same as the occurrence of only 1 fraction (Schröffel, Hojný 1962).

In order to extend the number of genetic systems of serum polymorphic characters suitable for practical determination of paternity in pigs, we used the thread proteins, described first in cattle (Ashton 1958b) and later in pigs (Ashton 1960a). For these special thread-like fractions, genetic control through two alleles has been suggested by the author, the corresponding alleles being called T^A and T^B. With this mode of inheritance, each allele would correspond to one fraction, heterozygotes would show two fractions. We have found an exception to this scheme; in some individuals this fraction could not be identified. And therefore we wanted to check the system of inheritance of these fractions. Our results together with a proposed system of genetic control of $S\alpha_2$ globulins are presented in the following report.

Material and Methods

Blood serum samples from pigs of the Large White, Black and White-Přeštice and Landrace breeds were used for the analyses. The tested set of animals comprised 278 offspring, originating from 34 different matings, including the parents and a group of further 37 animals. Prior to analysis, the serum was kept frozen at $-20°C$ and was not thawed more than three times. The sample was inserted in the insertion line by means of a double piece of saturated filter paper Whatman 3 which was removed from the gel after 30 minutes.

We used a modification of the vertical electrophoresis (Smithies 1959) with water cooling from the bottom of the gel and a continuous phosphate buffer system according to Ashton (1960b).

The gel was prepared by boiling our own hydrolyzed potato starch in phosphate buffer diluted 1 : 50 and in a concentration of 16 g. starch per 100 ml. buffer. After inserting the samples and covering the insertion line with a mass of boiled starch, the surface of the gel was covered with an even layer of paraffin.

The voltage gradient amounted to 7—8 V/cm. at a 6 mm. strength of the gel and a seven-hour period of electrophoresis. In one gel ten samples were tested simultaneously, two of which usually were standard sera of known type. The usual staining technique with Naphthalene Black 10 B was used.

Results

Sα₂-globulins

The electrophoretogram on Fig. 1 shows the 6 possible phenotypes of the $S\alpha_2$ globulins and the different phenotypes are given in the scheme in Fig. 2.

By analyzing the progeny of parents of different mating types it was proved that the differences between the single phenotypes were really an expression of individuality of the given animal and were relatively stable throughout the life; this was also proved in 37 pigs, mostly of the Landrace breed, originating from 8 litters. In 27 pigs the samples were taken at the interval of 1—2 month at the age of 1—8 months, in 4 pigs at 7—14-day intervals, at the age of 6—13 months, and in 6 pigs at varying intervals from birth to 24 months. The type of $S\alpha_2$ globulins, as originally identified, remained unchanged throughout the observation, even at a reciprocal comparison of samples from several collections, as far as the sera were not denatured by prolonged storage.

The fractions of all types showed a slightly higher mobility in the gel in adult pigs than in piglets, while the relative distance within one type or between the different types remained the same. The $S\alpha_2$ globulin fractions are as a rule sharper and more regular in young pigs than in adult animals. We had no difficulties, however, in determining the different types when analyzing sera samples of adult pigs or piglets in the gel at the same time.

The occurrence of 6 phenotypes of $S\alpha_2$ globulins suggests that the individual types are genetically controlled by a simple autosomal locus with 3 different alleles. We have designated these alleles $S\alpha_2^A$, $S\alpha_2^B$ and $S\alpha_2^C$ with a corresponding expression of genotypes

$$S\alpha_2^A/S\alpha_2^A, \ S\alpha_2^B/S\alpha_2^B \quad \text{etc.}$$

The facts in support of the proposed hypothesis, obtained by analyzing 278 offspring from 34 litters of different types of mating, are summarized in table 1; there is a satisfactory agreement between expected and observed numbers of animals (X^2 2·17 5 d. f.). although not all types of mating were available.

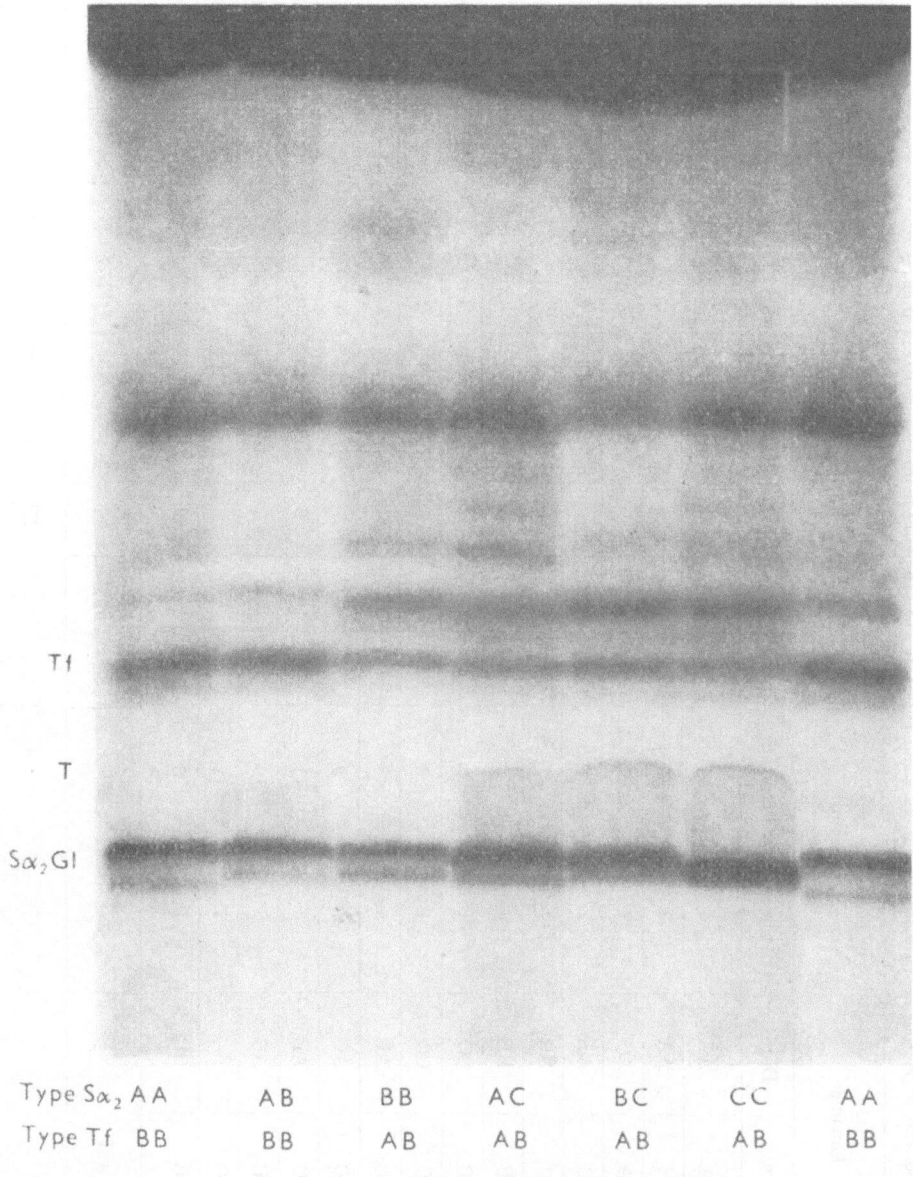

Fig. 1. Starch gel electrophoretic patterns of $S\alpha_2$ globulins, thread proteins (T) and transferrins (Tf).

Table 1

Observed and expected distribution of Sα_2-globulin types in 278 offspring from different matings

| Matings | | Number of | | Type of Slow α_2-globulins | | | | | | | | | | | |
Sire	Dam	litters	offspring	AA obs.	AA exp.	AB obs.	AB exp.	AC obs.	AC exp.	BB obs.	BB exp.	BC obs.	BC exp.	CC obs.	CC exp.
AA	AA	4	34	34	34·00										
AB	AA	3	25	11	12·50	14	12·50								
AA	AC	1	7	3	3·50			4	3·50						
AC	AA	2	18	8	9·00			10	9·00						
AC	BB	4	27			16	13·50					11	13·50		
AB	BB	3	22			8	11·00			14	11·00				
AB	CC	1	8					6	4·00			2	4·00		
AC	CC	1	3					1	1·50					2	1·50
AB	AB	5	48	12	12·00	24	24·00			12	12·00				
AC	AC	1	9	2	2·25			5	4·50					2	2·25
AB	AB	5	41	9	10·25	8	10·25	13	10·25			11	10·25		
AC	AC	2	19	2	4·75	7	4·75	4	4·75			6	4·75		
AB	BC	1	12			3	3·00	4	3·00			3	3·00	2	3·00
AC	BC	1	5			2	1·25	1	1·25	1	1·25	1	1·25		
Total		34	278	81	88·25	82	80·25	48	41·75	27	24·25	34	36·75	6	6·75

$X^2 = 2.17$ $0.90 > P > 0.80$ 5 d. f.

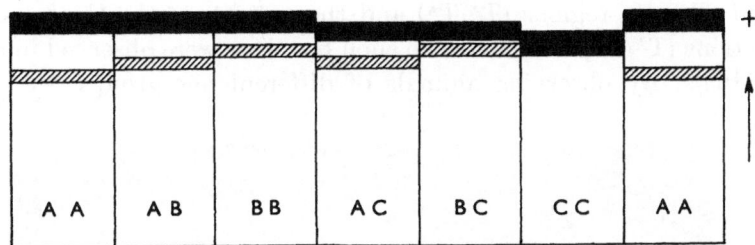

Fig. 2. Scheme of 6 possible phenotypes of $S\alpha_2$ globulins in pigs.

According to our results, the $S\alpha_2$ globulin system appears to be independent of other systems of serum protein polymorphic characters (transferrins, haptoglobins, prealbumins, and thread proteins) the types of which have been determined on all samples analyzed at parallel testing. No differences could be detected with regard to sex of the tested animals (Table 3).

Thread proteins

The thread protein fractions were determined together with the $S\alpha_2$ fractions on the same gel. In agreement with the data published by Ashton (1960a) the tested animals could be divided into 3 groups according to the types of these proteins (Fig. 3). The first shows only one arc-like fraction between $S\alpha_2$ globulins and transferrins (genotype T^B/T^B), in the second group

Table 2

Observed and expected distribution of thread-protein types in 159 offspring from different matings

| Matings | Number of | | Type of thread-proteins | | | | | |
| | | | AA | | AB | | BB | |
	litters	offspring	obs.	exp.	obs.	exp.	obs.	exp.
AA × AA	1	8	8	8·00				
AA × AB	2	12	7	6·00	5	6·00		
AA × BB	1	7			7	7·00		
AB × AB	2	15	2	3·75	11	7·50	2	3·75
AB × BB	5	50			23	25·00	27	25·00
BB × BB	8	67					67	67·00
Total	19	159	17	17·75	46	45·50	96	95·75

325

the same fraction is located approximately in the middle between the transferrin and albumin regions (T^A/T^A) and the animals of the third group show both fractions (T^A/T^B). Moreover, no such fractions were observed in a number of individuals. By observing animals of different age groups, we succeeded

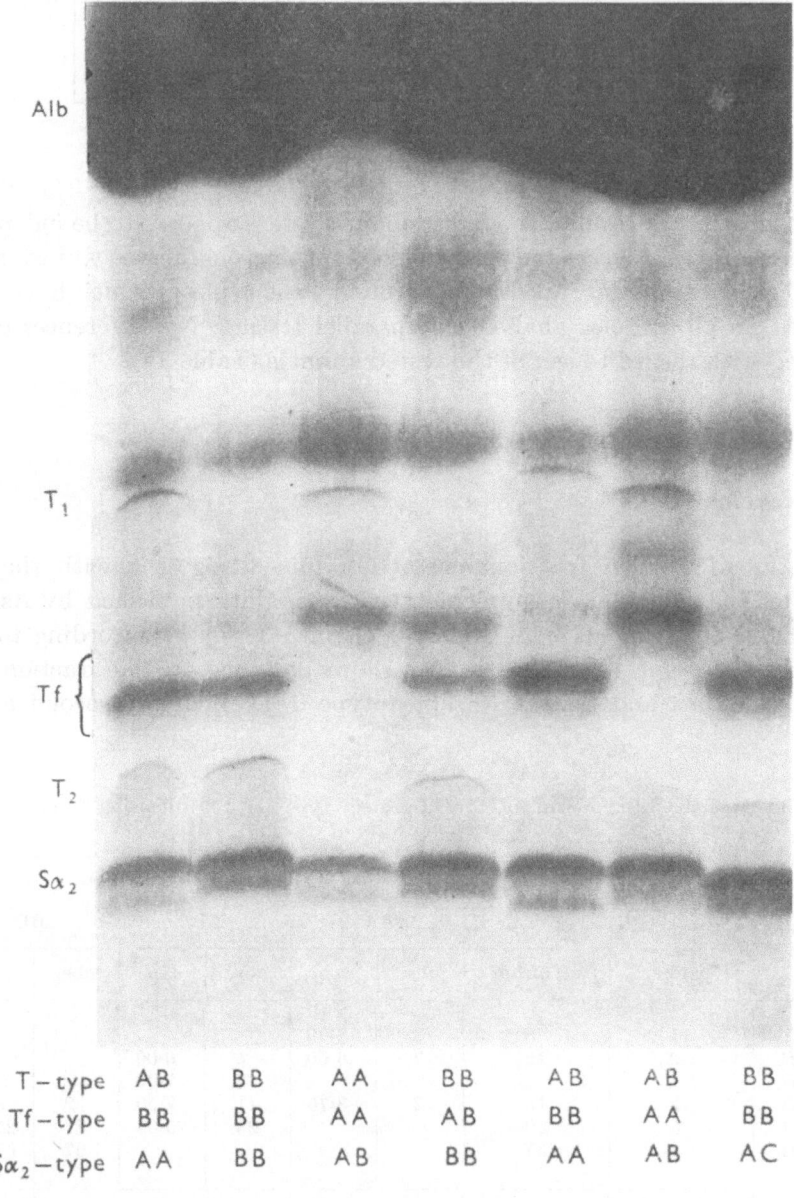

T – type	AB	BB	AA	BB	AB	AB	BB
Tf – type	BB	BB	AA	AB	BB	AA	BB
$S\alpha_2$ – type	AA	BB	AB	BB	AA	AB	AC

Fig. 3. Starch gel electrophoretic patterns of thread proteins (T), transferrins (Tf) and $S\alpha_2$ globulins in pigs.

to confirm our previous hypothesis that the occurrence of these fractions is dependent on the age of pigs (Schröffel, Hojný 1962) and the fractions can be proved in the gel by means of the mentioned method only in animals at the age of over 4—5 weeks, and there are even differences in the time when the fractions can be identified either in different litters or, at a smaller rate, among piglets from one litter. Tables 3 and 4 comprise only litters where the T type was determined in all piglets. There is a good agreement between the expected and observed separation and this type does not appear to be linked to sex (χ^2 1·81 1 d. f.).

Table 3

Slow α_2-globulins and thread-protein distribution between males and females

$S\alpha_2$-type	♂	♀	Total	T-type	♂	♀	Total
AA	40	32	72	AA	10	8	18
AB	48	38	86	AB	24	21	45
AC	24	24	48	BB	54	42	96
BB	12	12	24				
BC	17	23	40				
CC	4	2	6				
Total	145	131	276	Total	88	71	159
$X^2 = 0.71$ $0.50 > P > 0.30$ 1 d. f.				$X^2 = 1.81$ $0.30 > P > 0.10$ 1 d. f.			

Discussion

Tests of $S\alpha_2$ globulin types by means of the mentioned method require that standard conditions are maintained owing to the subtle differences between the phenotypes.

The proposed 3-allele system of genetic control of $S\alpha_2$ globulins in pigs increases the number of genetically controlled polymorphic characters in the blood serum. The mode of inheritance is somewhat different from the known systems, where one allele is responsible for the presence of one fraction (haptoglobins in pigs) or a group of fractions (transferrins in cattle) where the localization of the fractions remains unchanged in a heterozygote. The alleles $S\alpha_2^A$ and $S\alpha_2^B$ each control two fractions and homozygous animals differ from each other only in the relative distance of both fractions. In a heterotygote $S\alpha_2^A/S\alpha_2^B$ the distance between the fractions has an intermediate character. In both homozygotes (AA and BB) and in the heterozygote AB, the fraction which

327

is more distant from start shows the same migration, only the localization of the slower fraction changes.

The $S\alpha_2^C$ allele controls only one sharp fraction, the localization of which can be detected in the middle of the reciprocal distance of a homozygote AA. We cannot exclude the possibility of two fractions, with but a very slight difference, which could not be separated by the method used. The heterozygotes AC and BC show two fractions again and a common feature to both these phenotypes is a lower migration, a more distant fraction (with regard to start) as compared with animals of the type AA, AB and BB. The respective distance of the $S\alpha_2$ globulin fractions of the types BB and AC is approximately the same with the difference that the pair of fractions of the AC type is located nearer to start as compared with the type BB. Both fractions of the heterozygote BC lie close to each other but their identification, when using a standard serum, presents no difficulties.

The frequency of the $S\alpha_2^A$, $S\alpha_2^B$ and $S\alpha_2^C$ alleles appears to be very favourable in all three breeds studied (Large White, Black and White-Přeštice, Landrace) but it could not be calculated because representative sets of unrelated animals have not yet been available. The thread proteins are of much interest for further research because their substance has not yet been studied, as far as known from available literature. The fact itself that their determination is dependent on age of the animal is interesting, but we should like to point out a few further interesting features. In a gel, prepared in phosphate buffer, these fractions form a partition between the darker and the lighter parts of the electrophoretogramm of a certain animal. The darker gel, probably containing some spreading protein, lies between the fractions and start, which is particularly evident in adult animals. On some gels, a further pair of thread-like fractions, similar to the type of the proper thread proteins, could be identified quite close to start. Moreover, there is another interesting fact that, when interrupting the electrophoresis for a longer period of time (e.g. 1 hour), two sharp fractions instead of the original one (homozygotes T^A/T^A and T^B/T^B) or four fractions instead of the original 2 in heterozygotes could be identified on the stained gel. A short interruption as needed to remove the papers does not produce the same phenomenon. The fractions do not disappear nor are in other ways affected even when the papers are kept in the gel during the whole period of electrophoresis. None of the mentioned phenomena, however, produced a negative effect on the identification of the T types and therefore does not affect the genetic hypothesis proposed by Ashton (1960).

The number of genetic systems of serum protein characters known in pigs is not final yet, we were able to detect another polymorphism in the postalbumin region. So far, we have found 4 phenotypes and we will continue in our attempts to detect the mode of their genetic control.

Summary

By means of the vertical system of starch gel electrophoresis in a continuous phosphate buffer, it has been possible to detect polymorphism in another genetic system of serum proteins in pigs. Evidence presented in this paper suggests that the 6 phenotypes of $S\alpha_2$ globulins found, are genetically controlled by a simple autosomal locus with three interchanging alleles, designated $S\alpha_2^A$, $S\alpha_2^B$, $S\alpha_2^C$.

The occurrence of thread proteins in pigs was found to be influenced by the age of animals, the fractions being detectable 4—5 weeks after birth.

I am very grateful to Mrs Langerová for her technical assistance and indebted to Mr. Hojný and Mr. Ling for their help with collections of samples.

References

Ashton, G. C. (1958a). Nature 182 : 193.
Ashton, G. C. (1958b). Nature 182 : 650.
Ashton, G. C. (1960a). Nature 186 : 991.
Ashton, G. C. (1960b). J. Agr. Sci. 54 : 321.
Gahne, B. (1962). The 8th Animal Blood Group Conference, Ljubljana, Yugoslavia.
Schröffel, J., Hojný, J. (1962). Čs. fysiologie 11 : 277.
Smithies, O. (1959). Zone electrophoresis in starch gels and its application to studies of serum proteins. Advances in Protein Chemistry. Academic Press, New York, 1959.

INFLUENCE OF SOME GONADOTROPHIC AND ANDROGENIC HORMONES ON THE MALE SEXUAL FRACTION IN THE SERUM OF PIGS

J. MATOUŠEK and J. SCHRÖFFEL

Laboratory of Physiology and Genetics of Animals, Czechoslovak Academy of Sciences, Liběchov

In a previous paper (Matoušek, Schröffel 1964) we reported of the detection of a protein fraction in the serum of sexually mature boars by means of electrophoresis in starch gel. (Fig. 1.) It has been also found that the mentioned fraction cannot be identified 3 weeks after castration.

The absence of the fraction in sexually immature males and its disappearance in sexually mature animals after castration led us to the assumption that its synthesis and detectability may be dependent on hormones controlling sexual functions.

In the present report we submit a brief survey of the results obtained.

Material and Methods

FSH (Folicotropin Spofa), Urinal Choriogonadotrophin (Praedyn Spofa), a combination of both and PMS (pregnant mare serum) were used to elicit premature sexual maturity and biosynthesis of the male protein fraction in the serum.

The above hormone preparations were injected i. m. into three boars of the Landrace breed, originating from one litter, in the following combinations. The boar 48/3 was given Folicotropin Spofa, in amounts of 15 RU per injection, on alternate days from the age of 12 to 16 weeks, 200 I. U. per dose of PMS, also on alternate days, from the age of 16 to 26 weeks. The boar 48/4 was given Praedyn Spofa in amounts of 1,000 I. U. of Choriogonadotrophin per injection from the age of 12 to 26 weeks. The boar 48/1 was given, at the same age, and timing, as the boar 48/4 500 I. U. of choriogonadotrophin hormone and 10 R. U. of FSH. The boar 48/5 from the same litter was not injected with any of these preparations and served as a control. In all these boars the length of the testicles was measured by means of a vernier caliper.

Two other boars from another litter of the Large White Breed No. 14/1 and No. 14/3 were injected i.m. with testosterone propionate (Testosterone propionate Spofa). The hormone was administered on alternate days in

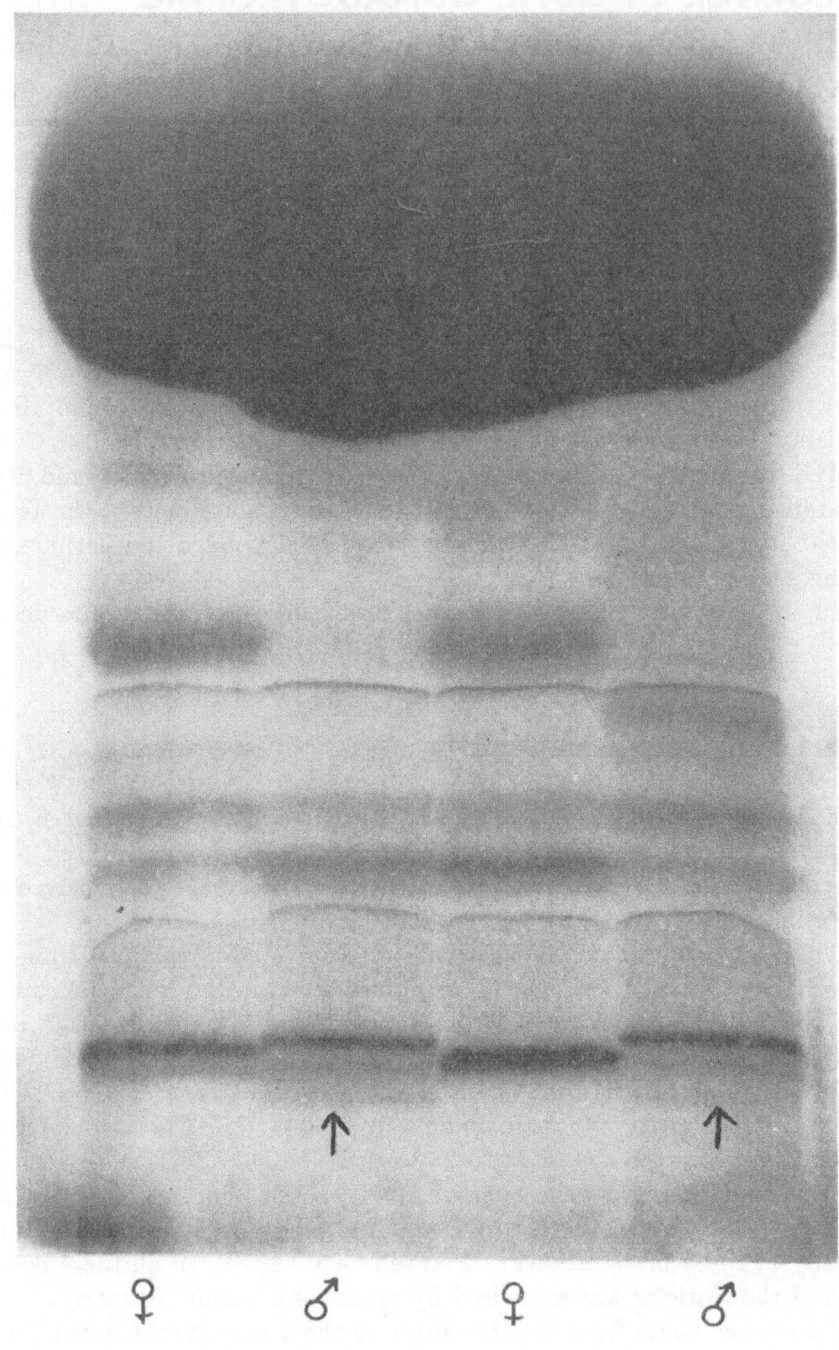

Fig. 1.

amounts of 30 mg. per injections. Another boar from the same litter No. 14/2 served as control animal. At the age of 39 weeks one of the treated animals, 14/1 and the control boar 14/2, were slaughtered, their genital organs brought to the laboratory, weighed and examined histologically.

After the detection of the male sexual fraction in the serum of the experimental boar 48/4 at the age of 27 weeks, this animal was castrated and beginning with the next day injected i. m. with 20 mg. of testosterone propionate every day.

In addition to the above animals, the boar 14/4 whose seminal vesicles had been removed at the age of 16 weeks, was observed from that age on, with regard to the male protein fraction in the serum.

The male protein fraction in the serum of experimental and control animals was determined by means of starch gel electrophoresis, in the vertical system and a modified method of the continuous phosphate buffer according to Ashton (1957) and Schröffel and Matoušek (1962).

Results and Discussion

By administering gonadotrophic hormones (FHS, choriogonadotrophins, PMS) to three boars at the age of 12—26 weeks, it has been impossible to promote a rapid growth of the testicles or an earlier detection of the male genital fraction in their serum. On the contrary, a considerable depression of the testicle growth could be observed in the boar 48/1 treated with combined FSH and choriogonadotrophin. The average length of his testicles at the age of 26 weeks reached 48 mm., whereas it ranged between 85 and 96 mm. in the other two animals and in the control boar. The male fraction in his serum was not detected even at the age of 9 months. After the slaughter at that age only primary and secondary spermatocytes were found in the tubuli seminiferi contorti. We assume, however, that this was due to an inborn defect, because a similar disturbance occurred in other related animals without hormonal treatment.

In the other two boars, injected with gonadotrophin preparations, and the control boar, the male sexual fraction was detected after 27—29 weeks. At the same age their spermatogenesis also was fully developed, because after the castration of one of the experimental boars (48/4) all stages of development of genital cells were found in the tubuli seminiferi contorti and even motile spermatozoa in the tail of the epididymis.

Similar, i.e. negative results with obtaining sexual maturity by means of gonadotrophic effect, were noted by Dörner and Deckart (1962) in immature male rats.

A more pronounced and more specific influence on sexual maturation and early detectability of the male sexual fraction in the serum was obtained with testosterone propionate.

Injections of this hormone to two boars caused a striking delay in growth of the testicles in both animals and also significantly delayed the appearance of the male protein fraction in their serum. In the control boar 14/2, not treated with hormones, the genital fraction was detected at the age of 25 weeks, and the size of the testes corresponded to normal conditions, whereas in the boar 14/1, treated with hormones, the male sexual fraction was not found even at the age of 9 months, when the two boars were taken to a slaughterhouse. A significant delay in the sexual development of the boar treated with testosterone propionate, though not in growth of the seminal vesicles, is shown in the following table.

Indices	Boars	
	14/1 injected with TP	14/2 control
Weight of both testicles	102 g	503 g
Weight of both epidydimes	45 g	90 g
Degree of spermatogenesis	secondary spermatocytes	spermatozoa
Weight of the bulbo-urethral glands	146 g	135 g
Weight of the seminal vesicles	379 g	158 g
Amount of fluid of seminal vesicles	203 ml	78 ml

The second of the two treated boars (14/3), also sexually immature according to the size of testicles and absence of the male sexual fraction in his serum, has been kept alive. When he was 39 weeks old, treatment with testosterone propionate was interrupted. From this time testicles began to grow rapidly. But the normal size of testicles, reached by the control boar at the age of 25 weeks, has been attained by this animal at the age of 51 weeks. The male protein in the serum was detected by means of electrophoresis also at the mentioned period (in the control boar at the age of 25 weeks).

From the above results it can be assumed that the male sexual fraction in the serum is not directly and exclusively dependent on gonadotrophinsalone or only on androgens. In spite of the fact that there were androgens in the organism so as to cause a hypertrophy of the seminal vesicles and even affect their fluid production, they were unable to induce biosynthesis of the male protein fraction. Neither has the presence and function of the seminal vesicles and other accessory genital glands (e.g. the bulbo-uretral gland) any significant influence on its biosynthesis, which can be seen not only from the mentioned results, but also from the normal presence of this fraction in the serum of boar whose seminal vesicles were removed before sexual maturity.

It was of interest to find out whether biosynthesis of the studied fraction could be maintained in the serum of a sexually mature boar, whose testes would be removed and the incretory function replaced by treatment with testosterone propionate. But this experiment also failed. Although the castrated boar was given 20 mg. of testosterone propionate daily from the first day after castration, the sexual protein fraction disappeared from the serum within 18 days of castration.

It seems therefore that biosynthesis of the male sexual fraction in the serum is strictly bound to the whole complex of metabolic functions. These are, we assume, under a considerable but balanced control of hormones, and quantitatively and qualitatively connected with complete spermatogenesis. As soon as some disturbance occurs, which is followed by an interruption in spermatogenesis, biosynthesis of the male protein fraction in the serum does not appear either. It is possible to give the following examples of such disturbances: an extreme increase in the amount of androgens in the organism and consequently decrease in hypophysis activity according to gonadotrophins, hypoplasis of the testicles (boar 48/1), removal of the spermatogenic epithelium by castration (boar 48/4).

Summary

The effect of FSH (Folicotropin Spofa), urinal choriogonadotrophin (Praedyn Spofa), combination of both and PMS (pregnant mare serum) on growth of testes and detectability of the male sexual fraction in the serum of immature boars were studied. None of the above hormones did accelerate growth of testes. The earlier occurrence of the male protein fraction in the serum was not observed, too.

The androgens, injected in the form of testosterone propionate (Testosterone propionate Spofa) inhibited growth of testes and also significantly delayed the appearance of the male protein fraction in serum of immature boars. On the other hand, growth of the seminal vesicles was stimulated by testosterone propionate. It seems therefore, that the male sexual fraction in serum is not only dependent on the amount of androgens in blood, but also on the degree of maturity of testes.

Testosterone propionate could not maintain biosynthesis of the male sexual fraction in the serum, not even in the boar castrated during sexual maturity.

References

Ashton, G. C. (1957). Nature 180 : 917.
Dörner, G. and Deckart, H. (1962). Acta biol. med. germ. 9 : 271.
Matoušek, J. and Schröffel, J. (1964). Folia Biologica (Prague) 10 : 30.
Schröffel, J. and Matoušek, J. (1962). Živočišná výroba 5 : 313.

SERUM PROTEINS IN CANIDAE: SPECIES, RACE AND INDIVIDUAL DIFFERENCES

M. KAMINSKI and H. BALBIERZ
with technical assistance of N. BRUNET
Laboratory of Histophysiology, University of France, Paris

Two species of pure-bred foxes from a fur-supplying farm were investigated by comparing proteinograms and zymograms obtained by starch and agar gel electrophoresis.

These two species are Alopex Lagopus L., called blue fox and Vulpes vulpes, a variety which gives the "silver fox" furs. Some litters of silver foxes yield a low frequency of a mutant, called platinum fox. No natural hybridization is known between these two species.

81 sera of blue, 22 of silver and 4 of platinum foxes were examined; the preliminary analysis using the tris-citrate buffer in starch gel and the borate buffer in electrode vessels showed definite differences between sera of two species, and in addition, differences between sera of silver and platinum animals (brothers and sisters from the same litter). Several individual variations were also noted in each group, thus extending the concept of serum protein polymorphism to one more family.

The buffer system used did not permit, however, a good resolution of β-globulin bands; therefore a reduced number of sera have been re-examined using a buffer system containing lithium hydroxide. The general migration was not changed on the whole, but more bands could be observed, and they were better defined.

Fox serum proteinograms are much different from those of horses or cattle, examined under the same conditions. Instead of being located at the anodic side of the plate, just behind the post-albumins, (β-globulins) the variable fractions in fox sera (6 bands altogether) have a much slower mobility, comparable to that of human haptoglobin.

For 24 sera of blue foxes the number of different phenotypes appears rather high, owing to many bands of intermediate mobilities; they were classed into two types (Table 1). The type I is characterized by the presence of the slower slowest band, called 1, and the band 3 being somewhat slower than in type II. Band 2 is absent from 3 sera out of 6, and band 4 is rather fast in 4 sera; in 4 sera band 3 is very intensely stained.

The type II has band 6 present in all sera examined, and band 7 is relatively frequent (it is present in only one serum of type I), but weak.

Table 1

Frequencies of individual protein bands in fox sera

Band	Present in	Total number of sera
1	6 sera	25
2	18	
2 rapid	4	
3 slow	6	
3	12	
3 rapid	6	
4	15	
4 rapid	4	
4 double	1	
5 slow	11	
5	17	
6 slow	6	
6	18	
7	9	

Frequencies of main patterns

I type bands						N° of sera	%
	1,	2,	3s,	4r,	5s	3	
	1,		3s,	4,	5	2	
	1,		3s,	4r,	5, 7	1	
pattern	1,		3 ,	4 ,	5	6	24
II type	2,	3,	4,	5,	6 —	10	
	2,	3,	4,	5,	6, 7	—3	
pattern	2,	3,	4,	5,	6, —	13	52
	2,		4,	5,	6	1	
	2r,	3r,		5s,	6s, 7	—5	

Only silver fox serum in this series of analyses was of type II, but some bands were slower, and the band 7 very intensely stained.

Generally, bands 1 and 2 are weak; band 4 also, except for two sera where it was faster than usual and very strong; band 5 is the most important in sera of type II; band 6 is strong in all sera lacking band 4. Type II accounts for 76% of analyzed sera, and the most frequent pattern is lacking only the band 1 : 2, 3, 4, 5, 6, — 52%.

It seems that all observed phenotypes are heterozygotes.

The esterases in fox sera, either blue or silver, are definitely more numerous and more differentiated than in Equidae (Kaminski, Gajos 1964). Depending on the substrate used, the obtained zymograms are different: with α-naphtyl or indoxyl acetates one strong and one weak bands are revealed, located immediately on the anodic side of the $S\alpha_2$ band, using buffer without lithium, and between bands 1—3 using the lithium-containing buffer. These two esterase spots are designated E_2 and E_3; they were both inhibited by prostigmine, by DFP and by p-chloro-mercuri-benzoate. The E_2 spot was the only one visible after incubation with carbonaphtoxycholine.

Using the β-naphtyl acetate 5 esterase spots were revealed: E_1, slower than $S\alpha_2$; E_2; E_4 and E_5: very intensely stained spots, located at the region of variable protein bands but not corresponding to either of them; a weak, fast esterase was situated immediately behind albumin; this reaction was rapidly inhibited by excess substrate — E_6.

Starch-gel proteinograms of:

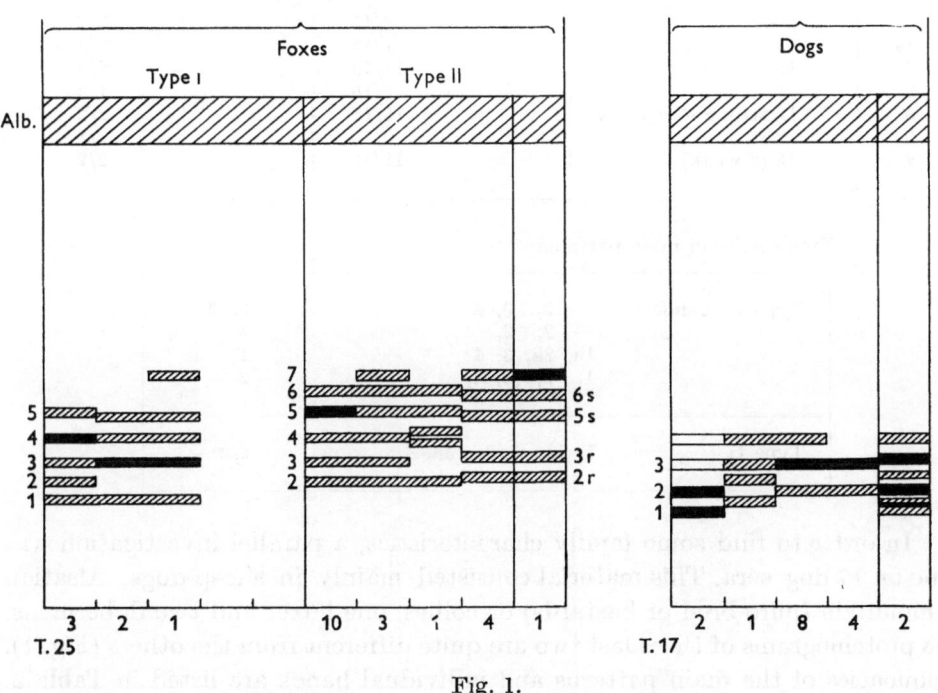

Fig. 1.

E_1 was inhibited by prostigmine only; E_4, E_5 and E_6 were inhibited by DFP completely and partly by prostigmine. In agar gel the esterases were detected using indoxyl or β-naphthyl acetates. The former gave two spots in the α_2-globulin region, and a weak, crescent-shaped spot on the cathodic side of the albumin area; the latter formed the same two slowly-migrating spots but the fast esterase was an intensely coloured spot covering the whole albumin area. This last component has not, by itself, the esterase activity: it was demonstrated by means of immuno-electrophoretic analysis, which showed four arcs coloured purple by β-naphtyl acetate diazo blue reaction; two of these were situated in the α-globulin region, and one was faster than albumin; the main arc had exactly the same mobility as the albumin, but was quite independent of the albumin line; this esterase arc was not visible as antigen-antibody precipitate, and was not stained by protein stains (an analogous observation was made concerning some bird sera esterases, (Kaminski 1964).

Table 2

Frequencies of individual protein bands in dog sera

Band 1	Present in	Total N° of sera (sheep-dogs + boxers) —15	Dobermans — 2
	4 sera	2/15 +	2/2
1a	3	1/15 +	2/2
2	16	14/15 +	2/2
2a	2	1/15 +	1/2
3	13 strong + 4 weak	15/15 +	2/2
3a	2	0/15 +	2/2
4	13 (2 weak)	11/15 +	2/2

Frequencies of main patterns

Type I bands	2, 3, 4	8/15
	2, 3,	4
	1a, 2a, 3, 4	1
	1, 2, (3, 4)	2
Type II	1, 1a, 2, 2a, 3a, 4	2/2

In order to find some family characteristics, a parallel investigation was done on 17 dog sera. This material consisted mainly in sheep-dogs, Alsatian or mountain (pure-bred or bastards) or collies, one boxer and two dobermans. The proteinograms of these last two are quite different from the others (Fig. 1). Frequencies of the main patterns and individual bands are listed in Table 2.

The general feature of these patterns is that the main variable system of bands consists of slowly-migrating proteins. The second feature is that, in comparison with foxes, the variable system is simpler, the number of bands being much lower.

Among sheep-dog and boxer serum proteinograms, only two have band 1, which is present simultaneously with band 2, more intense than in any other sera. 4 sera have bands 2 (weak) and 3 (very strong), one has abnormal bands 1a and 2a; all the others have bands 2, 3 and 4 — these are most probably heterozygous, but the previously described patterns do not seem to be homozygous with certainty. It is probable that the pattern 2, 3, 4 is a combination of 2, 3 and 3, 4; but the band 1 seems to be present in conjunction with 2, in these two patterns bands 3 and 4 are present in a very weak concentration.

Further investigations are necessary to determine whether other phenotypes will be detected in the same and in other races and breeds; for the moment the dog proteinograms do not resemble those of foxes.

References

Kaminski M. and Gajos E. (1964). Nature 201, 716.
Kaminski M. (1964). Protides in biological Fluids, in press.

DISCUSSION

M. Braend: I would like to ask M. Hesselholt whether they have made any Tf gene frequency studies in the Icelandic material. The reason why I ask is that the Tf^M allele was not found in Norwegian horses.

M. Helsselholt: We want to increase considerably our material of the Icelandic horses before we give a definite estimate of transferrin gene frequencies, but so much can be said that the frequencies of the Tf^F, Tf^C, and Tf^R genes are relatively high in the Icelandic horse, the component H is very rare.

The Tf^M homozygote has been observed.

C. Stormont: Mr. Hesselholt, have you studied the X substance only in pigs or in other species, too?

M. Hesselholt: We have investigated the X substance in horses. These preliminary studies showed only one X component in horse sera. But I can say that we have observed the X substance in several animals. In cattle 3 different phenotypes have been observed, a slow and a fast component occur in some individual sera and some individuals have both components. Work on the genetics of the X substance in cattle is in progress.

J. Moustgaard: The X-substance as we called it should be identical with that mentioned by Dr. Imlah yersterday. I would already like to have that discussed little more in detail. I don't feel quite convinced that there is an amylase. In case, if it is an amylase, its quite a funny amylase, so far as I cannot find it in pancreatic juice. It is also not exhausted out of the plasma by adding soluble starch to the plasma and incubated for 24 or 48 hours.

So we have the feeling or I have the feeling the whole time that it is a starch splitting enzyme, as was called then. But if it is truly saccharase or normal serum amylase I don't think we will be able to say at the time-being. I would like to hear also Dr. Imlah who was also looking for it in pancreatic juice.

P. Imlah: I have taken the pancreas and serum from pigs at the slaughterhouse. The amylase was extracted from the pancreas by grinding in trichloracetic acid. On running this extract and the serum in starch gel electrophoresis, it was found that the amylase from the pancreas remained at the insert line and digested the starch at the insert line, whereas the serum amylases migrated into the starch as described. The activity of the amylase at the insert line was much greater than the serum amylases. However, a component within the serum which does not move into the gel, and appears to be quite active has also been observed.

It may be a little presumptive to assume that this component is amylase. However, I have applied three distinct tests, which confirm the hypothesis that this component has activity similar to amylase.

They are:

1. The starch which is hydrolyzed by this enzyme is capable of reducing alkaline copper solution to cuprous oxide. This would suggest that the starch has been hydrolyzed to maltose, which is a disaccharide and a reducing sugar.

2. In the areas of activity by the enzyme the hydrolysis of the starch has also been confirmed by the absence of a positive blue staining reaction with iodine solution.

3. The activity of this component can be inhibited by incubation of the serum in a solution of urea.

On the basis of these tests, which can be found in any standard textbook, I would say that this component cannot be distinguished from the starch splitting enzyme called amylase. It is known that several amylases exist, for example the alpha- and beta-amylases, also the amylase of pigs is different from that of man. I consider that the enzyme found in pig serum which migrates in starch is a variant of amylase, which may be bound to a serum protein. I do not know, however, whether the binding of one protein to another like this is possible.

M. Hesselholt: With respect to amylase I should like to refer to some experiences we have had. When we make electrophoresis on purified amylase of pancreatic juice or saliva a digested zone in the gel situated near the origin appears. You don't see any X bands. The character of these digested zones of amylase activity is different from the character of the X bands in serum. But we must take into consideration the theory that an active part of a starch-splitting enzyme can be attached to a serum protein and migrate with this protein in the gel. This theory has to be investigated.

G. Efremov: Dr. Naik, did you use other techniques to study a new haemoglobin variant? Is the Hb X foetal haemoglobin which appears in adult life similar to Hb F in man? At the same time, a new Hb was reported in cattle which was found by myself and Dr. Braend. It migrates slightly slower than Hb A, and was found in 6 animals as homozygous and in 2 combinations with Hb A. We called it C and it can be explained by the occurrence of a third allele (Hb^C). It may be the same as that reported by Osterhoff.

S. N. Naik: The Hb-X is found to be different from that of foetal haemoglobin because foetal haemoglobin has slightly faster mobility when subjected to paper electrophoretic technique and this was also observed by Crockett et al. in USA. Besides, all the animals in which this Hb-X was detected were above the age of 5 years.

C. C. Osterlee: Concerning the paper of Dr. Osterhoff on transferrins, I can inform you that we found in the Netherlands a new transferrin type in serum which is, as far as I can see on the slides, parallel to the new G type of Dr. Osterhoff.

M. Braend: In what breed have you found the new allele?

C. C. Osterlee: This type has been found in the Black and White Friesian.

C. C. Osterlee: I was very much interested in the transferrins of sheep. In the Netherlands we did some research in this field this year. We typed animals of Dutch sheep breeds and thought first that the bands parallelled the types of English breeds described by Ashton.

Later we were so lucky to get some reference reagents from Ashton of Merino sheep in Australia and the types of the Dutch animals turned out to be parallel to the Merino bands.

There will be much confusion when each country gives different names to the transferrin bands of sheep.

Therefore I would like to suggest that we ask the Norwegian people that before publishing these results, the committee of the E. S. A. B. R. will organize a reference test for transferrin types in sheep.

M. Braend: We were not able to obtain reference samples. Therefore we named these preliminarily. I agree with Dr. Osterlee that we should have a sound nomenclature. I am going to meet Dr. Ashton in Italy in 14 days and will discuss the problem with him there.

M. Harboe: It is very striking that Hb variations in man are usually associated with disease, whereas the various Hb types here reported in animals occur in healthy individuals. Is there any information on the association of Hb variants with disease in animals?

G. Efremov: The first case of Hb N was found in a family suffering of a very strong parasitic invasion. In that time we thought that it is an abnormal haemoglobin. Later we examined about 700 blood samples from sheep, and in about 60% of animals having Hb^A allele we found less or more Hb N. Now we are inclined to believe that it is a normal haemoglobin.

E. M. Tucker: Evans and Blunt have described the appearance of a haemoglobin variant which occurs in the presence of anaemia in sheep which carry the haemoglobin A gene. We also have found this, and also, in agreement with Dr. Efremov, have noticed a haemoglobin variant, not foetal, which occurs in A type lambs.

G. Kovács: I wonder if all the 16 serum protein bands found by means of starch gel electrophoresis belong to the transferrins and moreover if this finding was supported by radioisotope investigations and by investigations performed on a family material.

M. Kaminski: Most of the 16 serum protein band belong to the beta-globulins. Radioisotope investigations were made only on parents. Family material will be investigated in the next season.

M. Hesselholt: Have you investigated the haptoglobins in dogs?

M. Kaminski: Yes, we have tried to investigate haptoglobins in dogs and we have preliminarily found 2 bands.

M. Hesselholt: Have you investigated the ceruloplasmins in dogs and foxes?

M. Kaminski: They were not investigated in dogs and foxes.

PROTEIN POLYMORPHISM IN SOME SEXUAL GLAND FLUIDS

POLYMORPHISM OF PROTEIN FRACTIONS IN THE FLUIDS OF ACCESSORY GENITAL GLANDS IN BULLS AND BOARS

M. VALENTA, J. MATOUŠEK, E. PETROVSKÝ, A. STRATIL

Laboratory of Physiology and Genetics of Animals, Czechoslovak Academy of Sciences, Liběchov

When studying the influence of the nutritional standards on proteins in the ejaculates of bulls (Valenta, Petrovský 1962b) differences were found, by means of electrophoresis in agar gel, in the number of protein fractions of the seminal plasma which corresponded to those found later in the fluids of the seminal vesicles of the same bulls (Valenta, Petrovský 1962a). During further investigations, the study of individual differences in the fluids of the seminal vesicles and other accessory genital glands in bulls and partly also in boars has been extended. Our results hitherto obtained are given in the present report.

Material and Methods

The genital glands were obtained from adult bulls, mostly of the Red and White Breed, and from boars of the Large White and Cornwall breeds, killed at the slaughterhouse. The glands were treated immediately, if possible, after the animal had been killed. The fluids of the seminal vesicles and from the ampullae were extracted by gently pressing the prepared glands. By inserting a hypodermic syringe into the seminal ducts, the fluids from the tail of the epididymis were expelled by air to the septa, in the vicinity of the body of the epididymis, and after perforating the back part of the tail of the epididymis, the drops were intercepted into centrifugal tubes. Spermatozoa were sedimented at 3,000 to 4,000 rpm. The fluids not directly used for electrophoretic analysis were kept in a refrigerator at $-20°C$.

Electrophoretic separation was made in a moist chamber in 2% agar gel (Difcoagar) on plates 12 × 20 × 0·3 cm. in size, using a modified veronal-acetate buffer, pH 8·6 (Wunderly 1953), prepared by diluting 5 g. veronal sodium, 3·2 g. sodium acetate trihydrate and 30 ml. N/10 hydrochloric acid in 1,000 ml. distilled water, or 1·6% sodium acetate/acetic acid buffer pH 6·6. The moist chamber was placed in a refrigerator or cooled by water. The time of separation was about 8 hours at 6 V/cm. length and 4 mA/cm. width of

Fig. 1. Phenotypes B, A, ab, bc of protein fractions of fluids of seminal vesicles in bulls.

the agar plate. The protein fractions were stained with a solution of Amido Black 10B (Uriel, Grabar 1956) or a similarly prepared solution of Nigrosin. 3% acetic acid was used for washing.

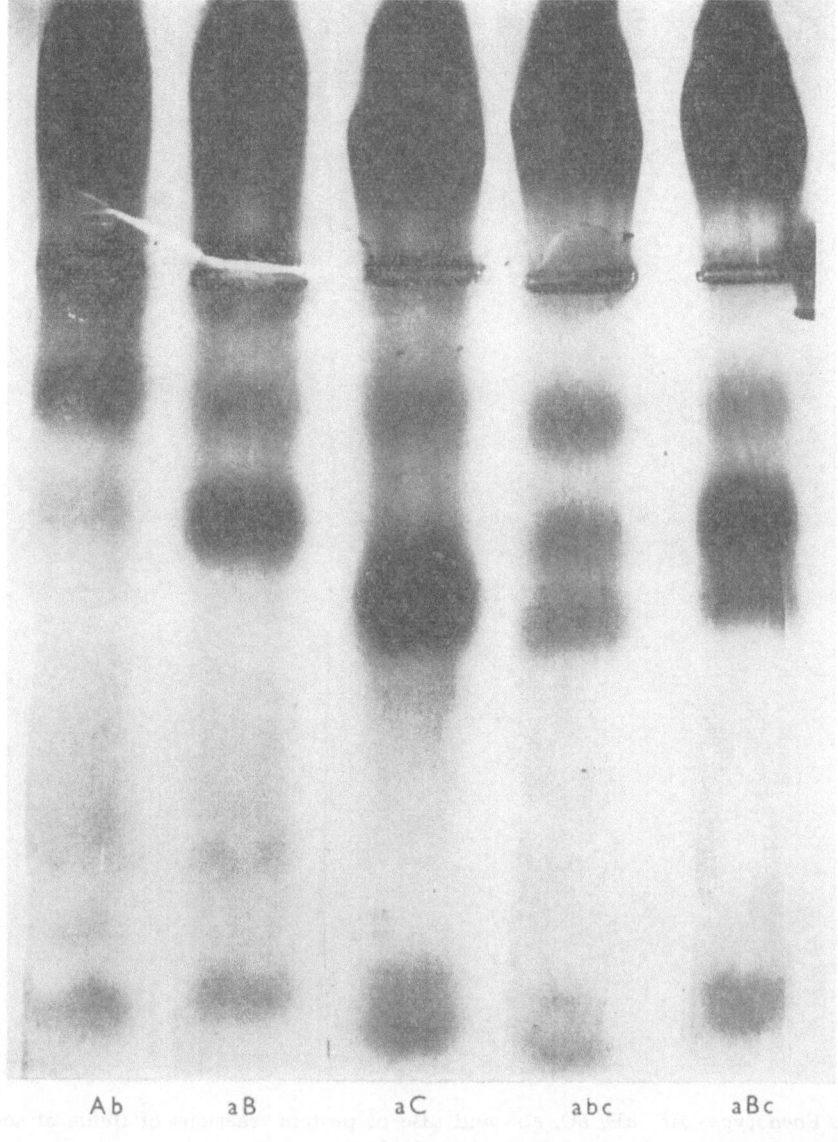

Fig. 3. Phenotypes aBC, ABc, ABC, AbD and ade of protein fractions of fluids of seminal vesicles in bulls.

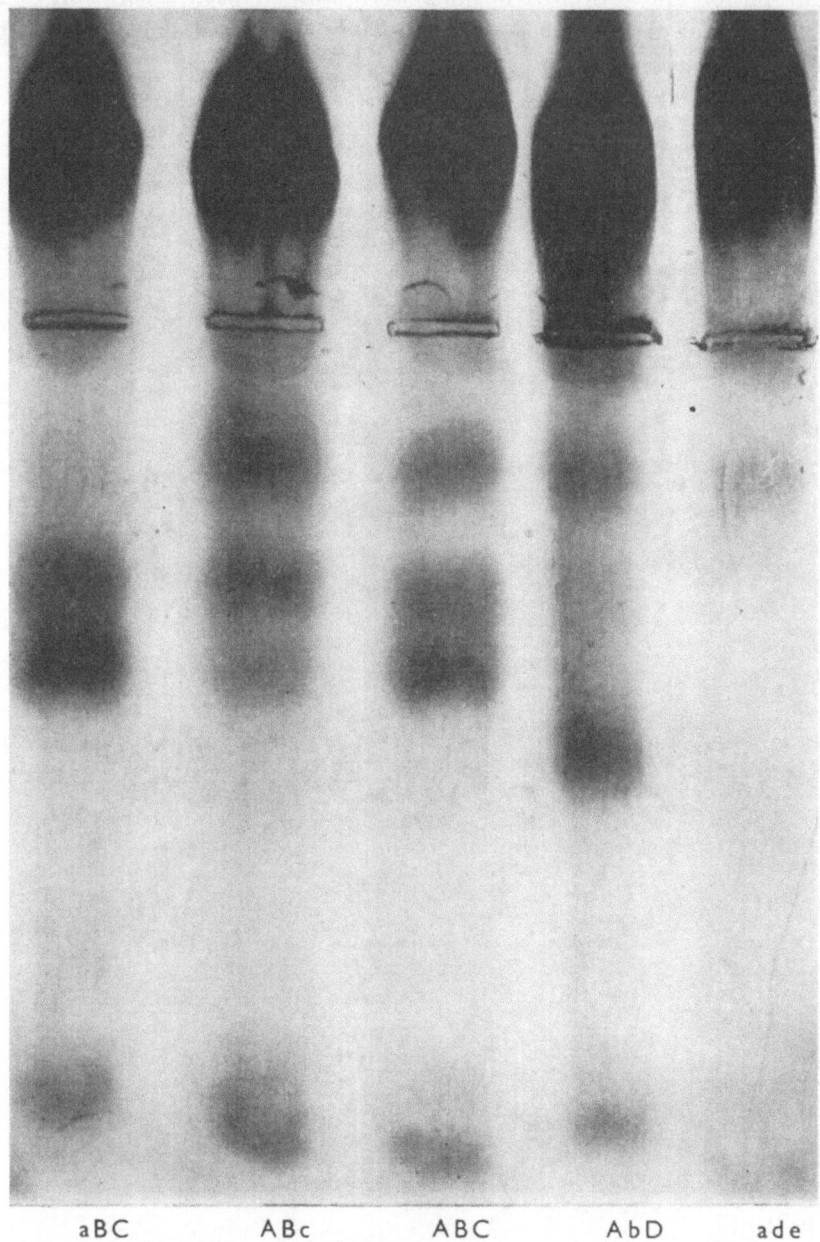

Fig. 2. Phenotypes Ab. aB, aC, abc and aBc of protein fractions of fluids of seminal vesicles in bulls.

Results and Discussion

In the fluids of the seminal vesicles of 402 bulls, differences in the number and intensity of 3 fractions migrating to the cathodic side of the electrophoretic pattern situated close behind the start, were found. The stronger fractions were designated by capital letters A, B, C, and the relatively weaker ones a, b, c. 12 phenotypes, differing in the number or intensity of fractions were found. The phenotypes B, A, ab, bc can be seen in fig. 1. In fig. 2 are the phenotypes Ab, aB, aC, abc and aBc. Fig. 3 shows the phenotypes aBC, ABc and ABC. As can be seen from fig. 3, the phenotype AbD has been also

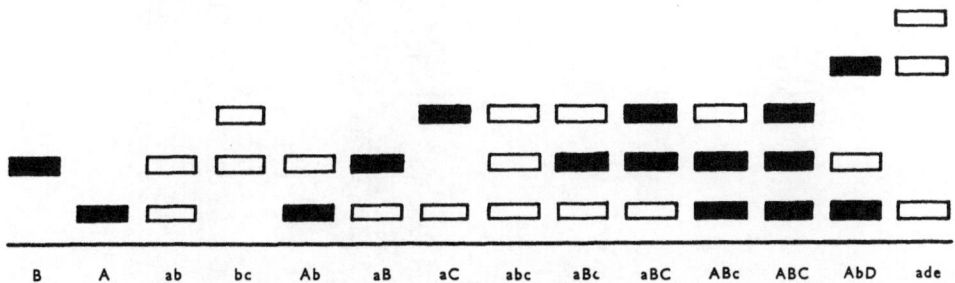

Fig. 4. A survey of all phenotypes of protein fractions of the fluids of seminal vesicles in bulls.

found in one case and the phenotype ade in two cases. A survey of all these phenotypes is presented in fig. 4. The frequencies of the mentioned phenotypes are given in table 1.

The combinations with the factor B (49%) occur most frequently, especially in phenotype aB (29·4%). In our further studies attempts will be made to determine other possible phenotypes of this system which we suppose to be a multi-allele system.

Table 1

Phenotypes of protein fractions of seminal vesicles in bulls

Phenotype	A	B	ab	bc	Ab	aB	aC	abc	aBc	aBC	ABc	ABC	AbD	ade	Total number of animals
Number of animals	4	9	129	2	27	118	4	28	24	8	35	4	1	2	402
Percentage of animals examined	1	2.2	32·1	0·5	6·7	29·4	1	6·9	6	2	8·7	1	0·2	0·5	100

In the fluids from the ampullae of some bulls, phenotypes similar to the types detected in the fluids of the seminal vesicles have been found. The remaining group of bulls has protein fractions similar to those of the fluid of the tail of the epididymis. Some cases are given in fig. 5.

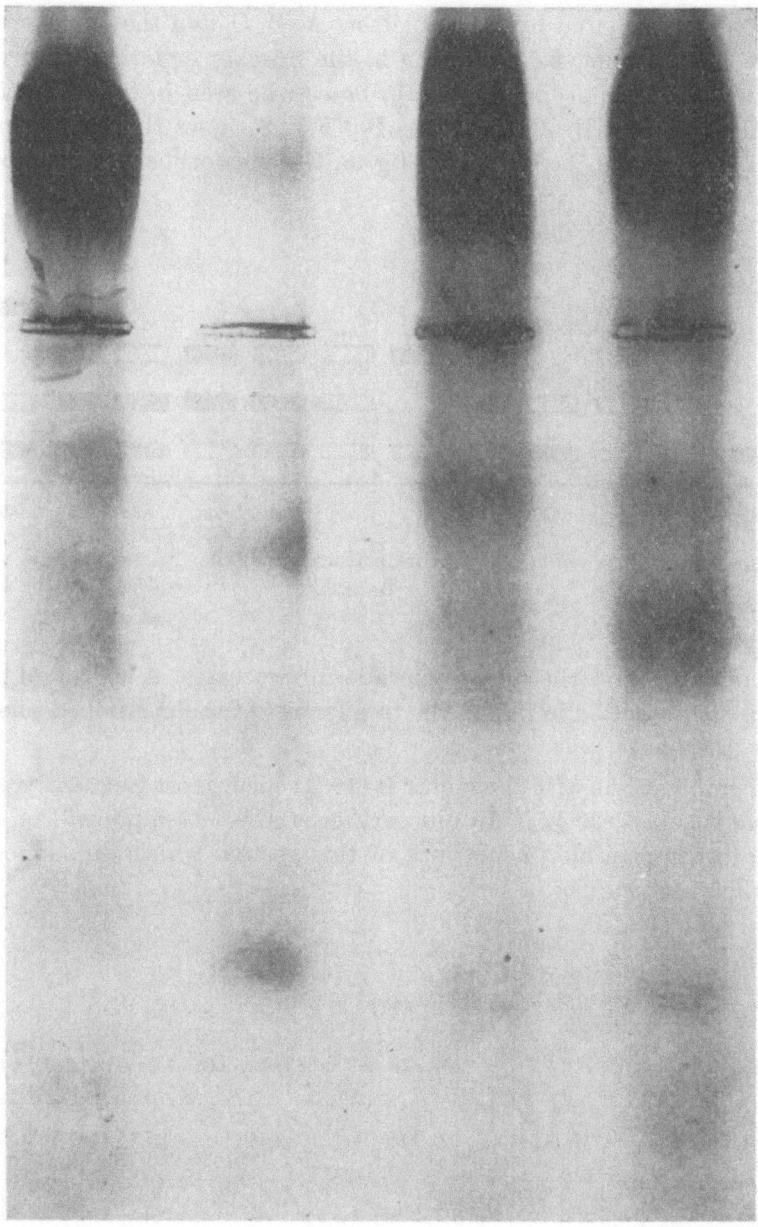

Fig. 5. Some phenotypes of protein fractions of the fluid from the ampullae in bulls.

In the fluids of the tail of the epididymis of bulls a system has been noted, which is expressed by the intensity of fractions located in the anodic side. As may be seen from fig. 6 there are 3 phenotypes A (strong fraction), a (weak fraction) and O (i. e. very weak fraction). The frequency of their occurrence

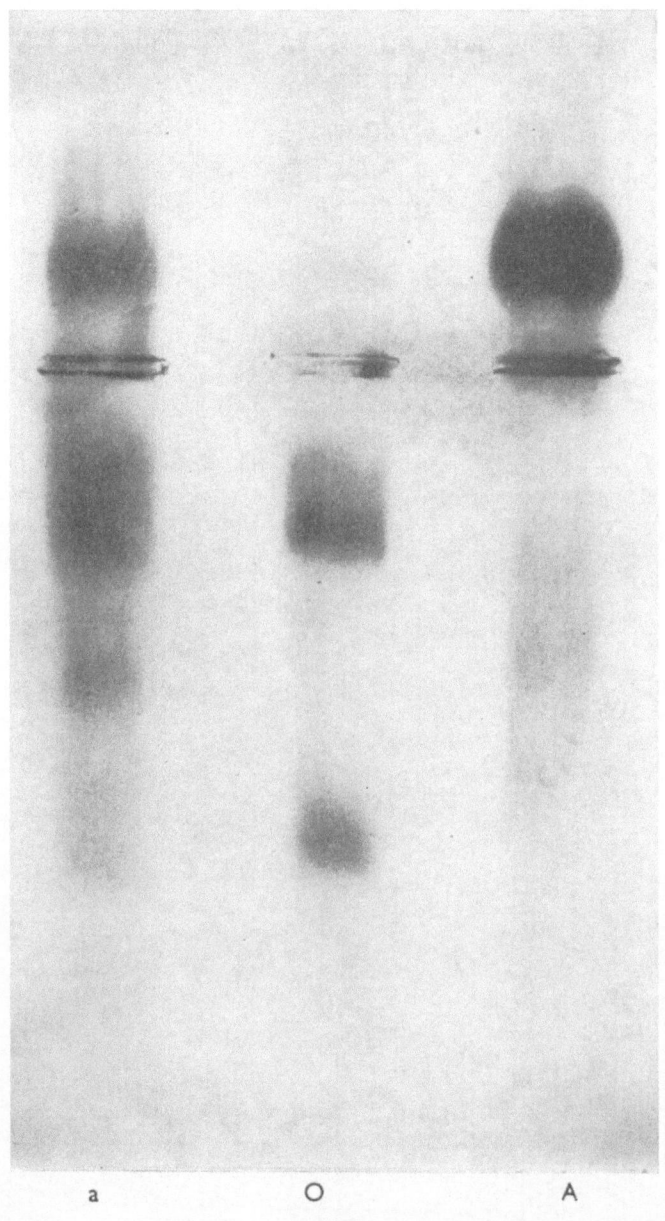

a   O   A

Fig. 6. Phenotypes A, a, O of protein fractions of the fluid of the tail of the bulls' epididymis.

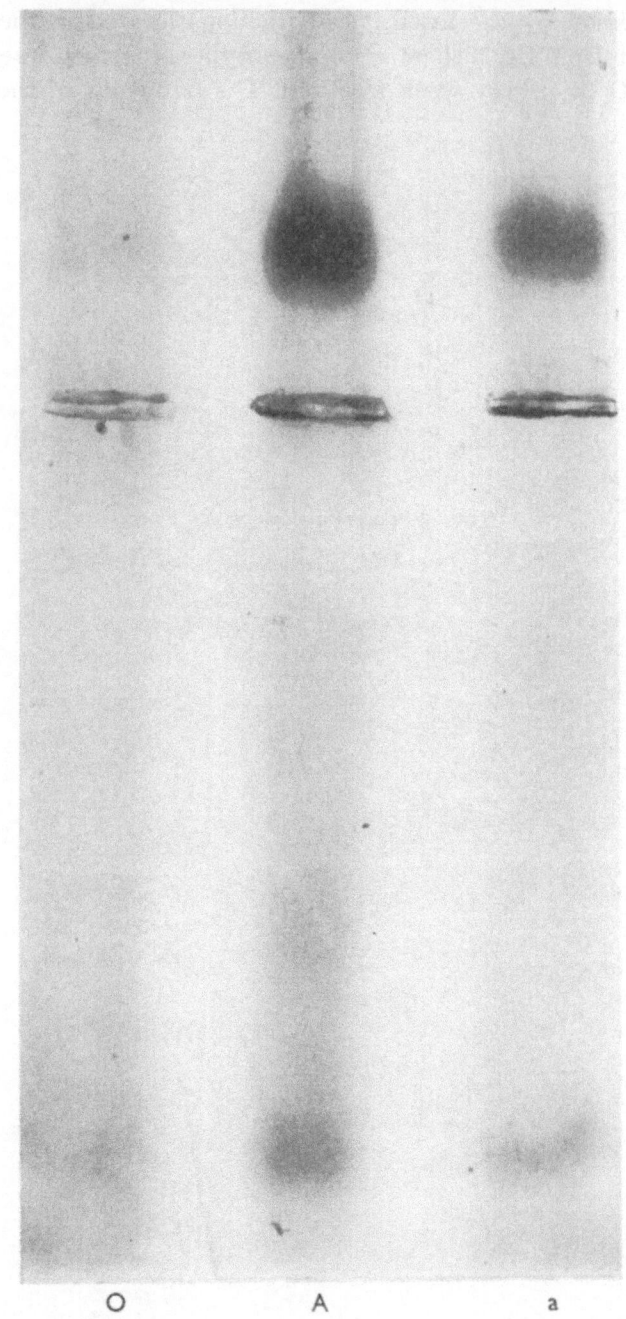

Fig. 7. Phenotypes A, a, O of protein fractions of the fluid of the tail of the boars' epididymis.

is given in table 2. The type O which practically has no proteins in the anodic side of the electrophoretic pattern, shows intense fractions in the cathodic side.

In preliminary studies the differences in the number and intensity of the protein fractions were also detected in the accessory genital glands of boars. In the fluids of the tail of the epididymis of boars a difference in the occurrence

Table 2

Phenotypes of protein fractions of the fluids of the tail of the bulls' epididymis

Phenotype	A	a	O	Total number of animals
Number of animals	29	58	16	103
Percentage of animals examined	28·3	56·5	15·2	100

and intensity of anodically migrating fractions was found. (Fig. 7.) Three phenotypes A, a, O, similar to those of the fluid of the tail of the epididymis in bulls, were detected. In type O, however, the cathodic protein fractions are not intense. The frequency of their occurrence is given in table 3.

Table 3

Phenotypes of protein fractions of the fluids of the tail of the boars' epididymis

Phenotype	A	a	O	Total number of animals
Number of animals	22	21	29	72
Percentage of animals examined	30·6	29·2	40·2	100

In the fluids of the seminal vesicle of boars differences in the occurrence and intensity of the fastest cathodic fractions except for one were found. Differences were also detected in the quickest cathodic fractions of the seminal vesicles and the fluids of the tail of the epididymis of bulls, and the epididymis of boars. Cathodic fractions of the fluids of the tail of the epididymis in boars, migrating at medium velocity, also showed individual differences. It will be necessary, however, to improve the electrophoretic method in order to separate these fractions more clearly.

From the given survey it may be seen that a relatively extensive polymorphism of protein fractions exists in the fluids of the accessory genital glands. At the present time we are studying the genetic control in the individual systems described here.

We are indebted to Mrs J. Novotná for her technical assistance.

Summary

By means of agar gel electrophoresis the extensive polymorphism of protein fractions in the fluids of the accessory genital glands in bulls and boars has been detected. 14 phenotypes have been hitherto proved in the fluids of the seminal vesicles of bulls, and 3 phenotypes in the fluid of the tail of the epididymis. Protein polymorphism has been also detected in the fluids from the ampullae. In boars 3 phenotypes have been preliminarily detected in the fluids of the tail of the epididymis. Further systems have been studied. Distribution of individual phenotypes were determined. At the present time the genetic control of the described systems is being investigated.

References

Uriel, J., Grabar, P. (1956). Ann. Inst. Pasteur 90 : 427; cit. Michalec, Kořínek, Musil, Růžička: Elektroforesa na papíře a v jiných nosičích. ČSAV, Praha 1959.

Valenta, M., Petrovský, E. (1962a). Čs. fyziologie 11 : 277.

Valenta, M., Petrovský, E. (1962b). Sborník ČSAZV, Živočišná výroba 35 : 481.

Wunderly, Ch. (1953). Chimia 7 : 145; cit. Lederer, M.: Paper electophoresis and related methods. Amsterdam 1957.

ANTIGENICITY AND POLYMORPHISM OF THE OVARIAN FOLLICLE FLUIDS IN COWS

J. MATOUŠEK

Laboratory of Physiology and Genetics of Animals, Czechoslovak Academy of Sciences, Libĕchov

The detection of polymorphism in some fluids of the genital tract in bulls and boars (Valenta et al. 1964) convinced me that certain individual differences must also exist in the fluids of the female genital tract. One of the most accessible fluids after the slaughter of the animal is undoubtedly the ovarian follicle fluid. The results hitherto obtained by the study of this fluid are submitted in the following report.

Material and Methods

The follicle fluid was obtained from follicles of both ovaries, at all stages of development, from cows and heifers of the Bohemia Red and White breed, killed at the slaughterhouse. The reasons for discarding the animals from breeding were not observed, only possible presence of the corpus luteum, ovarian cysts and pathological changes in the ovaries were noted.

The fluid from the single, differently developed follicles was obtained by inserting an injection needle into it and drawing off the fluid by means of a hypodermic syringe. The fluids were kept frozen at −15 to −20°C. Blood samples were also taken from all animals in order to prepare erythrocyte suspension and sera.

1. The presence of serologically active substances of the antigenic factor J in the serum and the ovarian follicle fluids were studied by means of the inhibition test (Stone, Irwin 1954).

2a. Preparation of antisera by immunization of rabbits and rams with follicle fluids of cows and some fluids of accessory genital glands of bulls. The fluids were injected into rabbits and rams in doses of 2 and 6—8 ml. once to twice a week by alternating intramuscular and subcutaneous injections. During one immunization period 4—5 injections were given. From the rabbits and the ram No. 1050 blood was taken 7 days after the last injection, from the remaining 3 rams 500 ml. blood were collected at the immunizations, beginning with the second, and the serum showing the most effective antibodies was used for analyses.

2b. Precipitation reaction on agar gel double diffusion plates — 1% agar Difco in borate buffer (boric acid — 4·04 g., tetraboric acid — 19·10 g., distilled water — 1,000 ml.).

2c. Microimmunoelectrophoresis — was based on the method of Grabar 1959. The same agar and buffer as in the current agar electrophoresis was used (Valenta et al. 1964). On slides of 50 × 50 mm. in size, 4 samples on small pieces of Whatman 3 filter paper were inserted into the insertion line. The period of electrophoresis was 40 min. at 10 V and 6 mA per cm. Undiluted serum was pippeted into the troughs.

The agar gel double diffusion plates and immuno-electrophoretic slides were stained with Amido Black.

3. The starch gel was prepared according to the method of Smithies (1955) and Schröffel, Matoušek (1962). Electrophoretic separation was effected by means of a discontinuous TRIS-borate buffer system (Poulik 1957, Kristjansson 1963) on plates of 220 × 105 × 6 mm. in size horizontally, without water cooling. Electrophoresis ran 4—5 hours at 250—300 V and 40 mA. Staining with Amido Black.

4. Electrophoresis on agar gel was performed according to the method described by Valenta et al. 1964.

Results and Discussion

The antigenic factor J in ovarian follicle fluids

The inhibition of natural antibodies of the anti J reagent 89/117 by the ovarian follicle fluids of 38 cows out of 94 gives evidence for the presence of this factor in this body fluids as well. (Table 1.)

Contrary to the seminal plasma of bulls (Matoušek 1961), the number of units of inhibitory J substances in the follicle fluids is considerably lower than in the sera. Consequently, it is impossible to prove this antigen in the follicle fluid of animals of the type Js with a low number of units of inhibitory J substances in the serum. Thus the group of animals of the Js phenotype is divided into two subgroups with regard to the presence of this factor in the ovarian follicle fluid, i. e. one where the antigenic factor is present in both the serum and the ovarian follicle fluids and the other, where the factor J is detectable only in the serum.

Table 1

Antigenic factor J in the ovarian follicle fluids and sera of 94 cows

Presence of antigenic factor J	Number of animals	Average number of units of inhibitory J substances	
		in the serum	in follicle fluids
On erythrocytes, in follicle fluids, in the serum	30	26·0 ± 12·8	8·3 ± 3·2
In follicle fluids, in the serum	8	5·5 ± 1·7	1·75 ± 0·4
In the serum	8	2·0 ± 0·0	0
Absent from all components	48	0	0

Table 2

Antisera used in precipitation tests of ovarian follicle fluids and sera of 82 to 94 cows

Designation of antiserum	Species and number of animal used for immunization	Body fluid used for immunization	Number of lines produced on agar double diffusion plates		Number of arcs produced at immunoelectrophoresis	
			with follicle fluid	with serum	with follicle fluid	with serum
B 58	Ram 58	follicle fluids from several cows	1—2	2	2	2
B 258	Ram 258	follicle fluids from one cow	1	1	1	2
K 14	Rabbit 14	follicle fluids from several cows	2	2	2	2
K 024	Rabbit 024	follicle fluids from one cow	4	3	7	7
B 8	Ram 8	seminal vesicle fluids from several bulls	3	3	7—8	6—7
B 1050	Ram 1050	seminal vesicle fluids from several bulls	4	4	6	5

☐ individual differences.

Antigenic individual differences proved by immune sera

It will be probably possible to detect another individual antigenicity in the studied fluid by means of some heteroimmune antisera. Of the 6 antisera used (Table 2), 2 are capable of detecting such individual differences. Both these sera were obtained by immunizing rams, i. e. a species phylogenetically comparatively closely related to cattle. First organ and species precipitins were formed, probably as a result of the mentioned affinity, but antibodies detecting individual characters were also produced.

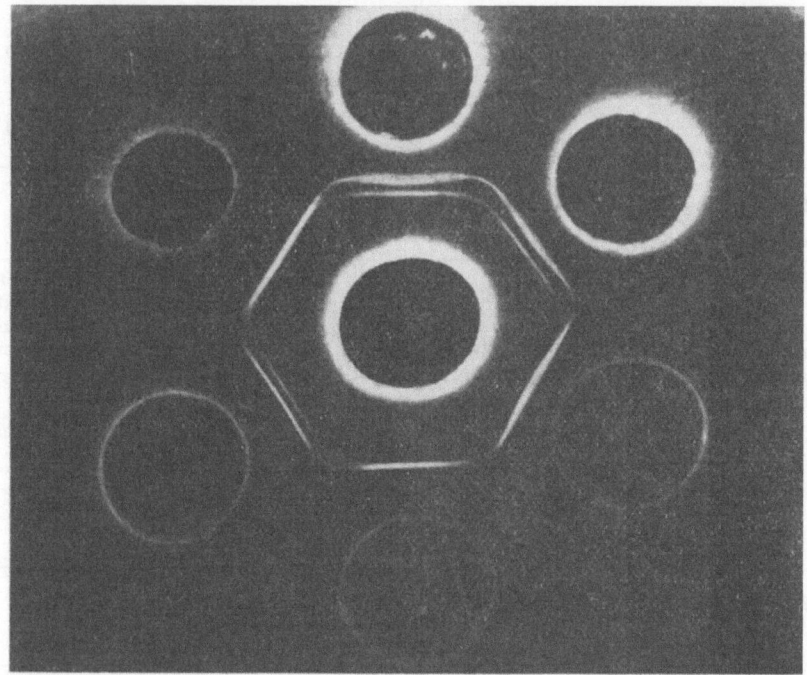

Fig. 1. Precipitation lines of ovarian follicle fluids with the antiserum B 58.

The precipitins of serum B 58 divide the ovarian follicle fluids into two groups. One group of follicle fluids produces one precipitation line on agar double diffusion plates with the serum B 58, the second group two lines (Fig. 1). We have designated A_1 the ovarian follicle fluids of the first group, A_2 those of the second group.

The immune antiserum B 8, obtained by immunizing a ram with seminal vesicle fluid of bulls shows antigenic individual differences at micro-immuno-electrophoresis. The difference between the fluids lies in the presence or absence of a pronounced arc, i. e. a certain protein component farthest from the start in the direction of the cathode (Fig. 2).

The follicle fluid of cows in which the reacting protein component is present, are designated B_1, without component BO. The first antiserum B 58 shows individual antigenic difference in the ovarian follicle fluids but not in the serum. The antiserum B 8 is also capable of differentiating sera of cows in a similar way (Fig. 3), but the differences proved in the ovarian fluids do not correspond to those in the serum.

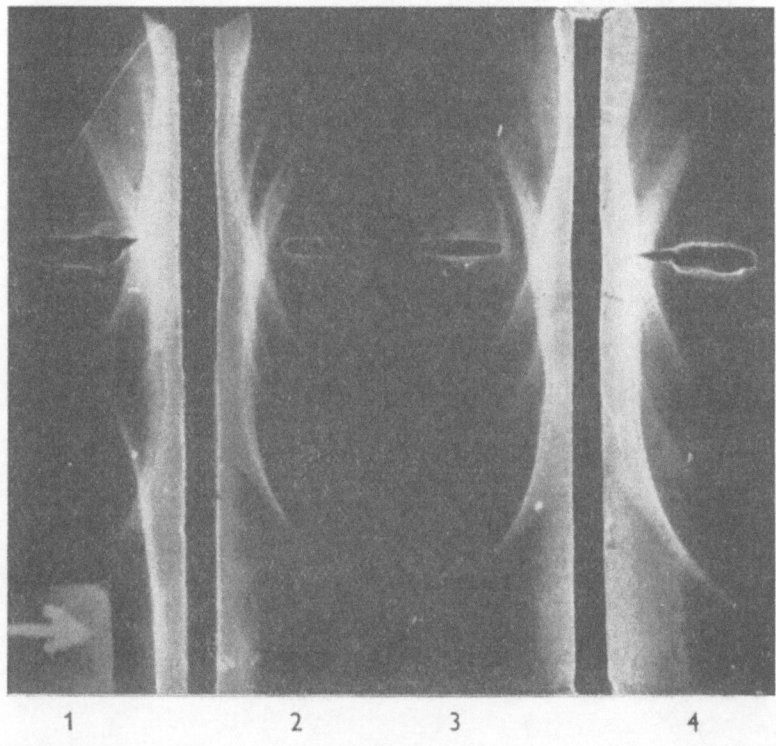

Fig. 2. Micro-immunoelectrophoretic precipitation reactions of antiserum B 8 with the follicle fluids of the type B 1 (denoted by figure 1 and an arrow) and the type B O (denoted by figures 2, 3, 4).

An individually reacting protein component is present in the ovarian follicle fluids of many cows, does not react at all with the serum or reacts only imperceptibly. The crossreacting protein seems to be quantitatively very irregular in the serum, whereas it is either present or absent from the ovarian follicle fluids or present at two different levels — which seems more probable — the higher always gives a clear precipitation reaction and the lower remains without response. The precipitation reaction types of both antisera with the ovarian follicle fluids seem to be independent of each other.

Preparation of specific antibodies against only one type of follicle fluids by absorption has not been successful in any of the individually reacting antisera. When e. g. the antiserum B 58 was absorbed by the follicle fluid of a cow, reacting only in one line, all antibodies have always been removed,

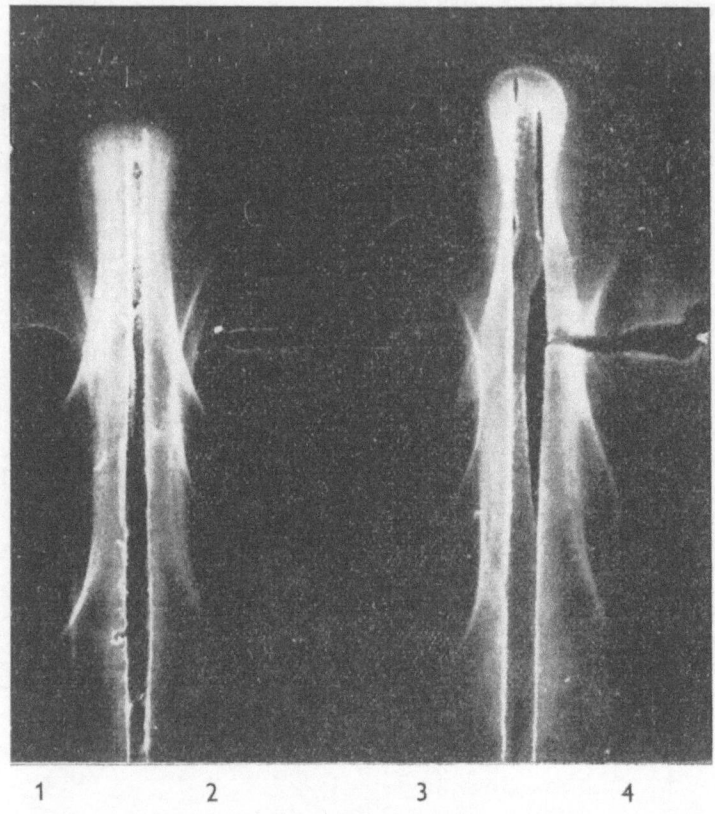

Fig. 3. Precipitation reactions of antiserum B 8 with the sera of cows whose follicle fluids were used fór analysis shown in Fig. 2.

against the second line as well. It seems therefore that closely related protein complexes are responsible for the formation of precipitation lines or there are quantitative differences in the proteins, reacting by precipitation, the same as suggested before for the reactions with antiserum B 8. Since the precipitation patterns of both "individual" antisera, except for the reaction of antiserum B 8 with the sera of cows, are fully reproducible, it can be assumed although the absorbing experiments were not successful, that the described

Table 3

Frequency of different types of individual antigenic differences of ovarian follicle fluids in cows

Type of individual antigenic difference	Number of reacting cows	Total number of cows observed	Frequency of individual difference in %
A_1	59	94	62·8
A_2	35	94	37·2
B_1	31	82	25·6
B 0	61	82	74·4

individual differences in the follicle fluids are specific. The frequencies of the different types are shown in Table 3.

The antibodies of all antisera in Table 2 form very similar precipitation patterns with the ovarian follicle fluids as well as with the sera of cows. (Compare e. g. fig. 2 and 3.) It seems therefore that these fluids are, with regard to the structure of their proteins, very close to each other.

Strong reactivity of the follicle fluids and sera with antisera B 8 and B 1050, obtained by immunizing rams with seminal vesicle fluids of bulls is striking. It seems that there will also be a considerable similarity in the structure of protein molecules in serum with that in the fluids, which are present only in one sex.

Organ specificity of precipitins in all antisera was also preliminarily studied using ovarian follicle fluids of two sheep and one goat. Except for antisera B 58 and B 258, the others reacted with the follicle fluids of the goat as well as with those of the sheep. However, only rabbit sera K 14 and K 024 gave reactions with the sera of these animals.

Electrophoretic analysis of individual differences

By means of starch gel electrophoresis, the presence of transferrins, exactly corresponding to the type of serum transferrins, has been proved (Fig. 4, Table 4). It is a further confirmation of a considerable structural similarity of these two fluids.

Table 4

Types of transferrins in the serum and follicle fluid of cows

Fluid	Transferrin type and number of animals						
	AA	DD	EE	AD	AE	DE	Total
Serum	5	51	0	34	2	7	99
Ovarian follicle fluid	5	51	0	34	2	7	99

In the electrophoretic pattern of starch gel, the ovarian follicle fluids differ from the sera only in the slow alpha fractions. These are either not present at all, or they are but little pronounced (Fig. 4).

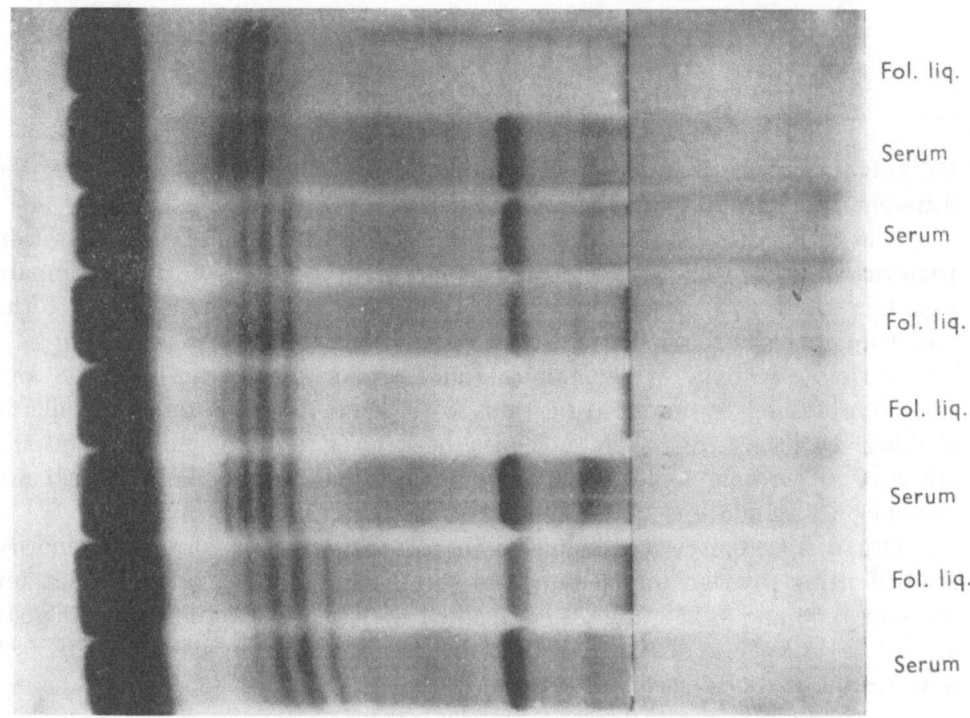

Fol. liq.

Serum

Serum

Fol. liq.

Fol. liq.

Serum

Fol. liq.

Serum

Fig. 4. Transferrins in the serum and follicle fluids.

Separation of protein fractions of follicle fluids and sera on agar gel showed no individual differences, but is also almost completely consistent (Fig. 5).

This suggests that in cattle as in man (Morgan, Watkins 1959) blood group substances soluble in body fluids (the antigenic factor J in cattle) are also present in ovarian follicle fluids.

The close structural similarity of serum proteins and the follicle fluid proteins is demonstrated not only by almost corresponding precipitation reactions but also by the same electrophoretic mobility of proteins on both starch gel and agar gel. Transferrins, in particular, were proved to be present in the follicle fluids in absolutely identical types as in the serum, and it may therefore be assumed that biosynthesis of these proteins is controlled by the same genes.

Individual antigenic differences, detected in the follicle fluids by some of the observed immune sera, provide some evidence for the possibility of extending the study of biochemical polymorphism in cattle also to the female genital tract.

Summary

The antigenic factor J present on the erythrocytes and in sera was also proved in ovarian follicle fluids in cows. It is possible to detect it in animals of the type J^{cs} even J^s.

The next possible individual differences were proved by immunoelectrophoresis and on Ouchterlony agar plates. Heteroimmune antisera acquired by the immunization of rams with cattle follicle fluids and the fluids of the seminal vesicles were used as suitable antisera.

Ovarian follicle fluid in cows is consistent with the sera of the same animals with regard to serum proteins divided by electrophoresis in starch gel. The pattern of the phenotype of transferrins in follicle fluids corresponds precisely to the phenotype in serum. No individual differences were found in ovarian follicle fluids by electrophoresis in agar.

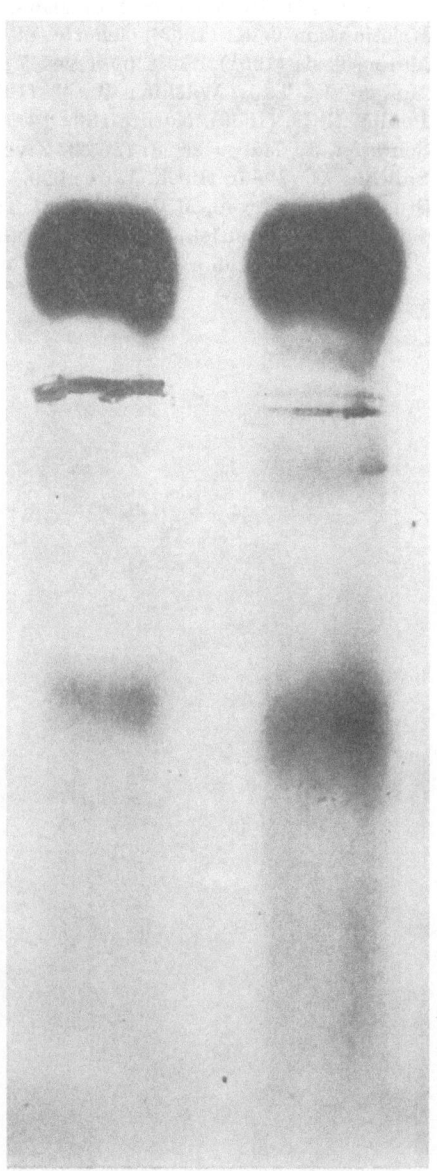

Fig. 5. Electrophoretic separation of ovarian follicle fluids and sera on agar gel.

Fol. liq.   Serum

References

Grabar P. (1959). Immunoelectrophoretic Analysis. Methods of Biochemical Analysis vol. VII. Interscience Publishers, Inc. New York.

Kristjansson F. K. (1963). Genetics 48/8 : 1059.

Matoušek J. (1961). Folia biologica 7 : 252.

Morgan, W. T. J., Watkins, W. M. (1959). British Medical Bulletin 15 : 109.

Poulik, N. D. (1959). Nature 180 : 1477.

Schröffel, J., Matoušek, J. (1962). Živočišná výroba 7 : 313.

Smithies O. (1955): Bioch. J. 61 : 629.

Stone, W. H., Irwin, M. R. (1954). J. Immunol. 73 : 397.

Valenta, M., Matoušek, J., Petrovský, E., Stratil, A. (1964). Blood Groups of Animals — Proceedings of a IX European Conference held in Prague 18—22 August.

THE POLAROGRAPHIC ANALYSIS OF SEMINAL FLUIDS AS A METHOD FOR THE STUDY OF THEIR ANTIGENICITY

E. PETROVSKÁ, E. PETROVSKÝ

Laboratory of Physiology and Genetics of Animals, Czechoslovak Academy of Sciences, Liběchov

Introduction

Polarography, a very sensitive analytic method, is used for its particularities only to a small extent for the evaluation of sperm quality of farm animals. We have found only little information about this problem in the available literature, as e. g. the work of Buruianu and Pavlu (1963).

The mentioned authors studied the question of the polarographic determination of seminal plasma proteins in farm animals and man in ammoniacal solution of Co^{3+}. In the course of these tests they found (hitherto only in bulls) a relationship between the sperm quality and the concentration of the seromucoid in plasmatic glycoproteins: a sperm rich in spermatozoa and having a high protein concentration contains a reduced amount of seromucoid.

In our work, attempts were made to reproduce some results of Buruianu and Pavlu by means of exhaustion tests, to determine the polarographic activity of fluids of accessory glands in bulls and especially to find out whether it is possible to use this method for the study of sperm antigenicity in farm animals. Some of the results obtained are presented in the present report.

Material and Method

The model LP-55 Heyrovský polarograph was used in the experiment; the galvanometer sensitivity was 1.7×10^9 A/mm. The polarographic curves were recorded from the outer tension 0.8 V. Novák's polarographic vessels were used with the reference electrode formed by the mercury on the bottom. Measurements were performed in an ammoniacal solution of divalent cobalt.

The fluids of the seminal vesicles and from the ampullae of bulls were obtained by carefully pressing these organs after their removal from the killed animals, the fluid from the tail of the epididymis by emptying this organ with a syringe. The polarographic measurement was carried out in the whole sperm

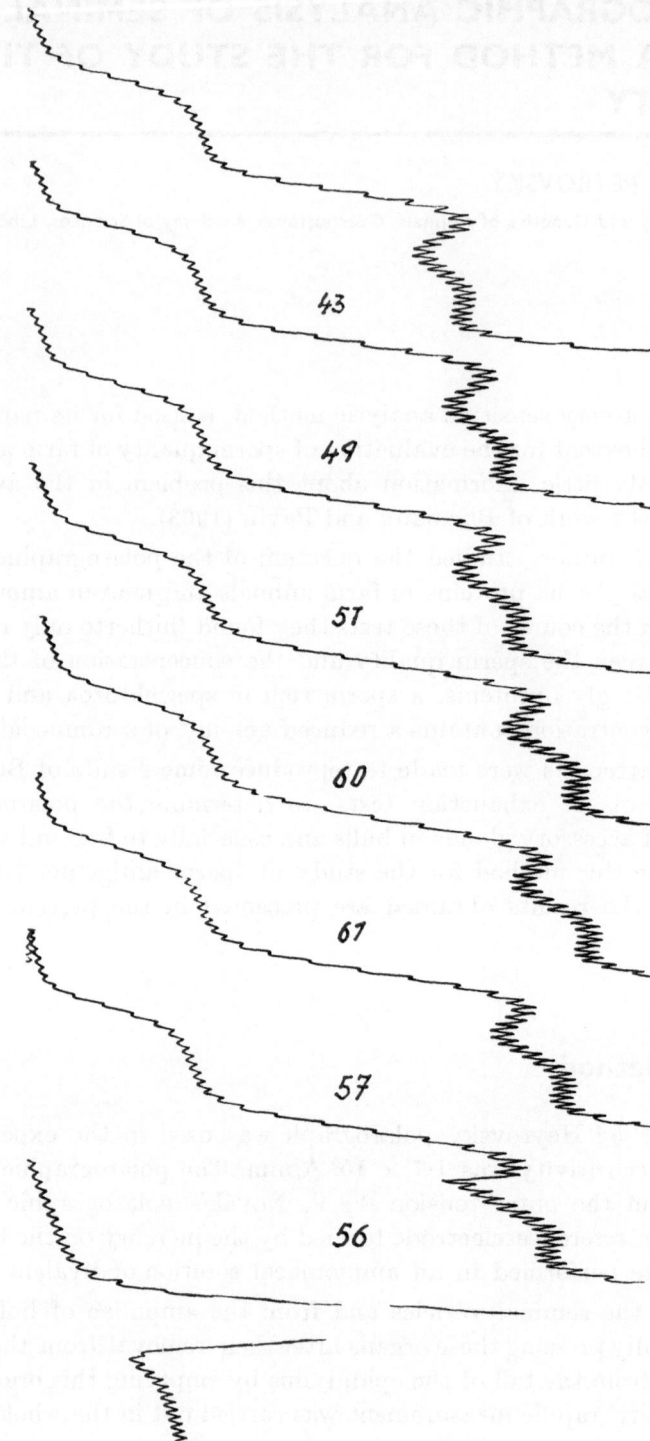

Fig. 1. The polarographic curves of 7 ejaculates of the bull Cesar. Curves 1—7: 1st to 7th ejaculate, curve 8 buffer; dilution: 12 µl. sperm/10 ml, sensitivity: 1 : 150, in the air, h = 60 cm.

or in the seminal plasma or fluid after separating spermatozoa by careful centrifugation in a cooled centrifuge.

The exhaustion tests were made twice in two bulls (at the interval of 11 weeks). The ejaculates were collected by means of an artificial vagina in the usual manner. Collections of ejaculates took usually 2 hours; they were finished when the bulls had stopped ejaculating or remonting. Immediately after the collection the motility of spermatozoa was determined in each ejaculate, the volume measured, pH determined, the dehydrogenation test carried out and samples were taken for the determination of the concentration of spermatozoa, fructose and for polarographic measurement. This was performed in all ejaculates always at the same intervals after collection. The concentration of spermatozoa was determined haemocytometrically, the fructose concentration by the procedure of Mann.

Results and Discussion

During the exhaustion tests we observed a strong oscillation in the level of waves, obtained by means of polarographic analysis of a series of successive bull ejaculates. In the second exhaustion test of the bull Cesar and Major, given as an example, the difference between the highest and lowest values amounted to 18 and 17 mm., respectively (with the analysis procedure — see

Table 1

Exhaustion test of the bull Cesar (23. 10. 1963)

Eja-culate No	Interval of collection	Volume of ejacu-late in ml.	Evaluation of motility*)	pH	Concen-tration of sperma-tozoa in mil./mm^3	Dehydroge-nation test	Fruc-tose in mg.%	Height of 2nd pol. wave in mm.
1	0	3·1	5	6·75	2·240	8′	408	43
2	15′	4·9	4	6·55	1·200	6′	555	49
3	27′	2·2	4	6·75	0·810	8′50″	558	51
4	47′	6·3	3	6·95	0·520	11′	628	60
5	54′	2·2	2+	7·05	0·470	14′	650	61
6	1 hour 27′	3·4	1	7·05	0·140	not decolor.	639	57
7	1 hour 38′	3·1	1	7·25	0·150	not decolor.	631	56
8	1 hour 49′	3·1	1	7·20	0·170	not tested	627	50
9	1 hour 53′	4·1	not tested	7·30	0·090	not tested	645	not test.
10	1 hour 56′	1·8	2	7·40	0·110	not tested	520	40
11	2 hour 09′	2·4	3+	7·35	0·310	21′	523	36
12	2 hour 16′	2·3	3+	7·35	0·280	15′	359	28

*) The motility evaluated by figures 1—5 according to % of motile spermatozoa.

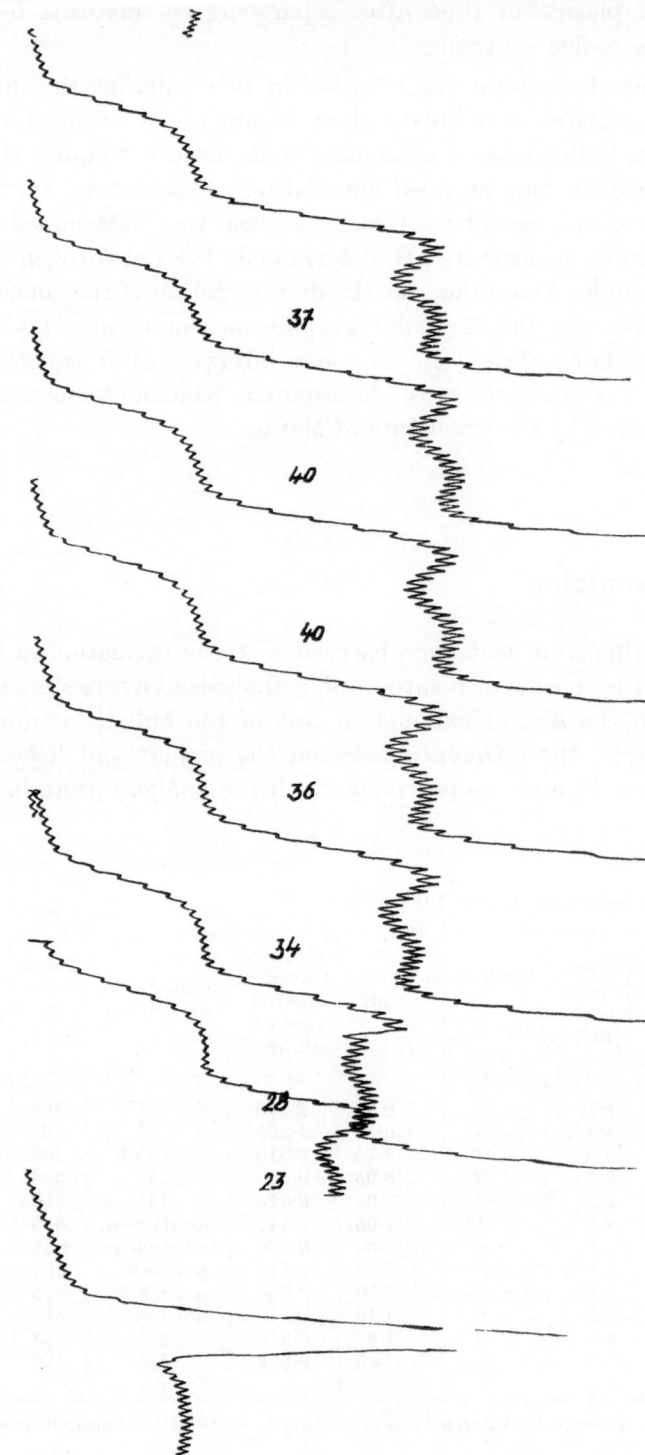

Fig. 2. The polarographic curves of 7 ejaculates of the bull Major. Curves 1—7: 1st to 7th ejaculates, curve 8 buffer. Dilution: 12 μl. sperm/10 ml., sensitivity: 1 : 150, in the air, h = 60 cm.

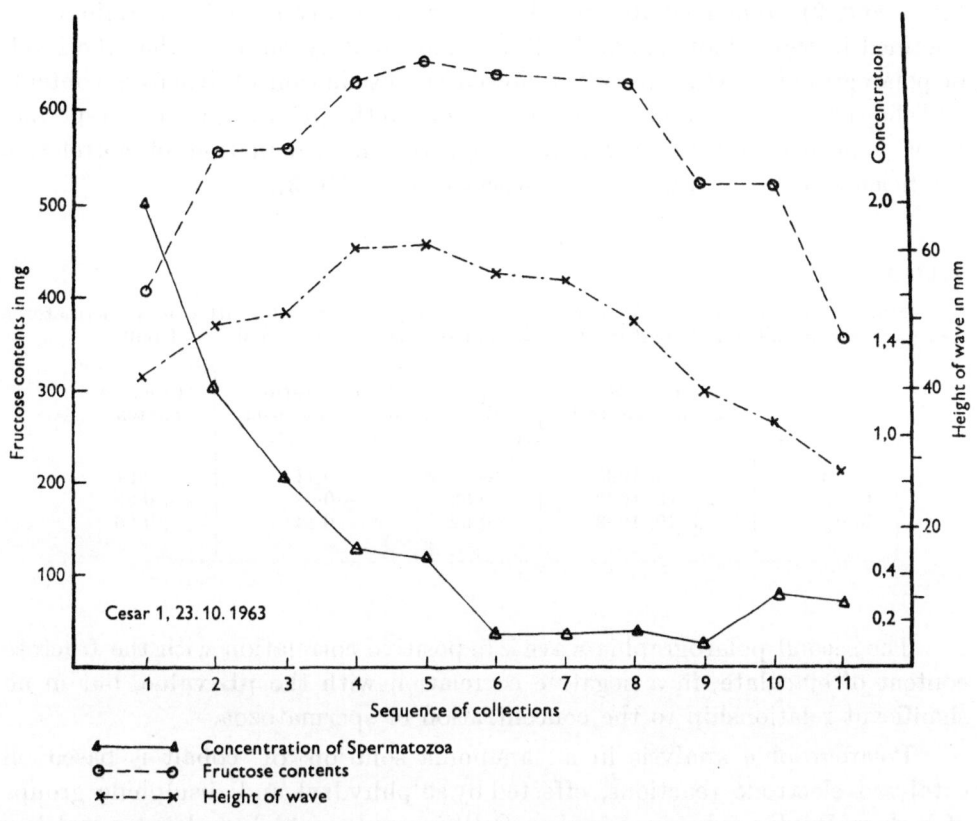

Fig. 3. Oscillation of height of the polarographic wawe (in mm), contents of fructose (in mg%) and concentration of spermatozoa in the sperm (10^6/mm³) during the exhaustion test.

Table 2

Exhaustion test of the bull Major (23. 10. 1963)

Eja-culate No.	Interval of collection	Volume of ejaculate in ml.	Evaluation* of motility	pH	Concentration of spermatozoa in mil./mm³	Dehydrogenation test	Fructose in mg. %	Height of 2nd pol. wave in mm.
1	0	3	4+	6·80	0·800	7'	615	37
2	27'	1·9	3	7·20	0·380	16'	652	40
3	37'	3·2	4	6·80	0·650	7'	567	40
4	47'	1·3	3	7·20	0·440	15'	551	36
5	1 hour 48'	1·4	4+	7·30	0·710	9·5'	509	34
6	2 hour 05'	1·4	4	7·35	0·680	10'	452	28
7	2 hour 23'	1·5	3+	7·30	0·590	not tested	362	23

*) The motility evaluated by figures 1—5 according to % of motile spermatozoa.

fig. 1 and 2). When comparing the values of analysis and determinations, obtained in the exhaustion tests, it was immediately obvious, that the levels of polarographic waves steadily followed the oscillation of fructose contents and showed, as a rule, a negative relationship to the pH-values and the spermatozoa concentration (table 1 and 2, fig. 3). The calculation of correlation coefficients partly confirmed this observation (table 3).

Table 3

Correlation of levels of the second polarographic wave (in mm.) with the pH-values, spermatozoa concentration (in mil./mm.[3]) and fructose content (in mg.) in the ejaculates of bulls

Bull	Date of exhaustion test	pH	Concentration of spermatozoa	Fructose content
Cesar	5. 8. 1963	—	−0·11	+0·94
Cesar	23. 10. 1963	−0·38	−0·03	+0·93
Major	23. 10. 1963	−0·62	−0·14	+0·95

The second polarographic wave is in positive correlation with the fructose content of ejaculate, in a negative correlation with the pH-value, but in no significant relationship to the concentration of spermatozoa.

Polarographic analysis in an ammonia solution of cobalt is based on catalysed electrode reactions, effected by sulphhydryl and disulphide groups of high molecular substances (when Co^{III} is used) or high molecular and low molecular substances (when Co^{II} is used). The polarographically active components of the sperm are therefore primarily proteins of seminal plasma. In this investigation, the buffer with Co^{II} was used in order not to overlook the low molecular weight substances of the ejaculate when searching for the relation of wave heights to sperm quality; the content of these substances in the sperm was low when compared with that of proteins, but they could play an important role in sperm metabolism and sperm evaluation as well. However, no such relation has been found.

The seminal plasma of bulls is a complex mixture of fluids of several glands of the male genital tract, which participate in different ways in its composition and thus in the polarographic behaviour. For this reason, we performed polarographic analyses of the fluids of the seminal vesicles, the ampullae and the epididymis and in one case also of the fluid of the urethal glands obtained by collecting the fluid secreted by the bull before ejaculation.

The urethal fluid in concentrations used for other fluids showed practically no waves in an ammonia solution of divalent cobalt, it merely suppressed the cobalt maximum. Very low polarographic waves have been observed in the

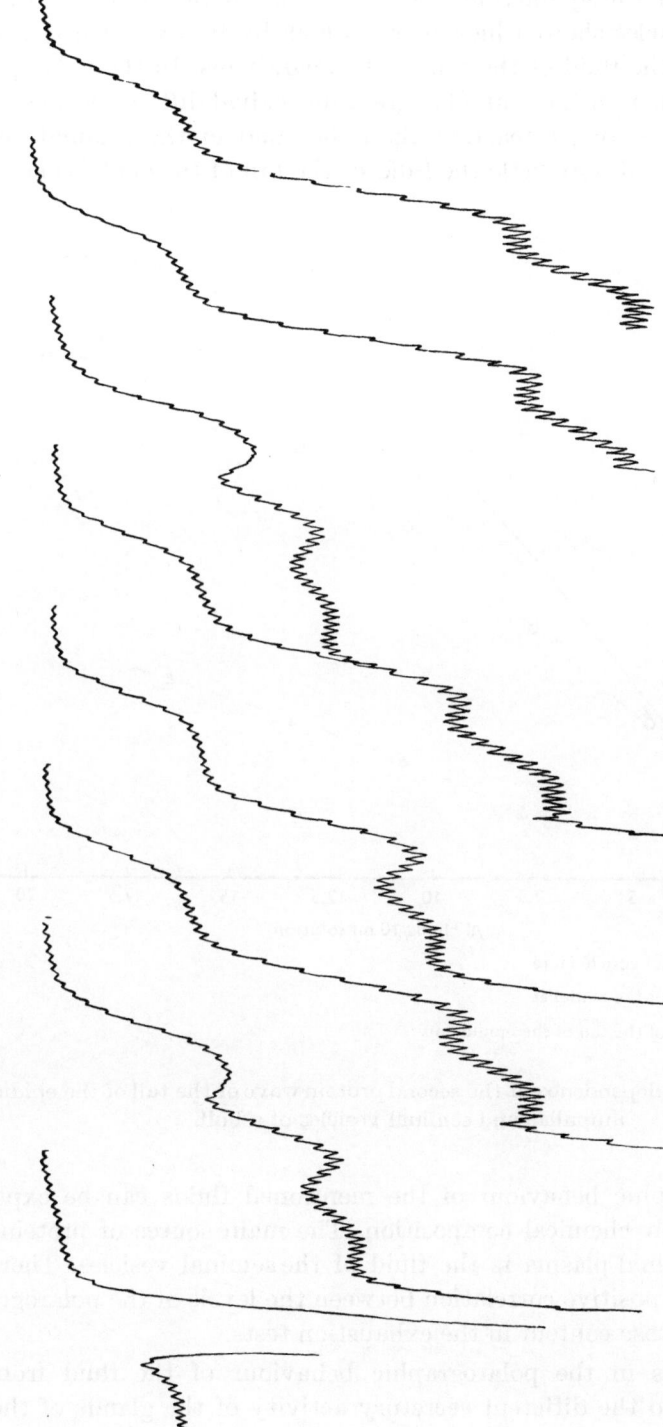

Fig. 4. The polarographic curves of the fluids of the seminal vesicles from the ampullae and the epididymis of 3 bulls. Curve 1, 4, 6: seminal vesicle fluid of bulls 1, 2, 3. Curve 2, 5: fluid from the ampullae of bulls 1, 2. Curve 3, 7: Fluid of the tail of the epididymis of bulls 1, 3. Curve 8: buffer. Dilution: 12 μl. seminal fluid/10 ml., sensitivity: 1 : 200, in the air, h = 60 cm.

fluid of the tail of the epididymis (fig. 4, fig. 5). On the contrary, the fluid of the seminal vesicles showed high waves which by far exceeded the polarographic curves of the fluid of the tail of the epididymis. In the polarographic picture of the fluid from the ampullae great individual differences were found: sometimes it gave stronger reaction than the fluid of the seminal vesicles, sometimes it exceeded only little the fluid of the tail of the epididymis.

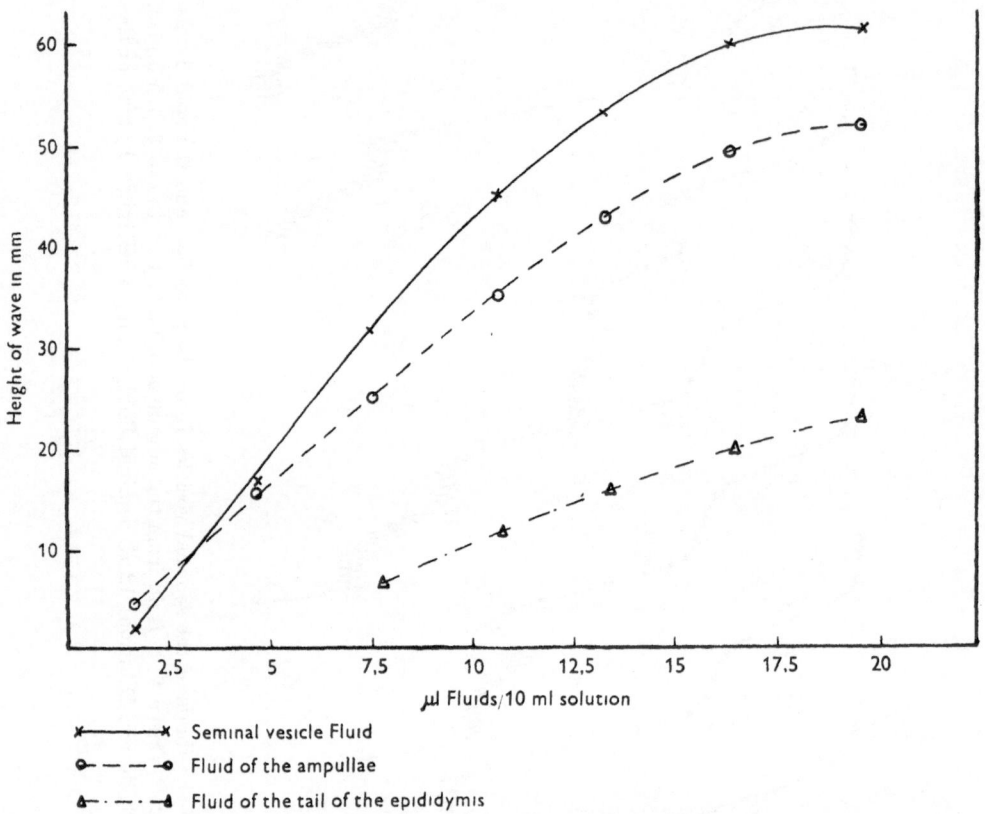

Fig. 5. Concentration dependence of the second protein wave of the tail of the epididymis, ampullae and seminal vesicles of a bull.

The polarographic behaviour of the mentioned fluids can be explained on the basis of their chemical composition. The main source of proteins and fructose of the seminal plasma is the fluid of the seminal vesicles. Therefore, we obtained a high positive correlation between the levels of the polarographic waves and the fructose content in the exhaustion tests.

The differences in the polarographic behaviour of the fluid from the ampullae are due to the different secretory activity of the glands of the am-

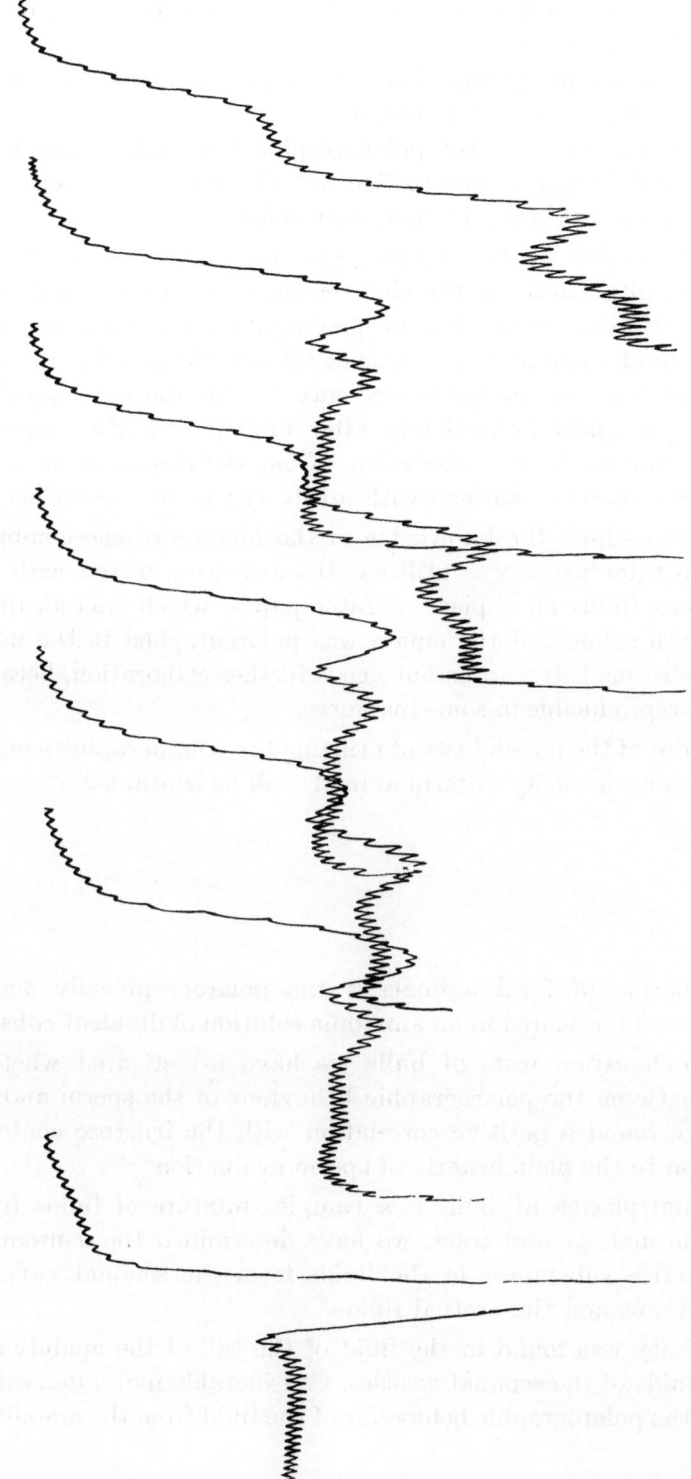

Fig. 6. The polarographic curves of the seminal vesicle fluid of a bull with the rabbit antiserum. Curve 1: 6 µl. seminal vesicle fluid + 6 µl antiserum/10 ml. immediately after mixing. Curve 2: 6 µl. seminal vesicle fluid + 6 µl. antiserum/10 ml. after removing precipitate by centrifugation. Curve 3: 3 µl. seminal vesicle fluid + 6 µl. antiserum 10 ml. immediately after mixing. Curve 4: 3 µl. seminal vesicle fluid + 6 µl. antiserum/10 ml. after removing precipitate by centrifugation. Curve 5: 1·5 ul. antiserum/10 ml. immediately after mixing. Curve 6: 1·5 µl. seminal vesicle fluid + 6 µl. antiserum/10 ml. after removing precipitate by centrifugation. Curve 7: buffer. Sensitivity: 1 : 150, in the air, h = 60 cm.

377

pullae or decrease in active substances of this fluid (proteins) as a results of their binding with spermatozoa.

The fluid of the tail of the epididymis has a low protein content and gives therefore small waves at polarographic analyses.

All the above observations on the polarographic behaviour of the sperm or the different seminal fluids are used to find out whether the polarographic method can be applied to the study of their antigenicity.

A direct polarography of the system: seminal antigen-corresponding antibody yields no results. However, the high sensitivity of this method could be utilized for quantitative evaluation of precipitation reactions of small amounts of seminal fluids. The procedure was as follows: the smallest measurable amount of titrated seminal fluids was mixed with the corresponding antiserum and polarographed immediately after mixing and after removal of the resulting precipitate by centrifugation. From the difference in wave levels strength of the reaction of antigen with antibody (fig. 6) was estimated.

In the second procedure, the knowledge of the binding of the complex: antigen-antibody on filter paper was utilized: the sera were mixed with the corresponding seminal fluids on a piece of filter paper, which was air-dried and then eluated with saline and the eluate was polarographed in the usual way. This procedure seems better to us but needs further elaboration, because the results were not reproducible in some instances.

Our investigations of the possibilities of utilizing the polarographic method in the study of sperm antigenicity in farm animals will be continued.

Summary

The seminal plasma of farm animals is the polarographically active component of the sperm (measured in an ammonia solution of divalent cobalt).

By means of exhaustion tests of bulls we have investigated whether there is a relation between the polarographic behaviour of the sperm and its evaluation. We have found a positive correlation with the fructose content, but no direct relation to the main criteria of sperm evaluation.

Since the seminal plasma of bulls is a complex mixture of fluids from several glands of the male genital tract, we have determined the content of polarographically active substances in the fluids from the seminal vesicles, ampullae and epididymis and the urethal fluids.

The lowest activity was found in the fluid of the tail of the epididymis, the highest in the fluids of the seminal vesicles. Considerable individual differences were noted in the polarographic behaviour of the fluid from the ampullae.

The urethral gland fluid in identical concentrations only suppressed the maximum of cobalt.

Attempts are made to utilize the knowledge of the polarographic behaviour of seminal fluids for an indirect quantitative determination of their reaction with the corresponding antisera.

References

Buruianu, L. M. and Pavlu, V. (1963). International Conference on the Reproductive Biology in Bucharest.

PROTEIN POLYMORPHISM OF THE SEMINAL VESICLES OF BULLS AND THE SENSITIVITY OF BULL SPERMATOZOA TO COLD SHOCK

J. FULKA, H. ŠULCOVÁ, M. VALENTA

Laboratory of Physiology and Genetics of Animals, Czechoslovak Academy of Sciences, Liběchov

Spermatozoa from the cauda epididymis and the spermatozoa from the ejaculate differ in a number of features. In addition to the commonly known difference in metabolic activity, they show a different sensitivity to a sudden decrease in temperature. Epididymal spermatozoa as noted by Lasley and Bogart (1944) are resistant to sudden cooling, whereas ejaculated spermatozoa show a considerable sensitivity (Milovanov 1940). In general, harmful changes produced by sudden cooling have been called cold shock, and a number of studies dealt with this subject (Mann 1954, Mann and Lutwak-Mann 1955, Blackshaw and Salisbury 1957).

Less knowledge is available about the reason for resistance of epididymal and sensitivity of ejaculated spermatozoa. Bialy and Smith (1959) noted that sensitivity to sudden cooling begins, in bull spermatozoa, during their passage through the ampullae of the seminal ducts. These authors assume that they are bound to some other substance which is no longer present in seminal plasma or is very instable and can be easily inactivated by other fluids of the accessory glands. Fulka, Šulcová and Valenta (1963) extended this knowledge by finding that the spermatozoa begin to be sensitive to cold shock in the ampullae of the seminal ducts but their full sensitivity manifests itself only after their contact with the fluids of the seminal vesicles.

The fluids of the seminal vesicles and the seminal plasma are not capable of eliciting an increased sensitivity of spermatozoa obtained from the cauda epididymis. Changes responsible for an increased sensitivity of spermatozoa begin to manifest themselves on the surface membrane in the ampullae of the seminal ducts. The fluid obtained after centrifugation of spermatozoa from the ampullae, is, however not, capable of producing changes in spermatozoa from in the cauda epididymis, which are characteristic for spermatozoa from the ampullae. This fact led us to suppose that spermatozoa might absorb some effective substances. Since the fluids of the seminal vesicles act in this respect, we considered it appropriate to investigate, whether and to what extent spermatozoa are capable of absorbing some substances from these fluids, and how absorption would influence their effectiveness. The gradual increase in the amount of antigens in spermatozoa during their passage through the male genital tract, as noted by Matoušek (1962) justified our supposition.

Furthermore, it was of interest to find out whether there is a relationship between the efectiveness of the seminal vesicle fluids and polymorphism of protein fractions, detected by Valenta et al. (1964).

Material and Methods

Genital organs of bulls, obtained at the abattoir, were used for tests. The organs were treated as soon as possible within 3 to 4 hours after death of the animal at the latest. Spermatozoa from the cauda epididymis were obtained by pressing air into the ductus deferens by means of a hypodermic syringe. After being filled with enough air, the cauda epididymis was perforated and the spermatozoa intercepted in test tubes. By gentle pressure it was possible to obtain spermatozoa from the ampullae of the seminal ducts. The fluids of the seminal vesicles were obtained by light pressure to prevent contamination with blood. Thereafter, spermatozoa were immediately centrifuged at 3,000 rpm. and washed twice. The washed and centrifuged spermatozoa were mixed with the fluids of the seminal vesicles 1 : 1 and incubated at 20°C for 30 minutes. After centrifugation the fluids of the seminal vesicles were examined as to their capacity to influence the sensitivity of spermatozoa from the ampullae to cold shock. Spermatozoa from the cauda epididymis and the ampullae were used for absorptions.

The number of dead spermatozoa after sudden cooling from room temperature to a temperature of 0°C for then minutes was taken as a criterion of effectiveness of seminal vesicle fluid. Unwashed spermatozoa from the ampullae mixed at an equal ratio with the treated fluids of the seminal vesicles, were used for preparation of samples for cooling. Untreated fluids of the seminal vesicles served as controls. 18 samples of spermatozoa from the ampullae from different bulls and the same number of samples of fluids of the seminal vesicles were evaluated. In order to evaluate spermatozoa before and after cooling, the Cyanine B dye was used, by means of which the living and dead spermatozoa could be discerned (Fulka et al. 1960).

Electrophoresis on agar gel, as described in detail in the paper of Valenta et al. (1964) was used to separate the protein fractions of the fluids of the seminal vesicles.

Results and Discussion

The results of the experiments are shown in table 1. The table shows that the number of living spermatozoa obtained from the ampullae is decreased after cooling from 81% to 45%. (P < 0·01). After the addition of the fluids of the seminal vesicles and cooling to 0°C the number of living spermatozoa decreased to 19%. The difference observed is statistically significant (P < 0·01). When using the fluids of the seminal vesicles which were absorbed by spermatozoa obtained from the ampullae, the number of living spermatozoa does not decrease to the previous values, but reaches nearly the level of spermatozoa from the ampullae. When comparing the semen vesicle fluids which were not absorbed with those which were absorbed by spermatozoa from the ampullae the difference is 21% (P < 0·01). Spermatozoa from the cauda epididymis seem to be more active for absorption. When used for absorption, the fluids of the seminal vesicles lose completely their capacity of increasing the sensitivity of spermatozoa to sudden cooling. With regard to high variability, the difference between activity of the fluids of the seminal vesicles, which were absorbed by epididymal spermatozoa or spermatozoa from the ampullae, is not statistically significant (P > 0·05).

Table 1

Number of determinations	Percentage of spermatozoa remaining unstained				
	without treatment	after cooling	after cooling in the presence of fluids of seminal vesicles	in the presence of seminal vesicle fluids absorbed by spermatozoa	
				from the ampullae	from the cauda epididymis
18	81 $s = 6·2$ $s_x = 1·4$	45 $s = 7·7$ $s_x = 1·8$	19 $s = 11·8$ $s_x = 2·8$	40 $s = 7·6$ $s_x = 1·8$	46 $s = 11·1$ $s_x = 2·3$

From these data it is evident that after absorption by epididymal spermatozoa or spermatozoa from the ampullae the fluids of the seminal vesicles lose to a certain extent or completely their capacity to increase the sensitivity of spermatozoa to sudden cooling.

At present it is difficult to explain what substances of the seminal vesicles, and in what manner, are responsible for the increase in sensitivity of the spermatozoa which disappears after absorption. There is relatively great resemblance between the haemolytic factors, described by Millar (1956) in seminal plasma of bulls. More detailed observations about the character of this factor are given by Hunter, Stevens and Hafs (1963). They noted that its source are the fluids of the seminal vesicles. It is not possible to remove it by dialysis. Evidently it is the question of a high molecular substance which cannot be inactivated even by heating to 56° and 70°C. Petrovský (1964) observed that the activity of the haemolytic factor in the fluids of the seminal vesicles could be reduced considerably or destroyed completely by absorption with

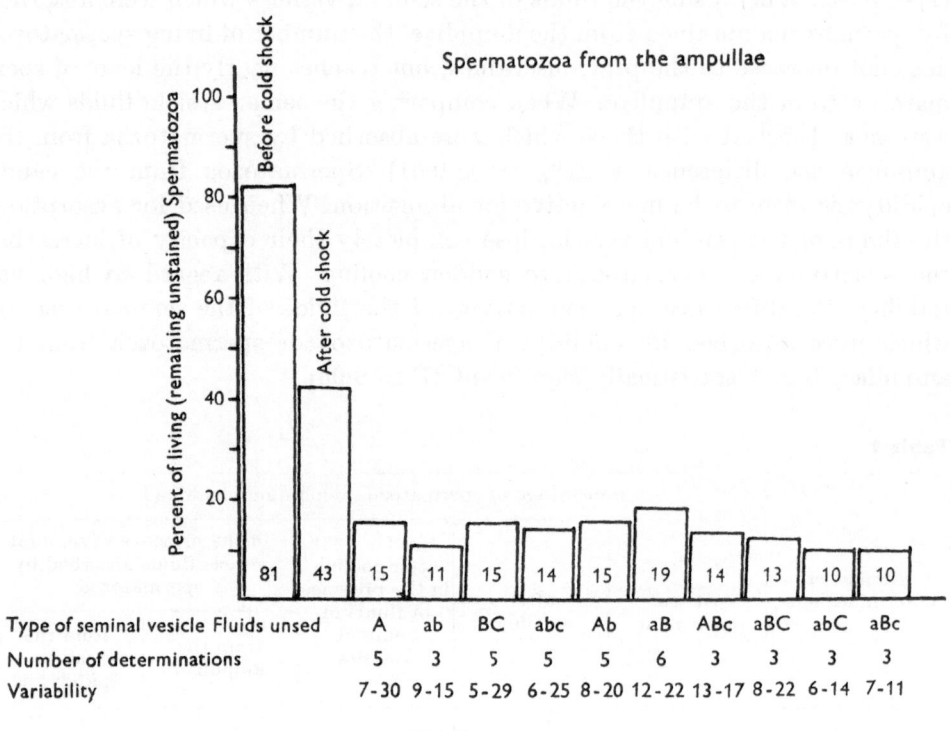

Fig. 1.

epididymal spermatozoa or spermatozoa from the ampullae. The addition of egg yolk in a low concentration reduced to a great extent the effectiveness of the factor. In this respect there is resemblance between the influence of the haemolytic factor and the shock effect of the fluids of the seminal vesicles. In some other respects, however, they are fundamentally different, as for example the possible inhibition of the haemolytic factor by bull or rabbit sera. The shock influence of the fluids of the seminal vesicle is not affected by the addition of serum.

In the course of tests it was also noted that the fluids of the seminal vesicles of different animals do not have the same capacity of increasing the sensitivity of spermatozoa to cold. We determined to investigate whether a relationship exists between this penhomenon and polymorphism of protein fractions, detected by Valenta et al. (1964).

Using the fluids of the seminal vesicles of almost all types described it was impossible to confirm the advanced supposition. The results obtained are shown in Fig. 1.

It may be seen from Fig. 1. that all types of fluids used have nearly conforming qualities in this respect, irrespective of which fractions and in what strength are represented in the fluids. Possible differences should manifest themselves primarily in the fluids in which some or all detected fractions (ABC) would be present in maximum intensity. From the graph we can also see that the capacity of the fluids of the seminal vesicles to elicit sensitivity of spermatozoa to cold is independent of the distribution and strength of the fractions. By adsorbing the fluids of the seminal vesicles with epididymal spermatozoa or spermatozoa from the ampullae they lose the shock effect. The number of protein fractions observed remains unchanged by this interference. Only their reciprocal relationship is changing.

Separation by means of agar electrophoresis permitted us to make a preliminary observation that the cathode fraction designated B is more absorbed by spermatozoa than other fractions. Absorption of substances from the seminal plasma or other fluids of the accessory genital glands was also found by other authors, e. g. Matoušek (1962) detected the binding of substances of antigenic character during passage through the male genital tract. It can be assumed that the binding of these substances is associated not only with changes bound to intensity of metabolism of epididymal spermatozoa and spermatozoa from the ampullae, but also with changes in the character of the cell membrane, responsible for an increased sensitivity to sudden cooling. For the time being sufficient evidence is not available to designate directly this substance or group of substances which are effective in this respect, and the possibility of their existence will be the object of further observations.

Summary

During the passage through the male genital tract sensitivity of spermatozoa to sudden cooling is increased. The capacity of the semen vesicle fluids to increase sensitivity of spermatozoa from the ampullae to cold shock can be limited by incubating them with washed spermatozoa from the epididymis or the ampullae. The epidididymal spermatozoa have a higher absorption capacity than those from the ampullae. Separation of the semen vesicle fluids by means of electrophoresis in agar shows that the number of fractions is not changed by absorption but their strength is visibly affected. The relation of this finding to polymorphism of protein fractions was not proved.

References

Bialy, G., Smith, V. R. (1959). J. Dairy Sci. 42 : 2002.

Blackshaw, A. W., Salisbury, G. W. (1957). J. Dairy Sci. 40 : 1099.

Fulka, J., Valenta, M., Pavlok, A., Icha, F. (1960). Sborník ČSAZV-Živočišná výroba 35(5) : 905.

Fulka, J., Šulcová, H., Valenta, M. (1963). III. International Conference on Biology and Pathology of Reproduction of Farm Animals, Bucarest.

Hunter, A. G., Stevens, K. R., Hafs, H. D. (1963). J. Dairy Sci. 46 : 618.

Lasley, J. F., Bogart, R. (1944). Am. J. Physiol. 141 : 619.

Mann, T. (1954). The Biochemistry of Semen. Methuen and Co., London.

Mann, T., Lutwak-Mann (1955). Arch. Sci. Biol. 39 : 578.

Matoušek, J. (1962). The VIII. European Conference of Blood Groups in Ljubljana, Yugoslavia.

Millar, P. G. (1956). Brit. Vet. J. 112 : 685.

Milovanov, V. K. (1940). Iskusstvennoje osemenenije s. ch. zhivotnych. Selchozgiz. Moscow.

Petrovský, E. (1964). Sborník živ. výroba (in press).

Valenta, M., Matoušek, J., Petrovský, E., Stratil, A. (1964). The 9th European Conference of Animal Blood Groups. Prague.

DISCUSSION

A. Eyquem: I wonder if during these studies, Dr. Matoušek has observed any damage associated with multinucleated cells which are probably spermatids and if immunofluorescence has been performed.

J. Matoušek: In rams immunized in order to obtain antibodies against antigenic components of the ovarian follicle fluids the process of spermatogenesis was not studied systematically. In rams, immunized by seminal vesicle fluids (ram No. 8 and 1050) the spermiogram was recorded only in ram 1,050. The ram No. 8 was not trained to ejaculate into the artificial vagina. The immunization of the ram 1,050 by seminal vesicle fluids from several bulls was accompanied by considerable disturbances of spermiohistogenesis. Four weeks after the end of immunization, the spermatozoa with bent and unterminated tails, moving in circle, began to appear in the ejaculate. The number of so degenerated spermatozoa amounted to 80% of all spermatozoa in the ejaculate. The total number of spermatozoa in the ejaculate was decreased.

Antigenic strength of the fluids of the genital organs without spermatozoa for the destruction of spermatogenesis is illustrated in the case of the bull 363/2, which was immunized by seminal vesicle fluids of boars. After the first immunization (a series of 4 injections) this bull ejaculated during 8 weeks only a fluid containing no spermatozoa.

As far as the second question of Dr. Eyquem is concerned, I must say that we didn't up-to-now use immunofluorescence. However, we will use this method in our further work, since we consider it very effective.

IMMUNOLOGICAL TOLERANCE
AND TRANSPLANTATION ANTIGENS

IMMUNOLOGICAL TOLERANCE AND BLOOD GROUPS

M. HAŠEK

Institute of Experimental Biology and Genetics, Czechoslovak Academy of Sciences, Prague

The antigen inducing antibody formation after the inoculation into the organism can, under suitable conditions, elicit quite the contrary reaction — specific immunological unresponsiveness or immunological tolerance. The critical factor deciding upon which of the reaction pathways will take place is the amount of antigen. Small doses lead to immunity, whereas large doses of appropriate antigen can induce tolerance. The second type of the immunological reaction to the introduction of antigen can be most easily shown in such an experimental model where the immunized animal is immunologically immature and the material used as antigen is not too foreign to the recipient. The third factor enhancing inducibility of tolerance is the use of antigen which is represented by cells capable of further multiplication in the recipient. Such conditions can be fulfilled if the isoantigen is used, i. e. the cells from another individual of the same zoological species, and the embryo or the newborn animal serves as recipient.

The cells can be introduced into the recipient either by means of embryonic parabiosis of two avian embryos or by intravenous injections into embryos or newborn animals. The embryonic parabiosis is nothing else than a laboratory imitiation of the experiment discovered by Ray D. Owen (1945) in natural dizygotic twin cattle. Vascular anastomoses between two embryos, which occur in twins by natural fusion of the placentas, are achieved in eggs by artificial fusion of chorioallantoic membranes (Hašek 1953a). Intravenous injections are performed in birds into the chorial vein of the embryo or into the vein in the wing of foot of the newborn bird.

After both embryonic parabiosis and intravenous injections of blood or spleen cell or bone marrow cell suspensions, cell chimaerism is produced in the recipients as in the case of dizygotic twins. In such animals, the presence of erythrocyte mosaicism can be easily detected, for example, serologically, in the peripheral blood. The erythrocyte mosaicism is the outcome of colonization in the recipient of primordial haematopoietic cells producing the erythrocytes of the donor type. It is of much interest that such graft hybrids often tolerate the donor tissue for their life-span and are incapable of destroying the foreign tissue by the immune mechanism.

The incapacity for the immune response was first demonstrated in tolerant chickens by their mutual immunization (Hašek 1953b). It is known that chickens are excellent producers of immune isoagglutinins. However, chicken parabionts form no isoagglutinins against the partner from the embryonic parabiosis, not even after repeated immunizations (three times 5 doses of 1—2 ml. blood). The tolerance is mostly specific. The chickens produce agglutinins after immunization with erythrocytes from other donors and are also capable of normal antibody formation against heterologous erythrocytes such as turkey erythrocytes (tab. 1). ^{51}Cr-labelled erythrocytes from the cell donor inoculated intravenously are not eliminated from the circulation in an immune fashion, but persist in the circulation as long as autologous erythrocytes. When the cell chimaerism spontaneously disappears in young tolerant animals, the capacity for antibody formation is quickly recovered. This dependence of persisting tolerance on the presence of antigen in the recipient's body was also experimentally verified by the abolition of cell chimaerism by means of passive transfer of hyperimmune sera and gamma globulin and this led to dissappearance of tolerance (Hašek et al. 1963). However, in old adult animals, a spontaneous disappearance of chimaerism leads no longer to such a rapid onset of antibody formation (Hašek and Hraba 1955). In these experiments, the presence of erythrocyte chimaerism was detected by absorption of serum (obtained by immunization of the third animal with blood from one of the parabionts) with erythrocytes from the other parabiont. Although no erythrocyte mosaicism was found in adult parabionts, they began to form agglutinins against the partner as late as 6 months after the follow-up had begun.

In general, it can be said that the antigen is necessary not only for the induction of tolerance, but also for its maintenance. This need not mean that

Table 1

Formation of immune agglutinins in adult parabionts and controls

Recipients	Donors of blood for immunization	Number of cases	Agglutinin titre
parabionts chickens	parabiont partner from embryonic parabiosis	12	0
	non-parabiont random bred chicken	17	2—128
	turkey	2	128 and 512
control random bred chickens	random bred chicken	94	16—128

the tolerant cell requires a further supply of antigen. First of all, it is possible that such a tolerant cell does not exist at all and specific tolerance is caused by the elimination of immunologically competent cells. However, if we accept the hypothesis that the tolerant cell exists, further supply of antigen might not be necessary, consequently tolerance at the cellular level is an irreversible state and a new antigen is required only for the induction of tolerance in newly differentiating immunologically competent cells (Hašek et al. 1961, Hašek 1962). A slower disappearance of tolerance after the interruption in antigen supply in adult animals as compared with young aninals may be explained by a decrease in cell-turnover of competent cells during ontogenesis of the organism (Mitchison 1962, Hašek 1963). There are pronounced differences even in inducibility of tolerance during ontogenesis. Tolerance is easiest to induce in immunologically immature animals, in adult animals many times larger dose of antigen is necessary (Hašek and Puza 1962). The greatest demand for the amount of antigen is during maximum growth of the organism, i. e. at the time of maturation (Iványi et al. 1964, Hašek 1964).

The most feasible explanation of the requirement for an enormous amount of antigen to induce tolerance in immunologically mature animals is that the antigen is necessary particularly for the exhaustion of already differentiated population of antibody-forming cells.

Although immunological tolerance represents an alternative reaction to antibody formation following the introduction of antigen, the spectrum of antigens inducing tolerance is much narrower than that eliciting immunity in small doses. The induction of tolerance to isoantigens, i. e. blood group antigens is very successful.

Immunological tolerance to heterologous erythrocytes, i. e. erythrocytes of different zoological species, is more difficult to induce (Hašek 1956). Although complete tolerance to various antigens such as soluble serum proteins, pneumococcal polysaccharides etc. (see review by Hašek et al. 1961, Smith 1961) was obtained, it seems to hold for the cells at least that with the increasing number of antigenic determinants foreign to the recipient's body, the success in the induction of immunological tolerance rapidly diminishes.

In the initial experiments on the induction of immunological tolerance the cells capable of further colonization and multiplication in the recipient's body were used as antigen. In further experiments we studied whether even pure erythrocytes deprived of leucocytes by repeated differential sedimentation were capable of inducing tolerance. Table 2 shows the results with the induction of tolerance in newborn chicks by injection of pure erythrocytes (Hašek et al. 1956).

These results show that the induction of tolerance does not require the animal to be colonized with donor cells, but even the introduction of cells incapable of further surviving and multiplying in the recipient's body leads

393

Table 2

Agglutinin formation in chickens intravenously injected with foreign erythrocytes, deprived of leucocytes on the 1st day after hatching

	Immunization with donor blood at the age			
	8 weeks		15 weeks	
	no antibody formation	antibody formation	no antibody formation	antibody formation
Chickens injected with 0·5 ml. erythrocytes	4*)	16	1	12
Chickens, control	0	8	0	8

*) Number of animals

to the induction of tolerance. However, in this case repeatedly injected antigen must be supplied in order to maintain tolerance permanently.

Although immunological tolerance is being studied to-date to various types of natural and purified antigens, one of the basic lines of research remains in the sphere of blood group research. It was found, for example, that tolerance related to both immune isoagglutinins and so-called natural agglutinins. In the same year as Landsteiner (1900) discovered the ABO system in man, Ehrlich and Morgenroth (1900) revealed the occurrence of immune isoantibodies in goat. However, the relationship between the two types of antibodies is not yet explained satisfactorily. The induction of immunological tolerance includes both these types of antibodies and complete tolerance can be induced even to erythrocytes, for example, from a different zoological species, against which natural antibodies are normally always present in the

Table 3

Suppression of the occurrence of "natural" heteroagglutinins in chimaeric parabionts

	A	B
Chicken parabiont with turkey	0/8	0 (0)
Chicken control	6/6	6·3 (2—16)
Chicken parabiont with guinea-fowl	0/5	0 (0)
Chicken control	9/9	7·1 (2—32)
Domestic duck parabiont with muscovy duck	0/4	0 (0)
Duck control	5/5	2·2 (1—4)

A = number of birds with heteroagglutinins against erythrocytes of the partner's species/total number of birds.
B = average titre; the figures in brackets indicate the extreme values of titre (both in reciprocal values of serum dilution).

recipient. The examples of suppression of formation of so-called natural antibodies in interspecific chimaeras-embryonic parabionts are given in table 3. These results support strongly the view that even natural antibodies are of active immune origin.

The study of immunological tolerance to isoantigens, in addition to it importance for research in therapeutical application of tissue or organ homografts, shows new pathways to the solution of immunological problems in the relation of mother to foetus including pathological conditions such as incompability of mother to foetus (Owen et al. 1954, Hašek et al. 1962).

The occurrence of natural chimaeras is of much interest for blood group typing. In addition to classical Owen's findings in cattle, natural chimaeras were described in sheep (Lampkin 1953, Stormont et al. 1953, Hraba et al. 1956), from two-yolk eggs in chicks (Billingham et al. 1956) and even in man (Dunsford et al. 1953, Booth et al. 1957, Nicholas et al. 1957, Chown et al. 1963).

Finally, immunological tolerance may also be useful in preparing the sera with narrower specificity as successfully shown in the preparation of anti-tumour sera (Zilber 1959, Levi et al. 1959).

This outline of the relationship between immunological tolerance and blood group research is far from showing all the perspectives and certainly will be overcome by further advancement and increasing knowledge. However, I have tried to show that this relationship is important for both sides, the blood groups provided us with an excellent model for the study of basic immunological problems on the one hand and this study can be worth while for blood group research on the other.

References

Billingham, R. E., Brent, L., Medawar, P. B. (1956). Phil. Trans. Roy. Soc. B 239 : 357.
Booth, P. B., Plaut, G., James, J. D., Ikin, E. W., Moores, P., Sanger, R., Race, R. R. (1957). Brit. Med. J. 1 : 1456.
Chown, B., Lewis, M., Bowman, J. M. (1963). Transfusion 3 : 494.
Dunsford, I., Bowley, C. C., Hutchinson, A. M., Thompson, J. S., Sanger, R., Race, P. R. (1953). Brit. Med. J. 81 : 409.
Ehrlich, P., Morgenroth, J. (1900). Berl. klin. Wschr. 37 : 453.
Hašek, M. (1953a). Čs. biologie 2 : 25.
Hašek, M. (1953b). Čs. biologie 2 : 265.
Hašek, M. (1956). Proc. Roy. Soc., B, 146 : 67.
Hašek, M. (1962). Folia biol. (Prague) 8 : 73.
Hašek, M. (1963). Immunopathology, IIIrd Int. Symp. La Jolla, p. 148.
Hašek, M. (1964). Immuntoleranz. Tagung der Deutschen Ges. f. Physiol. Chemie, Mosbach 1964 (in press).

Hašek, M., Hraba, T. (1955). Folia biol. (Prague) 1 : 1.

Hašek, M., Puza, A. (1962). In "Mechanisms of Immunological Tolerance", Prague, p. 257.

Hašek, M., Hraba, T., Esslová, M. (1956). Folia biol. (Prague) 2 : 54.

Hašek, M., Hraba, T., Lengerová, A. (1961). In "Advances in Immunology" 1 : 1.

Hašek, M., Hašková, V., Lengerová, A., Vojtíšková, M. (1962). In "Transplantation", London, p. 118.

Hašek, M., Hort, J., Lengerová, A., Vojtíšková, M. (1963). Folia biol. (Prague) 9 : 1.

Hraba, T., Hašek, M., Čumlivský, B. (1956). Čs. biologie 5 : 266.

Iványi, J., Hraba, T., Černý, J. (1694). Folia biol. (Prague) 10 : 198.

Lampkin, G. H. (1953). Nature 171 : 975.

Landsteiner, K. (1900). Zentr. Bakteriol., Parasitenk., Abt. I, 27 : 357.

Levi, E., Schechtman, A. M., Sherins, R. S., Tobies, S. (1959). Nature, 184 : 563.

Mitchison, N. A. (1962). In "Mechanisms of Immunological Tolerance" Prague, p. 245.

Nicholas, J. W., Jenkins, W. J., March, W. L. (1957). Brit. Med. J. 1 : 1458.

Owen, R. D. (1945). Science 102 : 400.

Owen, R. D., Wood, H. R., Foord, A. G., Sturgeon, P., Baldwin, L. G. (1954). Proc. Nat. Acad. Sci. 40 : 420.

Smith, R. T. (1961). In "Advances in Immunology" 1 : 67.

Stormont, V., Weir, W. C., Lane, L. L. (1953). Science 118 : 695.

Zilber, L. A. (1959). Vopr. onkol. 5 : 265.

THE RELATIONSHIP BETWEEN ERYTHROCYTES AND TRANSPLANTATION ANTIGENS IN CHICKS

F. KNÍŽETOVÁ, V. HAŠKOVÁ

Institute of Experimental Biology and Genetics, Czechoslovak Academy od Sciences, Prague

A number of authors were concerned with the relationship between erythrocyte and transplantation antigens in various species of animals.

In chicks, this relationship was investigated by Štark et al. (1960), Schierman and Nordskog (1962, 1963), Jaffe and McDermid (1962), Craig and McDermid (1963) and other authors.

The works of the above authors show that B locus and probably C locus of the blood groups influence the survival of skin homografts and possibly splenomegaly, whereas the differences at the A, D, and L loci have no effects on the graft survival (Jaffe and Payne 1962, Schierman and Nordskog 1961).

In our work, this relationsohip has been studied on inbred White Leghorn (C line) and on chicks of outbred population of White Leghorn to which grafts from C-line individuals were transferred.

Inbred C line was maintained by a strict brother x sister mating. During crosses the individuals were selected which tolerated a homograft from the partner in breeding for the longest period and tests were made using blood group reagents.

At the present time, segregation within the given line is maintained at the B locus and at that independent of the B which was preliminarily designated "404" locus.

Grafts were exchanged between members of the C-line tested by blood group reagents and the condition of the graft and the constitution of the

Table 1

Permanent survival of skin grafts depending on the B locus genotype of the donor and recipient of the C line

Recipient \ Donor	B_1B_1	B_1B_2	B_2B_2
B_1B_1	296*)/298	2/22	0/9
B_1B_2	31/32	34/34	11/12
B_2B_2	0/12	0/1	5/5

*) Permanently surviving grafts/total number of grafts

donor and recipient at the B locus (tab. 1) and "404" locus (tab. 2) were compared.

The tables show that the difference between the donor and recipient at the B locus results in skin graft rejection, whereas the constitution at the "404" locus does not affect the graft survival.

Using the same material, the relationship between graft survival and agglutinin formation was studied: the formation of serum antibodies against erythrocytes from graft donors 3—6 months after transplantation and after immunization with erythrocytes from the graft donors was followed up. We found that "posttransplantation" antibodies were formed in 15·7 % of cases, irrespective of whether the skin grafts had regressed or survived. Immunization of recipients with erythrocytes from the graft donors resulted in antibody formation in 29·4 % of cases (tab. 3).

Table 2

Permanent survival of skin grafts in dependence on the non B locus "404" of the donor and recipient of the C line

Donor / Recipient	404+	404−
404+	365*)/391	3/10
404−	16/22	20/23

*) Permanently surviving grafts/total number of grafts

Both transplantation and immune antibodies produced in the case of graft survival had anti-"404" specificity; in the case of graft rejection the antibodies possessed anti-B_1 or B_2 specificity.

Table 3

Relationship between the fate of graft and haemagglutinin formation in the birds of the C line

Graft survives permanently in 75/92 chicks		Graft was rejected in 17/92 chicks	
Haemagglutinins were present		Haemagglutinins were present	
after transplantation	after transplantation + immunization	after transplantation	after transplantation + immunization
6/75	12/75	8/17	17/17

In the outbred population WL we studied the effect of differences between the donor and recipient in the number of antigenic factors at the B locus on the survival time of the first and second grafts and haemagglutinin formation after transplantation (tab. 4.)

The table shows no significant differences in the survival time of the first grafts between the groups differing in the number of antigenic factors as

Table 4

Survival of 57 skin homografts in outbred Leghorn chicks and agglutinin formation after grafting

Antigenic difference between donor and recipient	Mean survival time of grafts in days (x)	Number of grafts (n)	Standard deviation (s)	Agglutinin formation
B_2	8	17	2·610	3*)/17
B_2 B_3	10	10	4·013	0/7
B_1 B_2 B_3	10	20	3·260	0/7
B_2 B_3 B_4	8	1		
B_1 B_2 B_3 B_4	8	2		0/1
B_3 B_4	8	1		
B_1 B_3	11	2		0/2
B_1 B_3 B_5	11	3	2·645	1/3
A_9 B_1 B_2 B_3	7	1		1/1

*) No. of chicks with agglutinins /total number of tested chicks.

confirmed even by statistical analysis. Neither was any interrelationship found between the condition of graft and agglutinin formation: after transfer of the second graft, a white-graft reaction was observed in a number of cases, although posttransplantation agglutinins were not formed. For example, 7 white grafts were found in the first group after the second transplantation, but they were not accompanied by agglutinin formation.

The present results confirm the relationship between the B locus of the blood groups and histocompatibility antigens, and moreover, evidence is again presented — as suggested before (Hort et al., 1961, Mitchison, 1962, Kinský, Mitchison, 1963) — that the spectrum of erythrocyte and transplantation antigens overlaps only in part.

References

Craig, G., McDermid, E. (1963). Transplantation 1 : 191.
Hort, J., Hašek, M., Knížetová, F. (1961). Fol. biol. (Praha) 7 : 301.
Jaffe, W., McDermid, E. (1962). Science 137 : 984.
Jaffe, W., Payne, L. (1962). XIIth World's Poultry Congress Papers, p. 1—6.
Kinský, R., Mitchison, N. A. (1963). Transplantation 1 : 224.
Mitchinson, N. A. (1962). Immunology 5 : 341.
Schierman, L., Nordskog, A. (1961). Science 134 : 1008.
Schierman, L., Nordskog, A. (1963). Nature 197 : 511.
Schierman, L., Nordskog, A. (1962). Science 137 : 620.
Štark, O., Křen, V., Frenzl, B. (1960). Fol. biol. (Praha) 6 : 64.

RELATION OF BLOOD GROUPS
TO TRANSPLANTATION ANTIGENS IN RABBITS

P. IVÁNYI

Institute of Experimental Biology and Genetics, Czechoslovak Academy of Sciences, Prague

I would like to review evidence that the "major" blood group system in rabbits, the H_c system can be regarded as a "real, rabbit blood group". The antigens of this system were not found in individuals of other species, are not transplantation antigens and are not present in body fluids or cells other than erythrocytes.

1. Interspecies relationships. The isoantigens of the system $H_c^{\text{ŠNO}}$ do not cross-react with human antigens ABO, Rh-Hr, MNS, P, Le, Lu, Kell, Fy (Boyd and Feldman 1934, Fischer and Kindler 1955, Sziszkowicz and Eyquem 1959, Joysey 1959). Neither were we successful in finding a specific reaction of strong anti-Š and anti-N sera with chicken, guinea pig, rat, mouse and human erythrocytes. Wild rabbits reacted with anti-N and anti-Š serum, the hares did not specifically react with anti-Š and anti-N sera (Tolarová-Koutková and Varga-personal communication).

2. Occurrence in body fluids and cells other than erythrocytes. The isoantigens of the system H_c were not found in body fluids and cells other than erythrocytes as shown by the examination of the saliva, gastric juice, blood serum, leucocytes, spleen, liver, muscles, heart and kidneys (Boyd and Feldman 1934, Kellner and Hedal 1953, Sziszkowicz and Eyquem 1959). We were not able to find the antigens Š and N by means of absorption test in leucocytes and epidermal cells in embryonic tissue and the amniotic fluid. The sera with high titre anti-Š and anti-N antibodies (capable of producing lethal haemolytic shock) exert no cytotoxic effects upon lymphoid cells homozygous for the respective antigen. The results were also negative, if a mixture of both sera against heterozygous lymphoid cells (ŠN) was used.

3. The relation to transplantation antigens.

a) Medawar (1946) was not able to induce transplantation immunity by means of rabbit erythrocytes. Kapitchnikov et al. (1962) showed that cytotoxic and haemagglutinating antibodies occurring after skin homotransplantation could be selectively absorbed.

b) The presence of immune isoagglutinins obtained by active or passive immunization does not influence the survival time of blood group incompatible graft. The inoculation of serum with high titre anti-Š antibodies into pregnant females produced "rejection" of incompatible offspring in the uterus but had no effects on incompatible skin graft ("split rejection") (Iványi 1962).

c) Split tolerance of erythrocytes and leucocytes was observed (Iványi et al. 1961).

d) We were not able to find any relationship between inducibility of tolerance to skin and the blood group of the system H_c. Blood group incompatible skin grafts transplanted to newborn animals survived for a long period (Iványi and Iványiová 1961, Iványi 1962, 1964).

e) We also failed to find a relationship between the survival time and the blood group during skin transplantation to adult partially inbred rabbits (Fabián et al. 1963). Blood group incompatible skin grafts survived for a long period of 2 to 3 months.

It can be concluded that all the findings support the view that the blood groups of the system H_c are not transplantation antigens. We have obtained no results in support of the assumption of Kapitchnikov et al. (1962) that this major blood group system is a weak transplantation antigen. In order to exclude definitively this possibility, inbred rabbits are, however, lacking in which blood group incompatible graft should survive permanently.

A puzzling question arises, however, considering the fact described primarily by Zotikov (1956) and confirmed by a number of other authors that relatively strong haemogglutinins are produced in rabbits after skin transplantation. The fact that haemagglutinins are produced after skin transplantation in rabbits was confirmed by all authors dealing with this question and Hancock and Mullan (1962), Kapitchnikov et al. (1962) and Iványi (1962) pointed out that these were blood group specific antibodies according to the major blood group system H_c. The observations of Grozdanovič et al. (1959) and Chutná et al. (1961) that haemagglutinins were produced after immunization with epidermal cells only could not be confirmed in our work. Repeated injections of pure epidermal cell suspensions or a dose of epidermal cells representing the pool obtained from two ears of a grown-up rabbit injected intramuscularly produced a high titre cytotoxic serum but no haemagglutinins. The difference may lie in purity of the epidermal suspension or in the method of immunization. It is not excluded, for example, that the injection of epidermal cells in Freund adjuvant (Chutná et al. 1961) could result as a non-specific stimulus for the increase of the titre of rarely occurring "natural" isoagglutinins in rabbits.

Nevertheless, a regular occurrence of blood group specific antibodies after skin transplantation is difficult to explain because the antigens against which these antibodies react are not transplantation antigens and are not present in cells other than erythrocytes.

The simple explanation offers itself that antibodies occur as a result of the presence of erythrocytes in skin graft. This assumption is also supported by the finding (Iványi and Ujhélyiová 1958) that even as small a dose of erythrocytes as seven injections of 0·5 μl. (approximately 2 . 10^6 erythrocytes) is sufficient to produce isoagglutinin formation. However, this simple explanation is in disagreement with the fact that rabbits produce immune isoagglutinins worse after active immunization with the amount of blood much larger than that in the strip of skin employed, than after skin transplantation. Maybe a different method of immunization (with erythrocytes in skin graft) supports especially non-specifically the formation of immune isoagglutinins. Another assumption remains, of course, that the antigens of the system H_c are present in an unknown, concealed form even in epidermal cells and their immunogenic capacity manifests itself only during immunization in the form of transplantation. We are even inclined to prefer the possibility that red cells present in the skin graft are responsible for the haemagglutinin production after homotransplantation in rabbits.

References

Boyd, W. C., Feldman, D. A. (1934). J. Immunol. 27 : 547.
Chutná, J., Rychlíková, M., Pokorná, Z. (1961). Fol. biol. (Praha) 7 : 107.
Fabián, Gy., Iványi, P., Széky, P. (1963). Fol. biol. (Praha) 9 : 440.
Fischer, K., Kindler, M. (1955). Zschr. Hyg. 141 : 122.
Grozdanovič, J., Chepov, P. M., Zotikov, E. A. (1959). Fol. biol. (Prana) 5 : 113.
Hancock, D. M., Mullan, F. A. (1962). Ann. N. Y. Acad. Sci. 99 : 534.
Iványi, P. (1962). Fol. biol. (Praha) 8 : 322.
Iványi, P. (1964). Fol. biol. (Praha), (in press).
Iványi, P., Ujhélyiová, M. (1958). Brat. lek listy 38 (2) : 593.
Iványi, P., Iványiová, D. (1961). Fol. biol. (Praha) 7 : 369.
Iványi, P., Czambelová, A., Dornetzhuber, V., Ujhélyiová, M. (1961). Fol. biol. (Praha) 7 : 337.
Joysey, V. Y. (1959). Brit. med. Bull. 15 : 158.
Kapitchnikov, M. M., Ballantyne, D. L., Stetson, C. A. (1962). Ann. N. Y. Acad. Sci. 99 : 497.
Kellner, A., Hedal, E. F. (1953). J. Exp. Med. 97 : 33.
Medawar, P. B. (1946). Brit. J. Exp. Path. 27 : 15.
Sziszkowicz, L., Fyquem, A. (1959). Ann. Inst. Past. 96 : 184.
Zotikov, E. A.: (1956). Byull. eksp. biol. med. 42 (7) : 58.

THE ONTOGENETIC DEVELOPMENT OF H-2 ANTIGENS *IN VIVO* AND *IN VITRO*

J. KLEIN

Institute of Experimental Biology and Genetics, Czechoslovak Academy of Sciences, Prague

At the present time, three blood group systems are known in mice: The systems H-2 (Gorer 1936) and the system H-5 and H-6 (Amos et al. 1963). In addition to erythrocytes and leucocytes, the antigens of the H-2 system are also present in the cells of other tissues such as the liver and spleen and represent strong transplantation antigens. The mouse belongs to the animals most frequently used for the study of transplantation problems and therefore it seemed useful to us to accumulate a body of information about H-2 antigens.

In the present paper, we pay attention to the ontogenetic development of H-2 antigens because of some discrepancies between different investigators: According to Gorer (1938), Mitchison (1953) and Pizarro et al. (1961) the H-2 antigens are differentiated only a few days after birth, whereas Billingham et al. (1956), Billingham and Silvers (quoted by Medawar, 1959), Chutná and Hašková (1959), Möller (1961a, b), Möller and Möller (1962), Möller (1963), Tyan and Cole (1962), Doria (1963), Simmons and Russell (1962) assume that the H-2 antigens are present in embryonic cells already.

The aim of this work has been to answer the question of whether or not the H-2 antigens are present in early embryonic tissues, and whether the differentiation of H-2 antigens takes place in vitro in the same way as in vivo.

Material and Methods

Mice. The mice of a coisogenic pair C57BL/10ScSn (further B10: H-2^b/H-2^b) and B10 . D2(H-2^a/H-2^a) were used in most experiments. Apart from H-2, there are no other differences in histocompatibility genes between these two strains of mice (Klein — in press). In one experiment the mice of A/J and C57BL/Ks strains were used.

The mice were mated for one night. The next day was regarded as the first day of pregnancy in pregnant mice. The embryo in the first 14 days of pregnancy was denoted as early embryo.

Cell suspensions were prepared by careful homogenization of cut tissue with a loose glass homogenizer in Krebs-Ringer's buffered saline. The homogenate was filtered through silon mesh. Either whole embryos or embryos with cut off head and limbs were used for homogenization.

Tumours. Solid forms of tumour SaI maintained in A/J strain or methylcholanthrene-induced tumour D23 in B10.D2 strain were employed. The tumours were inoculated subcutaneously by trocar. The tumour growth was regularly tested by palpation.

Transplantation of skin grafts 1·5 × 1 cm. in size was performed by the method of Billingham et al. (1954). By the 6th day after transplantation, the grafts were removed for histological examination and the percentage of the surviving epithelium was determined.

Titration of anti-H-2 haemagglutinins was made using the method of Gorer and Mikulska (1954).

Absorption of anti-H-2 haemagglutinins. The cell suspensions were washed two to three times with Krebs-Ringer's physiological saline and after the last centrifugation an equal amount of antiserum was added to the sediment. The cell suspension, occassionally stirred, was left in antiserum for 30 minutes at room temperature. After centrifugation the antiserum was titrated for the presence of haemagglutinins.

Lyophilization of tissues was carried out in an all-glass apparatus in vacuo (0·9 mm. Hg.). Lyophilized tissue was suspended in saline, using a glass homogenizer.

Irradiation. Theratron (^{60}Co; Atomic Energy of Canada Ltd.; 366.5 r/min.) was used as the source of gamma radiation.

Diffusion chambers were formed by a ring from plexiglass and two membranes with pores 0·3—0·5 μ. in size (Synthesia, Uhříněves). The tissue was injected into the chamber in the form of a suspension in normal isologous mouse serum. The chamber was implanted intraperitoneally to adult isologous mice using the same procedure as that of Amos (1961).

Tissue cultures. The liver was taken from the embryos removed under sterile conditions, washed with PBS with admixture of antibiotics and homogenized by aspirating into a syringe and sprinkled against the glass bottom of a plate (repeated several times) and after centrifugation resuspended in medium (Parker 189 modified according to Michl (Michl — not published). The adhered cells of growing cultures were released by trypsinization.

My thanks are due to Miss M. Bílková and Mrs. J. Martínková for valuable technical assistance. To Dr. J. Michl and his colleagues of the Institute of Sera and Vaccines in Prague, I am indebted for their help in growing the tissue cultures. To Dr. I. Hilgert, I owe many thanks for his help in the section on enhancement of sarcoma I.

Results

The presence of H-2 antigens in embryonic tissues was studied using several methods:

Induction of second-set reaction. The first method was used to determine whether an accelerated rejection of skin graft B10.D2 (the so-called second-set reaction) could be obtained in B10 mice by preimmunization with irradiated (15·000 r) embryonic tissue B10.D2 in amounts of 50 millions of embryonic cells per recipient. The table 1 gives the survival of skin epithelium on the 6th day after transplantation in the control, untreated group of mice and in mice preimmunized with 12-day-old embryonic tissues. It can be seen that embryonic tissues contain H-2 antigens capable of producing an accelerated skin graft rejection in a suitable donor.

Table 1

Skin graft rejection after preimunization with irradiated embryonic cells. (Donor: B10.D2, recip.: C57BL/10)

	Percentage of surviving epithelium
Experimental group (after pre-imunization)	20 30 30 0 10 30 0 0 10 0 30 40 0 10
Control group (without pre-imunization)	90 90 80 90 90 80 80

Induction of enhancement. The second method was an attempt to induce an enhanced tumour growth by means of embryonic tissue. Lyophilized 17-day-old embryonic A strain cells were injected intraperitoneally to C57BL/Ks mice (three doses of 15 mg. lyophilisate per mouse every fourth day). 10 days after the last immunization, the recipient mice received subcutaneously 50 mg. Sa I. Non-preimmunized C57BL/Ks mice treated only, with tumour Sa I served as controls. The results are given in table 2. It appeared that the tumour grew better in mice treated with embryonic tissue than in the controls.

Table 2

Effect of embryonic tissue on Sal tumour growth

Group of C57Ks mice	No. of experimental animals	No. of dead animals	Day of death
Mice treated with lyophilized embryonic tissue of A strain	7	6	38 40 43 49 49 51
Control group	7	0	— — — — — —

In order to exclude the possibility that the non-H-2 antigens play a role in the induction of enhancement by embryonic tissue, we performed a similar experiment using the combination of B10 and B10.D2 strains differing only at the H-2 locus. B10.D2 mice were given a total of 75 mg./mouse lyophilized

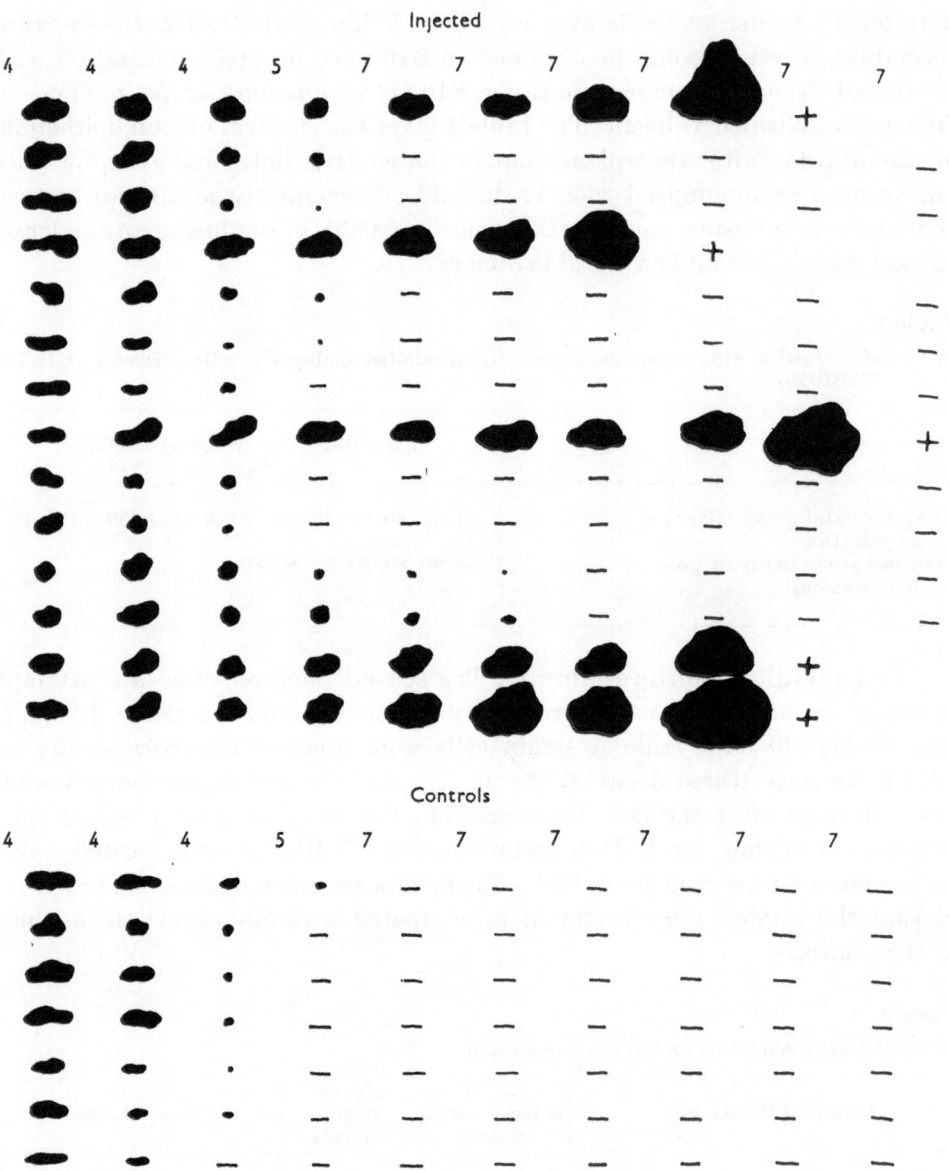

Fig. 1. Haemagglutinin absorption from serum C57BL/10 anti-B10.D2 by means of embryonic tissue B10.D2.

tissue from 12-day-old B10 embryo (12 doses, one dose every fourth day), and the tumor D23 8 days after the last immunization. Figure 1 shows tumour growth in the control and preimmunized mice. Even in this case, a much better tumour growth was obtained in mice treated with lyophilized tissue than in untreated controls, although to a much lesser extent (35% of deaths in the experimental group) than in the case of Sa I (85% of deaths).

However, both experiments indicate that the normal embryonic tissue contains the H-2 antigens capable of inducing an enhanced tumour growth.

Induction of anti-H-2 haemagglutinins. The third method consistent in finding out whether anti-H-2 haemagglutinins could be induced by immunization with embryonic tissues. The results of one experiment are presented in table 3. The tissue of 12-day-old embryo B10.D2 was irradiated with 15·000 r in order to prevent proliferation of cells and possible maturation of antigens in the recipient, and was injected intraperitoneally to B10 mice (the cells from 2 embryos per one adult mouse). Ten days later, the serum was collected from treated mice and titrated. The serum had a relatively high titre of anti-H-2 haemagglutinins suggesting the presence of H-2 antigens in the tissues of 12-day-old embryo.

Table 3

Hemagglutinin titre after one dose of irradiated embryonic cells.
(Donor: B10.D2 recip.: C57BL/10)

Serum dilution	2	4	8	16	32	64	128	256	512	1024	2048	4096
Serum 1	+++	+++	+++	+++	+++	+++	++	++	++	+	−	−
2	+++	+++	+++	+++	++	++	++	++	+	+	+	−
3	+++	+++	+++	+++	+++	+++	++	+	+	+	−	−
4	+	++	+++	+++	+++	+++	+++	+++	+++	++	++	+

Haemagglutinin absorption. The last method used was haemagglutinin absorption from specific anti-H-2 sera by means of embryonic tissues. The serum B10 anti-B10.D2 was used and the embryonic cells B10.D2 of different ages were employed for their absorption. The results in fig. 2 show that this method made it possible to detect the H-2 antigens only immediately before birth, but the amount of H-2 antigens increases rapidly after birth and in a few days reaches the level of H-2 antigens in adult individuals. In early embryos, the H-2 antigens were not detectable by this technique.

A second series of experiments were undertaken to determine the development of H-2 antigens in the cells which has lost the contact with the maternal organism.

Embryonic cell cultivation in diffusion chambers. In one case, a suspension of living embryonic liver cells B10.D2 from 14-day-old embryo in the normal B10.D2 serum was placed into the diffusion chamber (in amounts of about 2 million cells per chamber). The chamber was then implanted into the peritoneum of an adult B10.D2 mouse. Samples of the growing tissue were collected from the chambers at certain intervals and the presence of H-2 antigens was followed by means of antibody absorption from B10 anti B10.D2 antiserum. Figure 3 shows the results of these experiments. The degree of absorption achieved by embryonic cells is given on the axis of abscissa (1 part equals one value of absorption of antiserum diluted in the geometric progression), the age of embryonic cells calculated from the probable day of conception on the axis of ordinate.

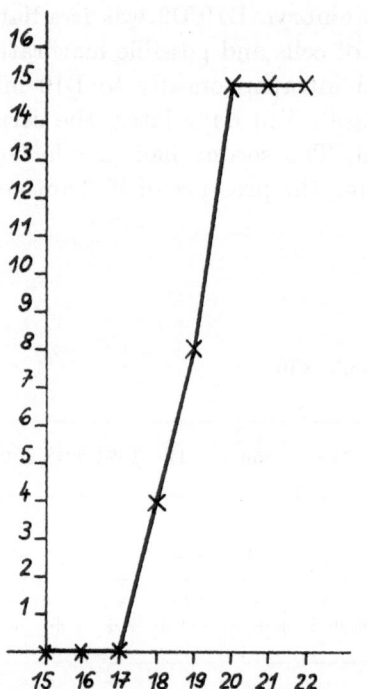

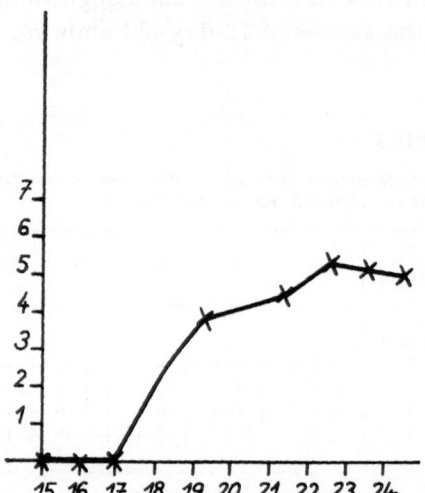

Fig. 2. Haemagglutinin absorption from C57BL/10 anti-B10.D2 serum by means of embryonic tissue cultured in diffusion chamber.

Fig. 3. Haemagglutinin absorption from serum C57BL/10 anti-B10.D2 by means of embryonic tissue cultured in tissue culture.

If this curve is compared with that obtained during antibody absorption by normal embryonic tissue (fig. 2), it can be seen that the amount of H-2 antigens increases in the same way as in the cells whose behaviour is no longer submitted to a coordination in the system of developing embryo.

Embryonic cell cultivation in tissue cultures. In a second case, the development of H-2 antigens in tissue culture, in vitro, was studied. A suspension of embryonic cells from 15—16-day-old B10.D2 embryos was transferred to tissue culture and samples were collected at definite intervals and tested for their capacity to absorb B10 anti-B10.D2 haemagglutinins. Figure 4 illustrates that even in this case the amount of H-2 antigens increases in the cells in the same way as during normal development in vivo.

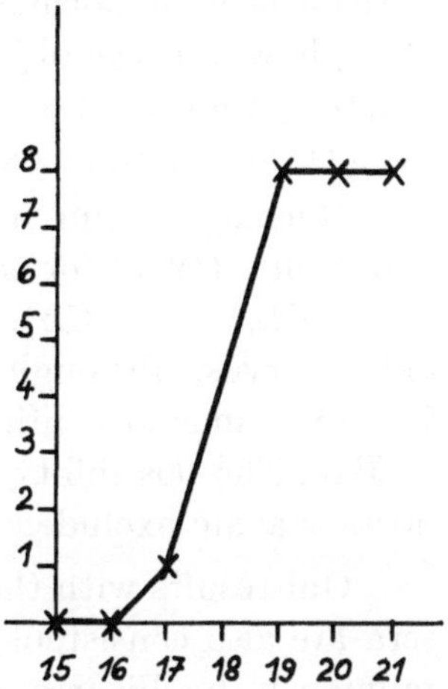

Fig. 4. Effect of injecting C57BL/10 mice with lyophilized 12-day-old embryos prior to inoculation of living tumour D23. Each horizontal line represents one mouse, each vertical line successive observations. The figures give the time interval (in days) between two observations. Treated mice were given a total of 75 mg./mouse lyophilized tissue and fresh tumour 10 days after the last injection.

However, in none of the two experiments did the concentration of H-2 antigens reach the level corresponding to that of cells maturing in a normal newborn animal. This is possibly due to the fact that the fibroblasts, having a lower concentration of H-2 antigens than other liver cells, do multiply in both diffusion chambers and tissue cultures.

Discussion

The method of accelerated rejection of skin grafts was first used by Billingham et al. (1956) and later by Chutná and Hašková (1959), Möller and Möller (1962) and Möller (1963) and other authors for the demonstration of the presence of H-2 antigens in embryonic tissues. All these authors noted the characteristic second-set reaction, but since they used the combination of strains differing in a number of other histocompatibility genes, in addition to the H-2 locus, it could be objected that the non-H-2 antigens were responsible for an accelerated rejection of grafts. (The presence of non-H-2 antigens in embryonic tissues has been, in fact, demonstrated — Klein, in press). The use of co-isogenic strains differing only at the H-2 locus, as it is the case in our experiment, refutes such objections. Since we did obtain an accelerated rejection of skin graft even in this case, it can be concluded that the H-2 antigens capable of inducing an accelerated skin graft rejection are already present in the 12-day-old embryo. The objection that the antigens mature rapidly in an adult organism was excluded by lethal irradiation of cells.

First attempts to induce an enhanced tumour growth after the pretreatment with normal embryonic tissue were done by Snell et al. (1948). Here, however, the objection could be raised, as in the case of the second-set reaction, that the non-H-2 antigens were involved in enhancement. The use of co-isogenic strains excluded again this possibility for the given case.

Haemagglutinin induction was also used by Möller and Möller (1962) and Möller (1963) for the detection of H-2 antigens in the combination (A $\times$ $\times$ C57BL/F_1 $\rightarrow$ A.CA). They found the H-2 antigens to be present in 12-day-old embryos, although their amount was lower than that in adult tissues. Our experiments confirm these results in the strain combination B10.D2 $\rightarrow$ $\rightarrow$ B10. The possibility of an accelerated maturation of H-2 antigens in adult hosts is again excluded by lethal irradiation of inoculated cells.

Our results with the absorption of haemagglutinins from specific anti-H-2 sera are also consistent with those obtained by Möller and Möller (1962) and previously by Pizarro et al. (1961), Mitchison (1953) and Gorer (1938) and show that the H-2 antigens cannot be demonstrated in early embryos by this method.

The answer to the question of the presence or absence of H-2 antigens in early embryonic tissues depends, therefore, on the method used for their detection: The method of induction of the second-set reaction, enhancement and haemagglutinins gives a positive answer, whereas the method of haemagglutinin absorption gives a negative answer. This discrepancy can be excluded as follows: Either there are the determinant groups of antigens involved in transplantation, enhancement and haemagglutinin induction differing from those involved in absorption, which seems improbable. Or, the amount of H-2 antigens in early embryonic tissue is lower than that in adult tissue and the discrepancy is due to a varying sensitivity of the methods used. The latter possibility is suggested by the experiments of E. Möller (1963) showing that the sensitivity of the absorption method depends on the concentration of antigens on the cell surface.

It can be therefore assumed that the H-2 antigens in a certain basal amount are already present in early embryonic tissues (Simmons and Russell (1962) detected them in 7-day-old embryos and Tyan and Cole (1962) in still younger embryos). Their concentration increases very slowly during pregnancy (as shown by the results of Möller 1961b, 1962, using the fluorescent technique), but their amount increases very rapidly immediately before and after delivery and reaches the level of adult individuals in a few days after birth.

The behaviour of H-2 antigens in embryonic cells prematurely deprived of the contact with the maternal organism has been previously studied by Hoecker (1959): The mice of A.CA strain were lethally irradiated and inoculated intraperitoneally with liver cells from A/Sn and C57BL embryos and then the appearance of homologous erythrocytes was determined in their circulation. Hoecker concluded that "the process of antigenic maturation occurs as if the H-2 genes of the transplanted embryonic cells had their own calendar, independent of the external environment". This conclusion was confirmed by our experiments on embryonic tissue culture in diffusion chambers. In embryonic cells present in the diffusion chamber in an isologous adult host, the period of a striking increase in the concentration of H-2 antigens also falls within the period of a hypothetical delivery of the foetus.

In order to exclude the possibility that the period of a rapid increase in the concentration of H-2 antigens is induced by some soluble factor circulating in the adult organism (this would be analogous to an enzymatic induction), we performed a similar experiment in vitro. Even here, with the influence of the adult organism being excluded, the embryonic cells retained "their own calendar" of antigenic maturation. The cause of this surprising phenomenon is not known.

References

Amos, D. B. (1961). In: "Transplantation of Tissues and Cells", The Wistar Institute Press, Philadelphia, p. 69.

Amos, D. B., Zumpft, M. and Armstrong, P. (1963). Transplantation 1 : 270.

Billingham, R. E., Brent, L. and Medawar, P. B. (1956). Nature 178 : 514.

Billingham, R. E. and Medawar, P. B. (1951). J. Exp. Biol. 28 : 385.

Chutná, J. and Hašková, V. (1959). Fol. biol. (Praha) 5 : 85.

Doria, G. (1963). Transplantation 1 : 311.

Gorer, P. A. (1936). Brit. J. Exp. Path. 17 : 42.

Gorer, P. A. (1938). J. Pathol. Bacteriol. 47 : 231.

Gorer, P. A. and Mikulska, Z. B. (1954). Cancer Res. 14 : 651.

Hoecker, G. (1959). Second Inter-American Symposium on Peaceful Application of Nuclear Energy, p. 211.

Medawar, P. B. (1959). In: "Biological Problems of Grafting", Les Congrès et Colloques de l'Université de Liège, 12 : 1.

Mitchison, N. A. (1953). J. Genetics 51 : 406.

Möller, E. (1963). Transplantation 1 : 165.

Möller, G. (1961a). J. Immunol. 86 : 56.

Möller, G. (1961b). J. Exp. Med. 114 : 415.

Möller, G. (1963). J. Immunol. 90 : 271.

Möller, G. and Möller, E. (1962). J. Cell. Comp. Physiol., Suppl. 1, 60 : 107.

Pizarro, O., Hoecker, G., Rubinstein, P. and Ramos, A. (1961). Proc. Nat. Acad. Sci. 47 : 1900.

Simmons, R. L. and Russell, P. S. (1962). Ann. N. Y. Acad. Sci. 99 : 717.

Snell, G. D., Cloudman, A. M. and Woodworth, E. (1948). Cancer Res. 8 : 429.

Tyan, M. L. and Cole, L. J. (1962). Transpl. Bull. 30 : 526.

CHIMAERISM IN SHEEP

E. M. TUCKER

Agriculture Research Council, Institute of Animal Physiology, Babraham, Cambridge

Freemartins in sheep have only been reported sporadically and evidence of vascular anastomosis during foetal life in sheep has been based upon one or several of the following criteria:

(a) Male sexual characteristics in the female member of unlike sexed twins. This intersex is known as a freemartin.

(b) Direct examination of placentae for vascular anastomosis.

(c) Acceptance of reciprocal skin grafts.

(d) Red cell mosaicism.

Moore & Rowson (1958) used criteria (a) and (c) for diagnosing foetal vascular anastomosis in a pair of sheep twins. Slee (1963) used criteria (a), (b) and (c) in diagnosing this condition in sheep. Slee also reported red cell mosaicism in one pair of twins in a quadruplet litter. The present paper reports the serological findings of an examination of the blood and saliva from Dr Rowson's pair of sheep and from 3 pairs of Dr Slee's sheep.

Table 1

Reaction of sheep red cells with sheep blood typing reagents using rabbit complement.
Degree of haemolysis = 1 − 5   5 = complete lysis
                         = no lysis

Sheep No.	Reagent						
	A	B′	$U_3 + I$	E′	N′	C	G
905	5	5	.	.	.	.	5
906	5	5	.	.	.	.	5
206	2	2	4	5	3	2	5
208	2	2	4	5	4	2	5
3F20	5	.	5	.	5	.	.
3F21	1	4	5	.	5	.	4
D24	5	1	5	5	5	5	1
D27	5	4	5	5	5	5	4

The serological test used was a haemolytic one employing rabbit complement. The haemoglobin variants were determined by both starch gel and agar electrophoresis. The red cells were washed in saline and then lysed with distilled water and the resulting haemoglobin solution was subjected to electrophoresis at pH 8·6.

Table 1 shows the results of testing red cell suspensions from 8 sheep with 7 immune blood typing reagents. Sheep Nos. 905 & 906 were Rowson's twin sheep which he found to be tolerant of each other's skin grafts. 905 was a

Table 2

Reactions of red cells from sheep Nos. 905 & 906 with South African reagents
Degree of haemolysis $= 1-5$
$\qquad$ 5 = complete haemolysis
$\qquad$ 0 = no haemolysis

| | | | | | | | | | | | | | Reagent |
Sheep No.	A	B′	I_x	N_2	N′	E′	S	U	O_x	$U_3 + I$	D	C	L
905	5	5	2	5	0	0	0	4	5	0	0	0	0
906	5	5	2	5	0	0	0	4	5	0	0	0	0

normal ram and 906 a freemartin. Complete haemolysis was obtained when the red cells from both 905 & 906 were tested against reagents A, B′ and G. No lysis was obtained with reagents $U_3 + I$, E′, N′, or C. There was therefore no evidence of mosaicism. Red cells from these 2 sheep were also tested against a number of sheep reagents from South Africa kindly provided by Professor Osterhoff. The specificity of some of these corresponded to that of my own reagents (A, B′, $U_3 + I$, E′, N′, and C, which I have renamed accordingly to fit his nomenclature). Others were different from mine and the remainder were unabsorbed immune sera. Table 2 shows the results. Since the characteristics of the individual antibodies in some of these sera were not known to me, it was not possible to be sure that incomplete haemolysis, where it occurred, was an indication of red cell mosaicism. However, one definite feature was that the reactions of the red cells from both sheep were identical with all the antisera. Sheep 905 and 906 were of haemoglobin type B.

Sheep 206 and 208 were the surviving male members of a triplet litter and this pair was shown by Slee to be tolerant of each other's skin grafts. When tested against my reagents there was evidence of mosaicism with A, B′, $U_3 + I$, N′, and C (Table 1). Incomplete lysis was obtained with each of these reagents even when stronger concentrations than the reagent test dose were used. Again, further treatment of the unhaemolysed red cells with the reagent which gave partial haemolysis failed to bring about further haemolysis. By differential lysis and testing the unhaemolysed red cells with other reagents, it was possible to separate the red cells of 206 and 208 into 2 populations in respect of their antigens. Table 3 shows the results. One population had $U_3 + I$ and N' positive red cells, and the other population had A, B', C positive red cells. Both populations shared G and E'.

When the haemoglobins were tested, it was found that both 206 and 208 had components A and B, but there was a distinctly stronger B component

Y	S8	S7	S27	K											
				90	48	38	188	130	154	132	192	16	28	13	33
0	5	5	0	5	5	0	4	0	5	0	4	5	4	0	0
0	5	5	0	5	5	0	5	0	5	0	4	5	4	0	0

than A suggesting that there was admixture of haemoglobin types as well as red cell antigens. By differential lysis and testing the haemoglobin of the un-haemolysed red cells, it could be shown that the $U_3 + I$, N' positive population of red cells in 206 and 208 had type B haemoglobin, whereas the A, B', C positive cells had haemoglobin type AB (Figure 1).

Table 3

Identification of red cell mosaicism

Sheep No.	Red cells	Reagents							Haemoglobin type
		A	B'	$U_3 + I$	E'	N'	C	G	
a) 206 & 208	One population	.	.	+	+	+	.	+	B
	Other population	+	+	.	+	.	+	+	AB
b) 3F21	One population	+	.	+	.	+	.	.	AB
	Other population	.	+	+	.	+	.	+	B?
c) D24 & D27	One population	+	.	+	+	+	+	.	AB
	Other population	+	+	+	+	+	+	+	AB

Similar tests were applied to the next pair of twins, 3F20 and 3F21, between which Slee has shown reciprocal skin grafts to be fully tolerant. These were female members of a triplet litter. Direct tests with the 7 reagents suggested admixture of red cell antigens to be present in 3F21 but not in 3F20

Electrophoretic separation of Haemoglobins after differential haemolysis

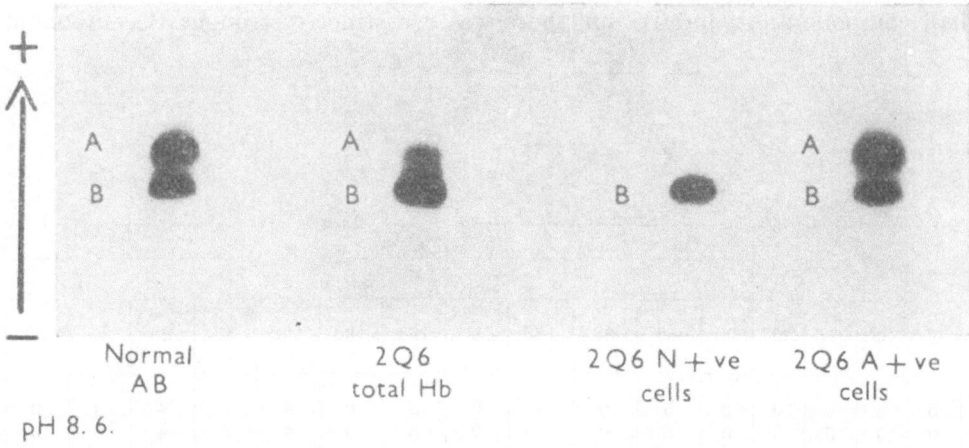

Electrophoretic separation of Haemoglobins after differential haemolysis

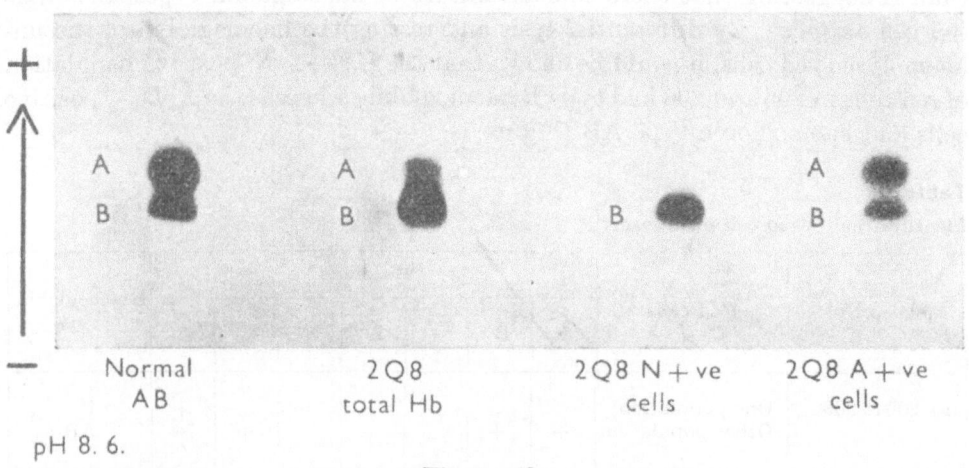

Fig. 1a, 1b.

(Table 1). The partial lysis of 3F21 cells with reagents AB′ and G was indicative of mosaicism and it was found that these red cells were able to completely absorb out anti-A, even though they only gave a haemolysis reading of 1 on the direct test. 3F20 red cells did not absorb out anti-B′ or anti-G so there was still no evidence of admixture in the red cell population of 3F20. By differential lysis it could be shown that in 3F21 there were 2 populations of red cells, one with A positive cells, the other with G, B' positive cells. Both shared $U_3 + I$ and N' (Table 3).

3F20 was of normal A B type haemoglobin. 3F21 was also of type A B but with a stronger B component than A and this again suggested admixture of the haemoglobins. Electrophoresis of the haemoglobins from red cells left after differential lysis showed that the A positive red cells in 3F21 were of normal haemoglobin type A B. However it was not possible to clearly demonstrate that the G positive cells contained only haemoglobin type B (Figure 2).

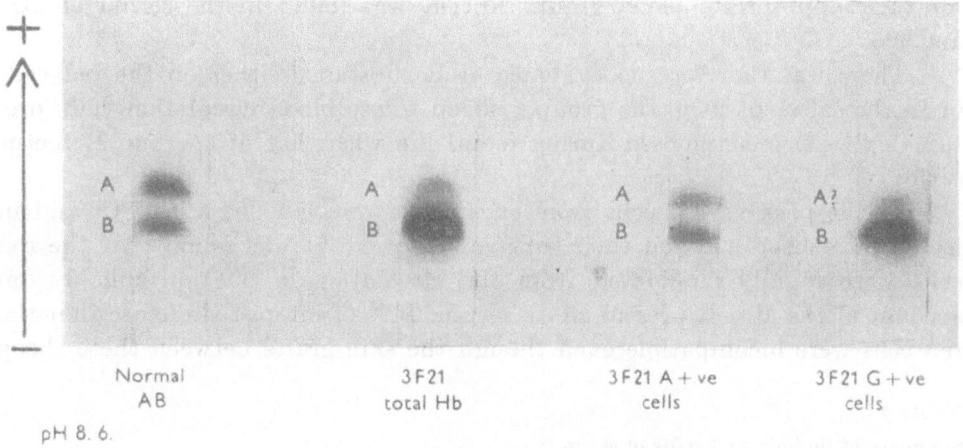

Fig. 2.

D 24 and D 27 were members of a quadruplet litter. D 24 was a male and D 27 a freemartin. Reciprocal skin grafts between D 24 and D 27 were tolerant. Direct tests with the reagents and absorptions again showed red cell mosaicism to be present in both sheep (Table 1). One population consisted of red cells which reacted with all 7 reagents, and the other population was of red cells which were G, B' negative. (Table 3.) No admixture of haemoglobins could be distinguished because both populations of red cells had type A B haemoglobin.

The R system was also investigated in these sheep. In agreement with the findings on the J system in certain dizygotic cattle twins (Stormont 1949) it was found that the red cell mosaicism was not reflected in the R system. Thus all the red cells in sheep 2Q6 were r positive whereas all those in 2Q8 were R positive. Similarly the red cells of 3F20 were r positive, whereas those of 3F21 were R positive. D 24 and D 27 both had R positive red cells. Sheep 905 had r positive cells and sheep 906 R positive cells.

Since R and r substances, like the cattle J substance are soluble substances which are adsorbed on to the red cells, it was not surprising to find that the foreign red cells in 2Q6 were r positive and those in 2Q8 and 3F21 were R positive. However, it was wondered whether perhaps if the factors controlling the production of R and r substances had been involved in the exchange in utero one might find evidence of these substances in fluids such as the saliva. All group R sheep have r substance in their saliva, but group r sheep do not have R substance in their saliva. The saliva of sheep 2Q6 was therefore examined for the presence of R substance by the inhibition test. No R substance was detected, however. The concentration of R and r substances in the salivas of sheep 905, 906 and 2Q6 was also determined and was found to be normal for their appropriate blood group. Anti-R was found in the serum of 2Q6 and 905.

There was therefore no evidence of R substance either on the red cells or in the saliva of 2Q6. the group r sheep whose blood circulation had presumably been anastomosed during foetal life with that of a group R foetus (2Q8).

The R positive red cells from sheep 906 were labelled with ^{51}Cr and injected into sheep 905 and their survival followed. It was found that the red cells were rapidly eliminated from the circulation of 905 presumably on account of the anti-R present in its serum. It is of interest therefore that the red cells were incompatible even though the skin grafts between these sheep

Table 4

Summary of findings in 4 pairs of sheep twins

Sheep No.	Source	Sex	Mutual skin grafts tolerant	Admixture of red cell antigens	Haemoglobin type	
					A	B
905	Rowson	Male	Yes	No, but identical cell		+
906	Rowson	Freemartin	Yes	types		+
206	Slee	Male	Yes	Yes	+	+ +
208	Slee	Male	Yes	Yes	+	+ +
3F20	Slee	Female	Yes	No	+	+
3F21	Slee	Female	Yes	Yes	+	+ +
D24	Slee	Male	Yes	Yes	+	+
D27	Slee	Freemartin	Yes	Yes	+	+

were tolerant. From this it could be concluded that the R system is not involved in the rejection of skin grafts.

In conclusion, Table 4 summarises all these results. Red cell mosaicism was found in 5 of the 8 sheep which had been shown previously to tolerate skin grafts from their co-twins. The mosaicism was identified by admixture of red cell antigens in all 5 sheep and also by admixture of haemoglobin types in 3 of them. The R system was not involved in the admixture and is apparently also not involved in the rejection of skin grafts.

References

Alexander, G., and Williams, D. (1964). Nature 201 : 1296.

Hraba, T., Hasek, M., and Cumlivsky, B. (1956). Folia Biol. Praha 2 : 276.

Lampkin, G. H. (1953). Nature 171 : 975.

Moore, N. W., and Rowson, L. E. A. (1958). Nature 182 : 1754 .

Roberts, J. A. F., and Greenwood, A. W. (1928). J. Anat. 63 : 87.

Slee, J. (1963). Nature 200 : 654.

Stormont, C., Weir, W. C., and Lane, L. L. (1953). Science 118 : 695.

Stormont, C. (1949). Proc. Nat. Acad. Sci. 35 : 232.

DISCUSSION

E. W. Tucker: I should like to point out that in sheep and I think in cattle also, the naturally occurring antibodies in the chimaeric animals are not suppressed. In sheep the natural antibody anti-R can be present in one member of a chimaeric pair and not in the other. This fact doesn't quite fit with your naturally occurring antibodies in your chicken species.

M. Hašek: What evidence of chimaerism was there in these sheep?

E. W. Tucker: Tolerance of mutual skin grafts and mosaicism of red cell antigens.

C. Stormont: It is true, as Dr. Tucker stated, that chimaerism in cattle twins has no apparent effect on the production of anti-J, and it is well to point this out in connection with the suppression of naturally occurring hetero-antibodies for red cells, as observed by Dr. Hašek, in artificially produced chimaeras between different species (turkey-chicken, etc.). However, we should keep in mind that chimaeric cattle twins are apparently never chimaeric for the tissues which produce J substance. Consequently, they remain unsuppressed for all antigens related in specificity to the J substance.

M. Braend: Dr. Hašek, could you advise us, please, the most appropriate number of injections, route of injections and amount of blood to inject to secure maximum antibody production?

C. Stormont: I would like to comment a bit on Dr. Braend's question which he directed to Dr. Hašek concerning advice about the isoimmunization of cattle. I recall the classic experiments of Felton on the production of immune paralysis in mice by simply giving them relatively large injections of pneumococcal polysaccharides. Such mice were not capable of producing antibodies against the polysaccharides and the paralysis persisted indefinitely. This suggests the possibility that certain types of antigens may be more effective than others in bringing about immune paralysis. I also recall that in the initial series of cattle isoimmunizations performed at the University of Wisconsin we used rather massive injections (2 to 3 liters per injection) which I personally regarded was far in excess of the dose necessary to bring about a good immune response. Furthermore, I was of the opinion that such massive doses could suppress the immune response. I think this view is now well supported by the experiments of Dr. Hašek and his colleagues on the induction of tolerance in adult animals by using large injections over long periods.

E. M. McDermid: To Professor Stormont's comments I would like to add that in my laboratory we have been able to make adult chickens tolerant by prolonged injections of blood over a period of several months. With some of these injections, the injected pairs differed at the B locus but the recipient animals after first responding by production of antibody eventually became tolerant in that such antibodies were no longer detectable. In other words, the immune mechanism was overwhelmed by antigen. I would also decree that small amounts of antigen should be used for immunization.

E. M. McDermid: Dr. Iványi, what agglutination technique do you use in your rabbit tests? Have you tried to detect incomplete antibodies using the antiglobulin test of Coombs?

P. Iványi: We use routinely the trypsin test.

F. Okerman: The breeds of rabbits which you were working with, do they belong to 1. a group of laboratory animals, 2. a strain of commercial meat rabbits, 3. fancy rabbits? What is the degree of inbreeding of these rabbits?

P. Iványi: Rabbits, (mentioned in your question,) were bred for commercial purposes. As to the last question, they were random-bred.

G. Efremov: 1. Dr. Tucker, do you know what is the reason that in sheep 9 there was variation in the quantity of haemoglobin bands, so that A band was thicker?
2. Did you test the total haemoglobin level in that sheep?

E. M. Tucker: The unequal amount of Hb A-B in the sheep 9 described was due to admixture of haemoglobin types. This was not the same phenomenon as that described by Huisman in anaemic sheep.

V. Tikhonov: Dr. Tucker has made very great and quite precise study, which clearly shows the presence of chimaerism in sheep on the erythrocyte and serum antigens. This phenomenon can be easily studied especially on multiparous animals, since it is frequent here. Chimaerism in sheep can be well studied, when using multiparous sheep, of e. g. the Romanov breed, which is kept in some regions of the Soviet Union. These sheep give up to 5—6 lambs per litter.

E. M. Tucker: Thank you for your remarks. If you have sheep in Russia producing large litters then I think you may well stand a good chance of finding chimaerism.

M. Harboe: Is there any evidence of why the red cell mosaicism (gradually) disappears in some of these sheep — if it has been present at all early in their life.

E. M. Tucker: I do not have any direct evidence that red cell mosaicism can disappear as sheep grow older. However, it has been found in cattle chimaeras that one population of red cells may well disappear as the animals age. Sheep 905 and 906 were at least 7 years old when tested and they both had identical cell types but no mosaicism. They may well have shown admixture of antigens earlier in life.

P. Imlah: Can I ask Dr. Tucker if she used her reagents at set dilutions? If she did not, then she might miss differential haemolysis, which could account for the absence of mosaicism for some of the loci, where it was not observed.

Secondly, I do not think it necessarily follows, that where you have multiple births, there should be an increase in anastomosis. If this were the case, then the incidence of "free-martins" would surely be higher in multiparous animals.

E. M. Tucker: Yes, I did titrate all my antisera from undiluted upwards and this did not produce any different results from just using the fixed test dilution. I also did many absorptions and other tests.

S. N. Naik: Dr. Tucker, may I know whether you have studied the blood smears particularly of the male co-twin of a free-martin to detect the sex-chromatin in poly-morpho-nuclear leucocytes?

E. M. Tucker: I have not yet done it but it shall be followed and used in future.

C. Stormont: Report of Dr. Tucker is an excellent report which independently confirms the occurrence of dizygotic mosaic or chimaeric twins in sheep and provides evidence that some of these twins were not only mosaic or chimaeric for blood factors but also for haemoglobin types. Although mosaicism for haemoglobin types was fully expected, it remained for Dr. Tucker to demonstrate this point. We have recently done the same for cattle twins (Stormont, Morris and Suzuki, 1964. Mosaic haemoglobin types in a pair of cattle twins. Science 145 : 600—601). I am completely in agreement with Dr. Tucker's conclusion that the one pair of twins (105—106?) with like blood types but with evidence for only one kind of red cells represents a case in which one of the two kinds of haemato-poietic tissue had regressed some time prior to the date these twins were first blood typed. In our laboratory, we have recently witnessed regression of one of two populations of red cells in two members of a set of Suffolk triplets. As a result of the regression, which took place over a period of about one year, the triplets ended up with like blood types. There is probably much more of this regression of cell types but, as in the case of the one pair of twins studied by Dr. Tucker, we will probably be "after-the-fact" observers of most of these events. Finally, I would like to return to another very satisfying aspect of Dr. Tucker's study, namely, that in the tests on two of the sets of twins she was able to demonstrate bilateral recovery of the two cell types. Our own experience to-date with chimaeric sheep twins has been unilateral recovery only.

Someone suggested that it would be best to search for erythrocyte chimaerism in sheep by limiting the studies to triplets, quadruplets and higher multiples simply because the crowding in utero would bring about a better chance for anastomoses. Comments: While such a view appears logical, there are reasons to believe that the incidence of anastomoses between ovine triplets, quadruplets, quintuplets and sextuplets is probably no higher than between twins. In the first place, the chorions of embryonal sheep twins invariably become fused regardless of their positions in utero. The trick is getting the blood vessels to cross connect. Here, in all probability, this occurs as a result of genetic compatibility rather than as an accident of juxtaposition.

ACTIVITY REPORTS FROM LABORATORIES

WORKING REPORT OF THE BLOOD GROUP LABORATORY OF THE INSTITUTE FOR ARTIFICIAL INSEMINATION AT SCHÖNOW NEAR BERNAU

J. PILZ

Institute for Artificial Insemination, Schönow near Bernau

So far, we have been able to produce the following monospecific sera by means of iso- and heteroimmunization:

A_1, A_2, B, G, I, I_2, O_1, P, Q_2, Y_2, D', E_2', O', Y', C_1, C_2, R, W, X_2, L', V, J, L, S_1, S_2, U_2, U_1', $Z/-$, Z/Z, G_{12}, G_{20}, G_{23}, Gn1, Gn2 .

At the time being the sera G_{12}, G_{20}, and G_{23} could not yet be classified and are thus bearing the designation given in our laboratory. The sera Gn1 and Gn2 are natural antisera which were found during a large-scale testing for natural antibodies using insemination bulls.

The above list of antisera developed by us shows that we are able to cover the factors completely or almost completely in some systems whereas there are still gaps in other systems, in particular, in the B system. In this connection it should be mentioned that the whole of our research work carried out up to the present time covered the Friesian breed. The blood spectrum of approximately 400 animals of Friesian cattle reveals that the factor Z', I, P, T_2, B', I', S', K', U_1 and U_2 are found very seldom or not at all. We are thus compelled to include other breeds into our investigations to ensure successful progress with a view to creating optimum immunization combinations on the one hand and obtaining a larger selection of bloods used for absorption. Moreover, the work with different breeds is also important for purely genetic considerations, since we must work in the future with all breeds which are distributed in the G. D. R. as follows:

	%
Friesian cattle	87·0
Brindled highland cattle	11·0
Franconians	1·3
German red cattle	0·4
F_1 animals and Danish red cattle	0·3

The work carried out in other laboratories showed that every cattle breed is capable of producing antibodies but that this capacity differs according to the breed. By including other breeds, in particular, of brindled highland cattle, we hope to contribute to a further completion of the spectrum of factors, especially in the B system.

Artificial insemination is being widely used in our country, too. Approximately 80% of all cows and heifers are artificially inseminated. This involves the higher probability that cows (heifers) are inseminated during two consecutive ruts with sperm from different bulls. For this reason, the future work in our Blood Group Laboratory will be focused on the determination of identity and issuance of paternity certificates for the herds already existing with a view to determine a reasonable breeding value of the sires used for insemination and to check the descent of young bulls to be selected for breeding. These measures are simultaneously supported by electrophoretic investigations.

In addition to these questions, we are working in the field of twin research since monozygotic and dizygotic twins can safely be recognized by means of blood group examination.

REPORT ON THE PRESENT STAGE OF INVESTIGATIONS IN THE PROBLEM OF CATTLE BLOOD GROUPS IN RUMANIA

V. DERLOGEA, I. GAVRILET, I. GRANCIU, Z. SIRBU, I. SOCEANU, S. RUSU

Institute of Zootechnical Research, Bucarest

In the Rumanian People's Republic the problem of blood groups was approached for the first time in 1957 by the Institute for Veterinary Research "Pasteur", in connection with the haemolytic disease in foals and piglets.

The study of blood groups from the point of view of the practical zootechnical requirements was initiated in cattle in 1962 by the Genetic and Selection Department of the Institute for Zootechnical Research, Bucarest. No special laboratory exists for the study of blood groups, and the investigations are carried out in the Laboratory for Genetics, within the programme of this department.

As our investigations on blood groups are only at their beginning, our chief aim is the preparation of our own test sera on the more representative cattle breeds of our country and the utilization of these sera in the zootechnical practice for the assessment of origin and other genetic purposes.

Taking into account the individual differences between blood factors, immunization pairs were formed; up to the present, two series of immunizations have been carried out each year, in spring and in summer-autumn, totalling seven immunization series which comprised 83 donor-recipient couples.

At the beginning, in order to obtain a greater number of antibodies, donors were chosen which had more numerous factors in their erythrocyte antigens as compared with the recipients. Soon it proved more advantageous to obtain sera with 1—2 antibodies, as the former technique needed too many absorptions, so the line of donor-recipient pairs with differences in 2—3 factors, was followed.

Immunization was carried by using 20 ml. whole blood inoculated subcutaneously at weekly intervals, for 4—7 weeks. In the immunization series of March—April 1963, the recipients were immunized with 20 ml. red cell suspension, washed 2—3 times with physiological saline, inoculated intramuscularly, using the same scheme of injections. The sera obtained by this technique had somewhat higher titres than those obtained by previous inoculations of whole blood. The number of cases was too small to draw any definite conclusions. Moreover, this technique required supplementary work so that subsequent immunizations were made only with whole blood.

From the results obtained in the 6 immunization series it can be seen that, out of the 83 recipients, 10 (12%) had not reacted to immunization, and of those who had reacted, 47% had low titres and only 41% had titres higher than 1/64.

Generally, it was established that for animals reacting from the very beginning, 4—5 inoculations are sufficient to obtain sera with satisfactory titres and the prolongation of the immunization period does not lead to a rise in the titre.

The reimmunizations carried out with the same donors to obtain sera with higher titres, or with other donors in order to produce other antibodies, have yielded variable results according to the individuality of the respective animals.

The period of time, during which the antibodies were maintained in the blood, varied considerably. In the majority of animals the antibodies disappeared after 1—3 months. In this respect we must mention the case of the cow Saftea 206, in which antibodies were found even after 16 months and her antibodies gave quite intense positive reactions. As a matter of fact, this recipient had high titres after the first immunization, which were at the same level even after reimmunization.

Parallel to isoimmunization in cattle, heteroimmunization in rabbits and wethers were done. In the raw sera obtained, which had titres much higher than those in cattle (over 1/1, 024), antibodies for the A and V factors were identified after absorptions.

Sampling of sera was made at a titre of at least 1/64. Following absorptions with a part of the raw sera, antibodies were identified for the following factors: A, B, G, I, O, P, A', I', O', C, R, V, L, S_1, S_2, U_1.

REPORT FROM GENETICS RESEARCH LABORATORY, DAIRY HUSBANDRY DIVISION, NATIONAL DAIRY RESEARCH INSTITUTE, KARNAL, PUNJAB, INDIA

P. G. NAIR

This laboratory, which was established in 1962, is at present interested in the antigens on the erythrocytes and in the milk of dairy animals of India. Some work on haemoglobin and transferrin types had been started, but had to be temporarily discontinued.

Studies on blood groups. Through the kind courtesy of Prof. Moustgaard, 219 blood samples from the herd of the Institute herd were blood typed between December 1962 and December 1963. Results indicated that most of the antigenic factors reported in the literature are present in Indian cattle (but animals negative for some of the factors like F were not observed) and interesting observations have been made on their distribution. However, since the samples were not drawn from animals taken at random, these observations may not be typical of the breeds involved.

Isoimmunizations were started in March 1962. So far, the following reagents have been produced: A, Z', B, G, O(?),*) Q, Y, B', I'(?), K', Y'(?), R, W(?), V_2, J(?), M(?), U and U'. In addition, several new factors have been recognized. In general, Red Sindhi and Sahiwal breeds appeared to be more responsive to immunizations and also more subject to occasional distress symptoms, when compared to the Tharparker breed.

While isoimmunizations are still continuing, heteroimmunizations (with rabbits now and with other species later) are being started. Genetic studies also are due to be started shortly.

Antigens in milk. The dairy industry in India was faced with the problem of detecting adulteration of cow milk with diluted milk from the water bufallo. Since bufallo milk, which constitutes more than 50% of the market milk in the country, is also richer in its solid constituents, this is a problem peculiar to India. By immunizing rabbits with buffalo skim milk, it was possible to produce an antiserum which, after suitable absorptions, is able to detect buffalo milk in proportions as small as one per cent (Nair, P. G. and K. K. Iya, 1962, Milchwissenschaft 17: 477—479). Similar sera have since been prepared specific for goat milk and cow milk. Absorption studies

---

*) Reagents marked thus (?) indicate that they are still under investigation.

have indivated that these sera are anti-casein sera. More recent observations have also indicated that there may be individual differences in the reactivity of milk from different animals with homologous sera (Nair et al., papers under publication).

Personnel. The laboratory was organized by P. G. Nair who has been in charge all along. However, several colleagues also have rendered valuable assistance. Two students have recently completed their dissertations for the M. Sc. degree of the Punjab University.

Acknowledgements. The work reported above would not have been possible but for the valuable suggestions and support from Prof. Dr. K. K. Iya, the Director of this Institute, and Prof. Dr. S. N. Ray, the Head of the Division of Dairy Husbandry.

THE ACTIVITY REPORT OF THE BLOOD GROUP LABORATORY

A. SCHINDLER

Institute of Animal Breeding of the Bern University, Bern

The determination of blood groups in cattle was started in our institute in the year 1958, after our collaborator Mr. Eric Müller had returned from Copenhagen, where he studied the principles and techniques of blood group typing. In the beginning, our Danish colleagues made it possible for us to start with the work since they kindly supplied us with a stock of all blood group antisera.

First of all, we made systematic investigations and frequency determinations on the Swiss breeds of cattle to have a basis for the practical application of the method.

The method of blood group determination, which opened new possibilities, evoked interest of both breeders and the breeding associations. To-day our principal activity comprises the following branches:

1. Investigations on twins: The blood group determination in dizygotic twins (200 to 250 pairs a year) is demanded frequently from the breeders to explain the potential fertility of the female twin-partners. In a recently published work we have summarized the results and obtained 5·9% of potentially fertile cow-calfs. Examinations of zygosity in monozygotic twins are not performed so frequently (in commission from the firms, which make fat- and other production experiments).

2. Origin resp. paternity controls: The Swiss cattle breeding associations have issued regulations, according to which the origin of the animal must be verified in certain cases, e. g.:

a) if a cow is inseminated or naturally mated to 2 different bulls within 15 weeks and the duration of pregnancy, referred to the both mating dates lies between 270 and 308 days, the paternity control must be made for the calf.

b) the origin of all bulls used for the artificial insemination must be proved by means of the blood group determination.

c) in the Freiburg black spotted cattle the crossings with the semen of the Holstein bulls from Germany are made. The origin of all male crossing products is examined.

d) many breeders use the blood group determination voluntarily in the high-value production animals to prove their origin.

e) the tendency now exists on the part of the cattle breeding associations to declare the origin control by means of blood group determination as obligatory at least for the male animals included in the breeding books.

In the year 1963, nearly 2,000 animals were examined with regard to the origin controls, in the year 1964 it will be over 3,000 animals.

3. Production of immune sera: Three experimental herds of the Simmental cows from the state enterprises are available for the preparation of cattle immune sera. A smaller herd of the brown cattle was newly included.

First, immunizations were made using the following technique: 100 to 200 cc of blood in a flask with the Rous-Turner-solution were taken (blood: anticoagulant 3 : 1) from the donor animal and in this form injected i. m. to the recipient (25 cc per injection). The injections were repeated 6 to 8 times at the interval of 6 to 7 days. The obtained results were unsatisfactory, since of 80 such immunizations antibodies were formed, frequently even with very weak titrés, only in approximately 55% of cases. We have then slightly modified the technique, so that we now wash the donor blood 4—5 times in saline prior to injection (precisely in the same way as during the preparation of erythrocytes for the haemolytic test) and then we add the physiological solution up to approximately 25 cc, so that practically pure erythrocytes, diluted in NaCl, are injected, We have thus made 2 series of 20 immunizations with 100% success, with antibody titres 1 : 8 to 1 : 256. For one immunization, it is necessary to bleed the donor animal only twice, each time approximately 100 to 150 cc of blood are taken. Storage in a refrigerator and the regular shaking of the sedimented erythrocytes make it possible to store the blood for 4 to 5 weeks. Both with the earlier and now employed technique we have never encountered any unpleasant accidents such as the anaphylactic reactions, decrease in milk production, abortions, etc. in the recipient animals.

4. Haemoglobin- and serum transferrin types in cattle. Our collaborator Mr. Krummen, after a short study-stay in Munich, began to determine haemoglobin and serum transferrin types in cattle by means of starch gel electrophoresis (1962). These additional examinations are since then made also in all blood samples for the blood group determination.

SOME RESULTS OF BLOOD GROUP STUDIES IN ANIMALS

V. N. TIKHONOV

Institute of Cytology and Genetics, Academy of Sciences, Novosibirsk

Investigations on animal immunogenetics, carried out at the Institute of Cytology and Genetics of the Siberian Branch of the Academy of Sciences of the U. S. S. R., are directed on two problems.

First, to obtain new sera and improve the available serum-reagents for determining both the known animal blood groups and those which have not yet been investigated. The investigations directed to find new suitable methods of purification and concentration of specific sera are also concerned with this problem.

Second, to study theoretical and practical methods of immunogenetic control of the most important questions of animal selection, for example, prognosis of the best heterosis in different crossing combinations, selection of parents providing the highest productivity at purebreeding, creating of inbred and stable outbred lines, a more rapid determination of sire's breeding value, study of mentoral effects of sperm and embryos and study of selectivity of fertilization and other questions. Monospecific serum-reagents and special polyvalent sera may play an important role in working out and using these methods.

While studying the above problems, some other laboratories in our country have begun investigations on blood groups in pigs according to our common programme. These laboratories are organized under our participation in the Ukraine, the North Caucasus, Siberia and other regions.

In order to obtain reagents, 10 series of immunizations were carried out, in which 165 pigs, mostly adult swine, and boars were used as recipients and 38 — as donors. Some animals were reimmunized in 2 and 3 series. The immunization was interbreed, intrabreed and intrafamilial. The best effect with respect to forming antibodies was observed by immunization of adult sows, especially when intravenous and intramuscular injections of 10—20 ml. of a 50% erythrocyte suspension were used simultaneously. It should be noted that when 3—5 intravenous injections of washed erythrocytes were given at weekly intervals, and the injections were given slowly, no anaphylactic shock was observed.

As a result of the work carried out for obtaining isoimmune sera, and of their appropriate absorptions, more than 30 sera were obtained which appear to be monovalent with different specificity; they may be used as reagents for testing pig blood groups according to the erythrocyte antigens. Among the mentioned sera, 21 sera could be identified with identical reagents having the same designation in Czechoslovakia, Denmark and Poland. It became possible owing to the kindness of the scientists of these countries, who sent us samples of sera. Moreover, the scientists of the laboratory in Liběchov (CSSR) kindly helped us in testing 56 pigs according to the available reagents and in the organization of the international Comparative testing of reagents in which we took part.

The results of the Comparative testing of reagents and our data, obtained during the study with them, enabled us to conclude that the following reagents are now available: A, Ac(S_4), Ba, Ea, Eb, Ed, Ee, Ef, Fa, Ga, Gb, Ha(S_3), Hb(S_5), Jb(Tl-9), Ka, Kb(S_7). Kd, La, Ma(KII_2), Mc, Na. Among the available reagents Ed, one of the variants of Ha and La have shown weak sensitivity. Perhaps, the weak sensitivity of these sera may be due to the use of old samples which were weakened by a long period of storage and strong dilution. On the other hand, it should be noted that the anti-La and anti-Ha sera had a very high titre and usually worked very well at 20 times dilution. Therefore we cannot exclude the probability that the mentioned sera may show certain subtypes of antigens Ha and La. It is supposed that sera S_9 and S_{11} correspond to new subtypes Ha. The final solution of these questions requires additional testing. The Ef and Na(TL-12) sera which were used in testing appear to be not pure enough and require additional absorption. We suppose that we shall be soon able to obtain these reagents in a purer form. Moreover, we obtained the sera S_1, S_2, S_8, S_{12}, CZ-2, TL-6, TL-1/144 and some other reagents which do not show any immunogenetic similarity with any of the sera already available. By immunogenetic analysis (by statistical and family analysis) the antigens found could not be identified as products of alleles of the genetic systems which have been already studied.

REPORT ON SOME WORKING METHODS DEVELOPED AND APPLIED IN OUR LABORATORY

R. EBERTUS

Institute for Artificial Insemination, Schönow near Bernau

1. The practice exerted by Tolle (1960) to take the blood from the auricular vein appeared to us to be the most considerate method. However, it will be necessary to have some patience in this instance until the quantity required is obtained. When taking blood we use a method ensuring that

1. a perfect complement is obtained;
2. the blood required for this purpose is obtained within a short time;
3. the rabbits can be bled 20 to 30 times; and
4. the method requires less skill and a small number of instruments only.

Fig. 1.

We use a suction glass designed by me with an appropriate opening to enable the glass to be placed over the rabbit's ear. The opening must be so adjusted that the blood vessels at the base of the ear cannot be compressed. The suction glass must fit the average size of the rabbit's ear and therefore each laboratory must have this device made by itself. The length of the suction glass must allow the rear end to remain free even if the rabbit's ear is longer than on the average. A socket which can be connected with a water-jet pump with the aid of a hose connection is fixed to the front part of the suction glass. The rear end is furnished with an outlet socket bent downwards. A perforated rubber stopper is to be placed into the outlet socket in such a way that the lower part of the socket protrudes beyond the rubber stopper to cause the blood to flow over glass only. A centrifugal tube is placed onto the rubber stopper. Thus a self-contained system will be obtained in which a partial vacuum can be produced.

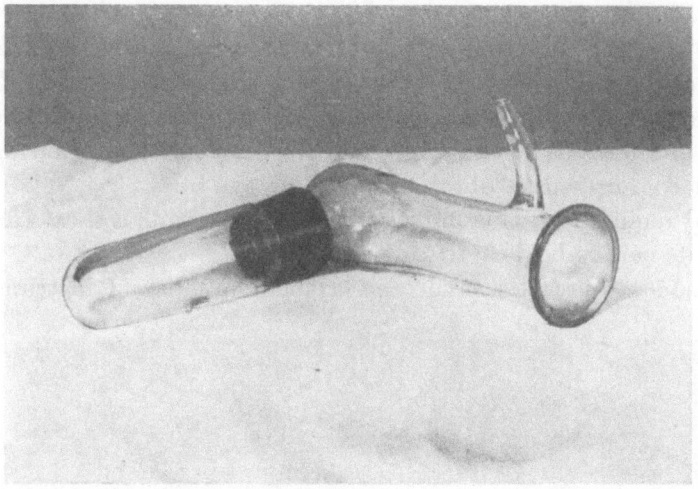

Fig. 2.

The edge of the ear is incised with scissors so that a marginal vein is opened. The ear is now introduced into the suction glass which is applied to the rabbit's head with moderately firm pressure. Physiological solution is applied around the ear to obtain a good sucking effect. After having started the water-jet pump, it is necessary to take care that the suction glass is slightly inclined towards the rear end to cause the blood to flow off quickly into the centrifugal tube. The blood flowing in a few moments out of the vein in a continuous jet will now pass directly to the centrifugal tube through the outlet socket. The required quantity of blood is obtained within approximately one minute and may amount to 30 to 50 ml. depending on the weight of the rabbit.

The blood thus obtained can now be processed in the usual way. Thus it will be possible to obtain a perfect complement from 10 to 15 rabbits without any loss by decanting within 60 minutes.

2. In the future we will use dry complement to a greater extent. Until now, lyophilized complement has maintained its efficiency for more than one

year. This fact has already become known by other authors like Tolle, Rendel, Neimann-Sörensen.

3. Experiments to preserve lyophilized test sera gave very good results so that our sera will be stored in this way in the future. We have started this work on the basis of the proposals submitted by Tolle et al. We hold that preservation of lyophilized test sera is the method of preference since

1. durability of dried sera is definitely superior to that of deep-frozen sera; and

2. storage does not depend on permanent technical equipment.

Lyophilized sera and complement can be preserved after dissolving at −20°C. This applied to quantities not used up which thus need not be destroyed.

4. Longer preservation of erythrocytes may take place in deep-frozen condition. Of the known methods we prefer the method of Tolle using sodium nitrate-ethylene glycol. However, our method includes an additional operation during thawing. After first centrifugation we give a 30% ethylene glycol solution and then increase the amount of ethylene glycol to the concentration stated by Tolle. We observed that suspensions of frozen erythrocytes could be preserved for a comparatively long time. This may be due to the fact that the less resistant cells are destroyed during freezing and thawing. A slight haemolysis in the suspension is of no importance. The reactions take place properly in the haemolysis test. This method is particularly suitable for preserving rare erythrocyte patterns for a longer period of time.

5. The yield of raw serum did not turn out to our satisfaction. So we tried to increase the quantity of serum per litre. The yield can be substantially increased when the following procedure is used:

2 g. sodium oxalate are added to each litre of blood to be collected in order to prevent coagulation. The blood is centrifuged immediately after its transport to the laboratory. The "blood plasma" is inactivated in small vessels (50 to 100 ml.) for half an hour at 56°C. The gelatinous coagulates of fibrinogen are separated from the serum by short and rapid centrifugation.

Comparative examinations of serum and "plasma serum" gave fully identical results both in the raw serum and later absorptions.

An advantage of this method resides in the fact that the erythrocytes thus obtained can be used for absorptions if required or preserved as test erythrocytes by deep-freezing.

RELATED QUESTIONS

COMPARISON OF COMPLEMENTS IN THE HAEMOLYTIC TEST

B. BUSCH

German Academy of Agricultural Sciences at Berlin-Institute of Animal Breeding Research, Dummerstorf

Production of sufficient amounts of rabbit complement in our laboratory is relatively laborious. And troubles may happen, if epidemic or other diseases destroy the rabbit stock. Therefore we investigated, whether sheep and pig sera are suitable as complement in the haemolytic test for the determination of cattle blood groups.

I will report of only the main results; our complete results will be published in detail elsewhere.

The investigated sera of 10 sheep and 75 pigs (from the slaughterhouse) were prepared as the rabbit complement used in our routine tests. Then these sera were tested using erythrocytes of 52 cattle and 4 monospecific and 1 polyvalent antisera; the results were compared with the reactions in which rabbit complement was used. The pig sera were tested against the erythrocytes to detect species-specific antibodies, but the results were negative in all cases.

The sheep sera had no complement activity and therefore the work with them was discontinued.

If pig serum was added to a mixture of test reagent and erythrocytes, the same reactions were observed as if rabbit serum had been added. But with one of the reagents, some pig sera showed no complement activity.

By inactivating the sera of pigs the complement should be destroyed. Our investigations show that serum of pigs may be used as complement in the haemolytic test, but it remains to clear by further investigations whether the effectiveness of pig serum is the same as that of rabbit serum when more reagents are used.

COMPARATIVE TESTS FOR OBTAINING ISOANTIBODIES AGAINST CATTLE ERYTHROCYTES

V. DIKOV

Institute of Biology and Pathology of Animal Reproduction, Sofia

In this preliminary report we give the results of a comparative test for obtaining isoantibodies against erythrocytes in the blood serum and in the milk of immunized cows. Seven cows of the Sofia Brown breed were involved. The immunization was carried out with citrated whole blood. The animals were injected intramuscularly at weekly intervals, in amounts of 2 ml. blood per 100 kg. live weight. A cross reaction prior to immunization was done for checking the biological compatibility between donors and recipients. Moreover, prior to each of the following injections both the blood serum and the milk of the recipients were checked for the presence of isoantibodies. The haemolytic test was used for this purpose.

Following the first immunization, antibodies against the erythrocytes of the donors with a titre 1 : 4 were detected in the blood serum of all the recipients. Antibodies in the milk were produced in only one recipient, with a titre 1 : 2. Data obtained during the second testing (prior to immunization III) show that the antibody titre considerably increased. Its maximum for the serum was 1 : 16. This time there was also a rise in milk antibodies in all of the cows involved. It must be stressed that their titre was very low — 1 : 2, and only in one of the cases — 1 : 32. It may also be of interest that at the final testing the antibodies of the blood serum were at a high level in all recipients, while those in the milk were increased in only one animal. The titre of antibodies in the serum of the same cow was 1 : 64, while that in the milk has reached its maximum — 1 : 512. In all other cows both titres remained unchanged.

The determination of isoantibodies in the milk of the recipients parallel to those in the serum throws more light on antibody formation in immunohaematology. Moreover, the great difference in the titre values of milk antibodies in one of the investigated cows rouses justified interest towards elucidating the phenomenon. The analysis of the physiological condition of the recipients showed that all cows were pregnant during the immunization course. Those with low titre of milk antibodies were in the second month of pregnancy, while the animal with a titre of 1 : 512 had completed the eighth month of pregnancy. It is well known that the udder plays an important biological role in providing ready antibodies for the still non-reacting newborn calf. However, the incidence of icterus gravis neonatorum is due exclusively to the presence of antibodies in the colostrum of the mother. These findings lead to the assumption that parallel to all other factors an advanced pregnancy and the forthcoming calving are of decisive importance not only for a well expressed immune response of the organism but also for the topographic site of the immunogenic process. In support of such an opinion we may well point to Ehrlich and Harris's findings (1942—1946) with regard to the leading role of lymphoid tissue in immunologic phenomena. Moreover, this was confirmed by some indisputable facts pointed out by Kenning et al. 1950, Dixon et al. 1967, Harris and Harris 1960, Hašek 1961. Physiologically, a rich lymphoid tissue is characteristic of the udder during this period.

These data show that the milk of cows immunized in advanced pregnancy may well serve for obtaining antisera. Our tests are still very insufficient to permit any generalized conclusions. However, we hope that further investigations will lead to positive results.

IMPORTANCE OF ANTIGLOBULIN REACTION FOR DETECTION OF NEW FACTORS IN CATTLE

P. MILLOT

Laboratory of Blood Groups in Cattle, Jouy en Josas

Antiglobulin sera have been obtained in three goats, each of which was immunized with one bovine plasma — Freund's adjuvant was used. After suitable absorption of the anti-red cell antibodies with a titre from 1/16 to 1/32, the precipitating titre of the three sera was in the ring test: $1°/\frac{1}{4}$, $2°/\frac{1}{8}$, $3°/\frac{1}{2}$. Precipitation in agar gave bad results (only a weak line of precipitation with the pure serum No 2).

We used the absorbed sera for the study of "incomplete" agglutinins in our bovine haemolytic reagents. Those, containing a single iso-haemolysin, are absorbed fractions of isoimmune sera, except for four fractions: two (anti-F, anti-V) prepared from a buffalo (Bubalius vulgaris) anti-bovine heteroimmune serum and two natural anti-J. The reagents had been previously absorbed by 1 to 4 doses of packed red cells in order to isolate the haemolysins.

The technique employed for antiglobulin reaction is an indirect test of Coombs modified as follows:

> serum reagent (not diluted) : 0·10 ml.
> red cell suspension (3%): 0·05 ml.
> shaking-incubation for 2 hours at 25°C
> three washings in saline
> draining on blotting paper
> addition of pure antiglobulin: 0·05 ml.
> shaking-incubation for 1 hour at 25°C
> centrifugation at 2,000 rpm
> reading with concave mirror.

Results

Of the 37 reagents, 24 have agglutinins and 13 are negative. Among the 13 negative reagents, 4 were obtained from 2 immune sera and 4 are fractions of various immune sera which also provide agglutinating haemolytic fractions.

The 24 positive fractions obtained from 19 immune sera (each of 5 immune sera gives two fractions) give with some suspensions strong agglutinations very easily read with a concave mirror. This indicates the presence of specific antibodies.

The 3 antiglobulins give similar results-2 are almost identical (less than 2% of error in the reactions), the third is weaker and gives the same reactions with most of the fractions but does not react with other fractions. Five reagents: anti-A, C_2, R and the two anti-J, have an "incomplete" agglutinin which reacts like haemolysin of the reagent. However, some J red cells do not react, especially with one of the anti-J.

The agglutinins of the 19 other fractions are different from all the haemolytic antibodies described until now. They give various reactions upon several series of suspensions.

Some agglutinating fractions detect rare agglutinogens but about the 2/3 of the fractions reveal agglutinogens whose frequency is more than 10% in several French breeds.

Control of purity of agglutinins by systematic cross absorption results in: 1. total absorption by all positive red cells and no absorption by all negative red cells; 2. total absorption by all positive red cells and by some negative ones. In the latter case, the agglutinin reveals the first subgroup of a more frequent antigen.

One of the reagents (anti-I') contains an agglutinin very strong in saline associated with an "incomplete" agglutinin. By these two pure antibodies, 2 rather rare and non identified factors are detected.

Each of the two anti-F and anti-V reagents obtained on the buffalo contain a different "incomplete" agglutinin. These correspond to agglutinogens of a rather high frequency. The study of the whole immune serum by the agglutination technique appears to be as complex as the analysis of the same serum by the haemolytic technique.

Investigations of the same haemolytic fractions are made just now by the following techniques: 1. agglutination in bovine serum albumin at 20% — (Poviet), 2. agglutination in saline of papainized red cells.

The first results show that these techniques also reveal specific antibodies in various fractions. These antibodies and the haemolysins do not usually correspond to the same antigens.

For instance, our anti-H reagent contains in addition to this haemolysin 1. an agglutinin reacting in saline, 2. an agglutinin of wider activity detected by antiglobulin serum, 3. an agglutinin detected by albumin technique, 4. several agglutinins detected by papainized cells. All these agglutinins seem to be different.

From the genetic point of view, our works are at the beginning stage. Four pure agglutinins have been tested on the red cells of 26 calves from a single Norman bull and on the cells of the respective mothers. Unfortunately, the bull which was tested for another purpose, did not possess the agglutinogens. However, this test gave the following results:

The agglutinogen detected by the anti-H serum (with antiglobulin serum) is present in 6 mothers — 2 of them transmitted it to their calves.

The agglutinogen detected by the anti-X_2 serum is present in 6 mothers and is not transmitted. Three of these mothers also have the preceding agglutinogen.

The two rare agglutinogens detected by I' reagent are found, one in 2 cows, the other in one of these 2 cows. A female possesses the four agglutinating factors but does not transmit them to her calf.

These first results show a possible existence of genetic relationships between the studied antigens. On the other hand, no link with B phenogroups seemed to be evident.

Summary

The agglutinating reaction with antiglobulin serum (from goat) makes various agglutinins appear in bovine reagents. These agglutinins are not usually homologous to haemolysins of the reagents. They probably detect several new factors.

BLOOD TYPING OF CATTLE OF TWO INDIGENOUS BREEDS: PODOLIC AND RED SPOTTED BREED OF VOJVODINA (YUGOSLAVIA), AND OF THE ORIGINAL SIMMENTAL BREED

V. JOVANOVIČ and L. KONČAR

University of Novi Sad, Faculty of Agriculture in Novi Sad
(Blood typing performed in the Veterinary Institute of Slovenia by O. Böhm and A. Gliha)

The occurrence of blood group factors and gene frequency for some of the factors has been studied in the population of three cattle breeds in Vojvodina (Autonomic region of S. R. Serbia, Yugoslavia), e.g. the Podolic Cattle and Red Spotted Cattle of Vojvodina and the strain of "Simmental Cattle" which has been imported and served as an important source in developing the ind:genous Red Spotted breed.

The Podolic cattle belongs to the Eastern group of primigenial cattle, originating directly from the European bison (aurochs). The native home of this breed is Podolia and Volhinia in the U. S. S. R. and from there the breed spread to Hungary, Rumania, Ukraine and Yugoslavia.

The native Vojvodina Red Spotted cattle have been developed using a group of Red Spotted Simmental Cattle in crosses with the Podolic breed.

Table 1

The occurrence of different blood group factors in cattle of three breeds expressed in numbers and percentage

Blood group factor	Podolic n = 76		Simmental n = 40		Red Spotted of Vojvodina n = 94	
	+	%	+	%	+	%
A	63	82·89	24	60·00	60	63·83
D	61	80·26	35	87·50	78	82·98
H	30	39·47	25	62·50	46	48·94
Z'	2	2·63	—	—	1	1·06
B	59	77·63	6	15·00	31	32·98
G	34	44·74	14	35·00	53	56·38
K	28	36·84	2	5·00	15	15·96
I	6	7·89	2	5·00	9	9·57
O	33	43·42	23	57·50	46	48·94
P	23	30·26	5	12·50	6	6·38
Q	18	23·68	5	12·50	17	18·09
T	24	31·58	6	15·00	8	8·51
Y	34	44·74	13	32·50	41	43·62
A'	37	48·68	21	52·50	39	41·49
B'	8	10·53	10	25·00	7	7·45
D'	25	32·89	9	22·50	17	18·09

Table 1 (continued)

Blood group factor	Podolic n = 76		Simmental n = 40		Red Spotted of Vojvodina n = 94	
	+	%	+	%	+	%
E'	55	72·37	17	42·50	40	42·55
G'	6	7·89	7	17·50	30	31·91
I'	13	17·11	20	50·00	29	30·85
J'	12	15·79	7	17·50	1	1·06
K'	21	27·63	7	17·50	6	6·38
O'	29	38·16	8	20·00	31	32·98
Y'	6	7·89	1	2·50	3	3·19
C_1	58	76·32	15	37·50	31	32·98
C_2	10	13·16	8	20·00	28	29·79
E	55	72·37	20	50·00	54	57·45
R	4	5·26	1	2·50	7	7·45
W	45	59·21	34	85·00	88	93·62
X_1	12	15·79	2	5·00	6	6·38
X_2	30	39·47	27	67·50	39	41·49
F_1	67	88·16	40	100·00	92	97·87
F_2	9	11·84	—	—	2	2·13
V	49	64·47	16	40·00	32	34·04
J	48	63·16	12	30·00	44	46·81
L	19	25·00	17	42·50	24	25·53
M	3	3·95	4	10·00	2	2·13
S_1	12	15·79	13	32·50	29	30·85
S_2	63	82·89	27	67·50	62	65·96
U'	14	18·42	6	15·00	14	14·89
Z	74	97·37	37	92·50	85	90·43
R'	56	73·68	13	32·50	27	28·72

The native Red Spotted Cattle of Vojvodina are still highly heterogeneous in their physical and functional characteristics.

The Podolic cattle that were bled for our investigations originated from peasant holdings in three villages in the Fruška Gora part of Srem and one village in Bačka. The investigated animals of the native Red Spotted breed originate from the herd of the experimental farm Rimski Šančevi (Institute for Agricultural Experimentations) and from the herd of the experimental farm Kamendin (Livestock Breeding Institute in Novi Sad).

Both herds were formed from the animals that had been bought as single animals from different private peasant holdings over a large part of the territory of Bačka.

The investigated cattle of Simmental breed represent a group of animals from the herd of imported cattle at Sirig, a farm of the Agricultural Combinate Novi Sad.

The sample of the Podolic cattle consists of 76 animals. Such a small number is due to the fact that the purebred Podolic cattle are very rare to be found in the territory. 94 animals represent the sample of the native Red Spotted breed and only 40 animals the purebred Simmental Cattle.

In Table 1 the occurrence of the different blood group factors for each of the three breeds is given, expressed in numbers and percentages.

Table 2

Gene frequencies for some blood group factors

Factor	Podolic	Simmental	Red Spotted of Vojvodina
A	0·586	0·367	0·399
D	0·556	0·646	0·587
H	0·222	0·388	0·285
Z′	0·013	—	0·005
F	0·678	0·800	0·830
V	0·322	0·200	0·170
J	0·393	0·163	0·271
L	0·134	0·242	0·137
M	0·020	0·051	0·011
S_1	0·325	0·230	0·339
S_2	0·648	0·731	0·646
R′	0·487	0·178	0·156

In Table 2 the gene frequencies for some blood group factors in each of the three breeds are presented.

INVESTIGATIONS OF THE SUBGROUPS IN THE BLOOD GROUP ANTIGENS P AND J' OF CATTLE

L. ERHARD and D. O. SCHMID

Institute for Animal Blood Group and Resistance Research —
Livestock Breeding Research Organization, Munich

The reference test 1963 showed two different groups of reactions in factors P and J'. The P-reagents of N, NL, D-Mü and F haemolyze the cells 13, 14, 18 of D-Mü and 20 of F, while the reagents of DK, PL, CSR and SA do not react with these cells.

The J'-reagents of D-Gö, PL, CSR, F and B reacted with the cells 2 and 15 of D-Mü and the cells 7 and 8 of F. On the contrary, the reagents from DK, S and NL did not react with these cells.

The ESABR-committee concluded that subgroups of P and J' probably exist.

Because the deviations from the reaction-pattern are only in blood samples of F and D-Mü, these two laboratories have performed the demanded absorptions. In Munich we have had the following results.

1. Anti-P

The absorption experiment was made with two anti-P-reagents (91/59 and 58/62) of the own production. These reagents were absorbed with 8 P-positive cells, including those with deviations in reactivity (cells 13, 14, 18) in the reference test. Thereby resulted, that the reagent 91/59 contained two kinds of P-antibodies. According to the nomenclature of other antigens with subtypes, these antigens are marked previously with P_1 and P_2.

Further absorption experiments in connection with the reference test 1964 and above all the results of this reference test confirmed the existence of the subgroup P_2.

The study of the records of the reference test 1964 shows that the reactions for P_2 as well in F as in D-Mü appear only in highland breeds (Montbelliard, Fleckvieh, Brown Swiss and Murnau-Werdenfels-cattle).

With the P_2-positive reactions of the blood samples of H-Vet we don't know the breed.

The phenogroups $P_2E_2'G'$, $P_1E_1'I'$ and $P_1QE'I'$ can be regarded as proved in the German highland breeds.

2. Anti-J'

In order to clarify the J'-subgroup problem, absorptions with two our own, one French and one reagent of the Netherlands were performed in the years 1963 and 1964. The results show that the J' (28/62) from Munich and the J' from France are identical. All 6 J'-positive cells absorbed the whole J'-specificity. These reagents contain only one kind of antibodies. On the contrary, the J' (11/61) seems to contain, as the P 91/59, two kinds of antibodies. The absorption records show that the cells 10 and 11 decrease strongly the activity of at least the reagent J' (11/61). In the beginning, we believed that the absorption power had not been sufficient and all three tested reagents were identical. Now we know that the reagent, analogous to the P 91/59 $(P_1 + P_2)$, certainly contains both J' and J'2 antibodies.

The J'-reagents of France and Munich (28/62) contain certainly only J'$_2$ antibodies.

The J'-reagent 19/9 of the Netherlands reacts differently from the other tested J' reagents. It reacted positively in the lysis with all J'$_1$-positive cells and very weakly, partly only after 24 hours, with the J'$_2$ cells. Absorptions with 4 J'$_1$-cells and 3 J'$_2$-cells show that all J'-cells absorbed the J'-reactivity. This reagent is evidently a J'$_2$-serum, which reacts in the lysis nearly exclusively with the J'$_1$-positive cells.

According to our experiments and the records of the reference test it can be stated that the subgroup J'$_2$ was observed only in phenogroup $O_3E'J'$.

It is striking that this phenogroup can be found only in highland breeds. This phenogroup has been proved so far in the French Montbelliard and German highland breeds Fleckvieh, Brown Swiss and Murnau Werdenfels cattle.

The investigation of the subgroups of P and J' was only possible by a close cooperation between the European blood group laboratories and the ESABR.

We have done this investigation in collaboration with the French laboratory and we had an interesting contact with Dr. Grosclaude and the Secretary of our ESABR Dr. Bouw.

BLOOD GROUPS IN CHICKEN OF SPANISH STRAINS AND BREEDS

A. JOVER and A. RODERO
Veterinary Faculty, Cordoba

In a work of ours before this report, studies of poultry of Spanish strains and races were begun; and herein we continue the course of investigations.

The animals investigated belonged to one line of the Black Castilian breed and two of the White Leghorn breed that had been bred in Spain for more than ten generations in closed pens. Studies have also been initiated on the race Utrerana Franciscana.

Experiments were started with reagents supplied by Dr. McDermid. The following reagents at the dilutions used were tested:
B2 (1 : 80), B2 (1 : 20), B2 (1 : 5), B3 (1 : 20), B6 (1 : 96), B7 (1 : 20), B7 B8 (1 : 10), B9 (1 : 5), B11 (1 : 40), B12 (1 : 20), B13 (1 : 10), B14 (1 : 20), B19 (1 : 20), B19 (1 : 40), B21 (1 : 10), B21 (1 : 20), A2 (1 : 20), A6 (1 : 40), E9 (1 : 40), E10 (1 : 20), E12 (1 : 20).

Isoimmune reagents are being produced within Black Castilians. There is an inbreeding coefficient of 59%. Three alleles have been identified and their reactions with McDermid's antisera are shown below:

	B7	B7+8	B9	B11	B21	B14	E10
BI	−	+	−	+	+	+	−
BII	−	−	+	−	−	+	−
BIII	+	−	−	−	+	−	+

The BI most probably corresponds to B8 of McDermid, BII to B9 and BIII to B21.

The t test realized among the homozygotes and heterozygotes in the alleles B_8, B_9, and B_{11}, for the characters: mature sexual, winter laying, the egg weight and total laying was as follows:

For sexually mature, $t = 1\cdot45$; not significant.

For winter laying, $t = 1\cdot12$; not significant.

For weight of egg, $t = 1\cdot86$; not significant.

For total laying, $t = 2\cdot42$; significant at the level of 5%.

Therefore it can be inferred that in the Black Castilian birds studied, the heterozygotes for the alleles cited are superior in the total laying to the homozygotes for the same alleles. The laying increased from 170 eggs for the homozygotes to 198 eggs being the average for the heterozygotes. The heterosis did not affect the rest of the studied characters.

In the breed White Leghorn we worked with two distinct strains with inbreeding coefficients of 48% and 57%, respectively, and observed that the individuals of the strain with lower inbreeding were more uniform in their genotype with respect to blood groups than the other. The characters studied were: weight of the fowl at three weeks and twelve weeks of age. Another three alleles were detected which seemed to correspond to the B_1, B_{11}, and B_7. Variance between homozygotes for each of the three alleles with respect to their weight at three weeks was analyzed and the value of $F = 3 \cdot 93$ was obtained which is significant at the level of 5%. The t test applied thereafter demonstrated that the homozygotes for the allele B_2 were less heavy than those of the other allele, differences between these last being not observed.

No differences in weight were noted among the homozygotes at twelve weeks of age.

The test for the first character between the homozygotes and heterozygotes demonstrated heterosis with regard to homozygosity in the allele B_2 in the individuals B_2, B_{11}, but not the B_2, B_{17}. The remaining heterozygotes when compared with the homozygotes do not reveal differences. The same tests were applied to the other character, i.e. weight at twelve weeks. No differences were noted between the homozygotes and heterozygotes.

The frequencies of the alleles detected are: $57 \cdot 14\%$ for the B_{11}; $25 \cdot 01\%$ for the B_7; and $17 \cdot 85\%$ for the B_2.

ON THE RELATION BETWEEN BLOOD GROUP GENES AND A LETHAL GENE FOR HAIRLESSNESS AND PROLONGED GESTATION

K. MAIJALA and G. LINDSTRÖM

Livestock Breeding Research Organization, Helsinki

The present report deals with the occurrence of hairless calves in connection with prolonged gestation periods in the isle Hailuoto in the Northern Finland (65th latitude), about 30 km. from the continent. A total of 12 hairless calves were observed in inbred matings of a bull Saku, brought from the continent. Because this occurrence of hairless calves in a geographically isolated area was co clear-cut, it was investigated more closely, and the possible relationship of the presumably recessive lethal gene to the blood group genes was studied. The finding of an association between a recessive lethal gene and a dominant marker gene would be of practical importance for future detection of lethal carriers and of theoretical interest from the viewpoint of mapping bovine chromosomes.

The bull Saku belonged to the Finncattle breed, where very few lethals are known to occur. In the isolated and primitive conditions Saku was used for several years and many of his own daughters were mated to him. From these matings more than 10 hairless calves were born. In most cases the calf was stillborn, but in some cases it died a few hours after birth. The gestation period varied from $10\frac{1}{2}$ to 12 month in length, and in some cases the cow was slaughtered before parturition. In every such case the foetus was hairless. Some hairs could only be observed on the back line around the eyes and on the edges of hoofs.

When data from milk-recorded herds only were considered, the ratio of hairless calves to normal calves from the back-cross matings was 4 : 16, while the ratio 2·5 : 17·5 was expected on the basis of the assumption of autosomal recessive inheritance. The χ_2 value of this deviation is 1·03, which is far from significant. Taking into account that some of the daughters which gave a hairless calf lost their chance to calve, it can be concluded rather safely, that recessive lethal gene has been involved and that Saku has been heterozygous for this gene.

Blood samples were obtained from 7 carrier daughters and from the dams of three of them. Samples were also obtained from Saku's sire and dam, and from an unrelated cow who had given hairless calves when mated to Saku. In order to determine the genotype of Saku, 17 other daughters of Saku were sampled, 11 of them with their dams. An unrelated cow who had not given hairless calves after 5 successive conceptions with Saku was also sampled. The erythrocytes of blood samples were tested with the conventional haemolytic technique against 42 different specific antibodies. The reconstructed blood type of Saku was A/–BGKE$_2'$/Y$_2$A'L'/L'F/F J/– –/– (L system) –/– (M-system) S$_2$/– Z/–.

In systems A, B, J and SU, the two Saku's alleles were represented among the carrier daugters with equal frequencies, which indicates that there cannot be a close linkage between these loci and the lethal gene. Similarly, the possibility that some of the alleles at these loci would cause the lethal effect as a pleiotropic effect can be ruled out.

In the C- and FV-loci homozygosity of Saku eliminates the possibility of a general linkage or pleiotropy for the whole breed, but it is still possible that a mutation causing the lethal effect has recently taken place in the other L'- or F-allele of Saku or in the vicinity of the C- or FV-loci.

In the L- and M-loci, Saku can be regarded as homozygous for the ,,empty" allele, whence reasoning is about the same as in the C- and FV-loci.

In the Z-system some of the carrier daughters lacked the Z-factor, while some of them did have it. Thus, there cannot exist any general association between the Z-locus and the gene for hairlessness. No conclusions are possible with regard to the special case of Saku's family.

On the basis of the information available it is thus very likely that a recessive lethal gene is involved, which is not closely connected with any of the blood group loci A, B, J and SU. Further it appears, that the blood group loci C, FV, L, M, and Z cannot generally be used as markers of the lethal gene for hairlessness in the Finncattle breed.

(A more detailed report will be published in Annales Agriculturae Fenniae, Seria Animalia Domestica.)

RELATIONSHIP BETWEEN BLOOD GROUPS AND BEEF PRODUCTION IN CHIANA BREED CATTLE*)

ALBERTO SALERNO

Institute of Animal Production of the University of Naples, Portici

We hope these investigations may provide a further contribution to the knowledge of the relationship between blood groups and beef production in some Italian breeds of cattle.

We considered the "live weight" of cattle at some typical ages (at birth, at 6, 12, 18 and 24 months of age) and we connected it with the presence of some particular blood group genes.

Table 1

Pleiotropic effects of blood group genes on body weight of Chiana breed calves at birth

Genetic system	Gene	Males				Females			
		n_1	n_2	D	t	n_1	n_2	D	t
A	A	19	6	—4·13	1·0325	75	19	2·77	1·6620
FV	F	—	—	—	—	47	24	—6·73	2·0863*)
	V	—	—	—	—	24	47	6·73	2·0863*)
J	J	14	11	1·94	0·5626	41	53	2·20	1·7600
L	L	9	16	0·65	0·1430	46	48	—2·18	1·4652
M	M	4	21	0·43	0·1034	5	89	—2·60	0·8840
S	S_1	5	20	—3·95	0·9480	15	79	—4·38	4·1610†)
	S_2	5	20	—1·70	0·3910	47	47	2·58	1·9350
	S_2U_1	20	5	1·70	0·4080	48	46	—2·78	0·7228
Z	Z	7	4	—1·2	0·2640	38	10	—4·00	1·7600

†) P < .001
*) P < .05.

Up to one year the "live weight" is, more or less, a mother character; after this age it becomes a highly individual character so that the value of heritability is 0·86 at 15 months, 0·34—0·53 at birth and 0·28 at weaning.

The genetic structure of production traits makes it difficult to study the relationship between blood groups, but considering that the values of heritability of beef production are rather high and the formation of beef Italian breeds occurred through a period of inbreeding, we cannot exclude the relation to

*) This paper was not read and discussed.

some blood group genes. Consequently, combined characters of productivity with some blood groups would be formed in cattle. To prove the existence of a pleiotropic relationship between blood group character and beef production, the bull's progeny was divided into two classes: (a) one with a particular blood group allele; (b) the other without this allele.

Table 2

Pleiotropic effects of blood group genes on body weight of Chiana breed calves at six months of age

Genetic system	Gene	Males				Females			
		n_1	n_2	D	t	n_1	n_2	D	t
A	A	19	6	−17·43	1·0458	75	19	4·64	0·6542
FV	F	—	—	—	—	47	24	−89·77	6·4634†)
	V	—	—	—	—	24	47	89·77	6·4634†)
J	J	14	11	−10·01	0·8001	41	53	− 1·88	0·3271
L	L	9	16	−21·37	0·1385	46	48	15·58	2·4552**)
M	M	4	21	4·26	0·2080	5	89	−12·89	0·9925
S	S_1	5	20	− 2·70	0·1512	15	78	−11·63	1·2793
	S_2	5	20	3·30	0·0561	47	47	− 2·95	0·5015
	$S_2 U_1$	20	5	− 3·30	1·8480	48	46	2·50	0·4250
Z	Z	7	4	−30·70	1·2824	38	10	− 0·90	0·0882

†) P < .001
**) P < .01.

In order to eliminate the influence of the genetic differences between bulls, all analyses were made within groups of paternal half-brothers and half-sisters.

The differences between the two classes were weighted considering the number of animals in the classes and variance within classes as follows:

$$D = \frac{\sum (\bar{x}_1 - \bar{x}_2) \dfrac{n_1 \cdot n_2}{n_t \cdot s^2}}{\sum \dfrac{n_1 \cdot n_2}{n_t \cdot s^2}}$$

where D = the averages weighted difference between the two classes
$\bar{x}_1$ and $\bar{x}_2$ = average of the two classes within the respective progeny groups
n_1 = number of animals in class 1
n_2 = number of animals in class 2
n_t = total number of animals in the respective progeny groups
s^2 = mean square within classes within the respective progeny groups

$$t = D \sqrt{\sum \frac{n_1 \cdot n_2}{n_t \cdot s^2}} .$$

Table 3

Pleiotropic effects of blood group genes on body weight of Chiana breed cattle at one year of age

Genetic system	Gene	Males				Females			
		n_1	n_2	D	t	n_1	n_2	D	t
A	A	19	6	−20·90	0·8987	75	19	2·41	0·2892
FV	F	—	—	—	—	47	24	17·56	1·2292
	V	—	—	—	—	24	47	−17·56	1·2292
J	J	14	11	−9·36	0·0842	41	53	− 5·01	0·7014
L	L	9	16	− 2·15	0·2042	46	48	− 9·11	1·0900
M	M	4	21	−15·61	0·5775	5	89	− 5·19	0·4619
S	S_1	5	20	− 9·15	1·0980	15	79	−14·73	1·4730
	S_2	5	20	13·35	0·4672	47	47	0·29	0·0133
	S_2U_1	20	5	−13·35	1·7355	48	46	1·30	0·5070
Z	Z	7	4	−22·20	0·2899	38	10	4·60	0·3634

Table 4

Pleiotropic effects of blood group genes on body weight of Chiana breed cattle at eighteen months of age

Genetic system	Gene	Males				Females			
		n_1	n_2	D	t	n_1	n_2	D	t
A	A	13	4	− 52·45	0·9965	69	16	7·66	0·6894
FV	F	—	—	—	—	41	24	11·34	0·5896
	V	—	—	—	—	24	41	−11·34	0·5896
J	J	10	7	− 6·02	0·0602	37	48	3·00	1·0800
L	L	6	11	− 65·31	0·6531	43	42	10·50	0·7350
M	M	7	10	−154·86	2·1835*)	3	82	−28·10	1·1240
S	S_1	4	13	− 48·43	0·8233	15	70	−10·36	0·9116
	S_2	3	14	23·50	0·3407	39	46	− 9·65	1·0615
	S_2U_1	14	3	− 23·50	0·3995	47	38	5·52	0·6624
Z	Z	6	4	− 18·50	0·1165	33	7	− 8·60	0·3354

*) $P < .05$

The investigation refers to 130 Chiana cattle reared in the "provincia" of Siena tested for 40 blood group factors by reagents that were kindly supplied to our Institute by dr. Jan Rendel of Uppsala University.

The results were as follows: At birth females with allele V weight 6·73 kg. more than those with allele F. On the other hand, females with allele S_1 weigh 4·38 kg. less than cattle without it. At six months females with allele V weigh 89·77 kg. more than those having allele F. ($P < .001$). At the same age, the females with allele L weigh 15·58 kg. more than L-negative cattle ($P < .01$).

On the other hand, allele M seems to have an unfavourable effect on the weight of males at 18 and 24 months of age. Their weight appears to be 154·86 kg. and 212·88 kg. respectively, less than cattle without allele M.

The results of these investigations are reported in five tables attached to this study.

Table 5

Pleiotropic effects of blood group genes on body weight of Chiana breed cattle at two years of age

Genetic system	Gene	Males				Females			
		n_1	n_2	D	t	n_1	n_2	D	t
A	A	11	4	− 46·17	1·2983	37	8	9·75	0·9750
FV	F	—	—	—	—	26	11	−10·91	0·6436
	V	—	—	—	—	11	26	10·91	0·6436
J	J	10	5	− 63·90	0·9009	18	27	4·14	0·4140
L	L	5	10	−139·64	1·3964	18	27	3·65	0·5110
M	M	7	8	−212·88	2·1835*	2	43	− 1·36	0·0061
S	S_1	2	13	19·16	0·2299	8	37	7·36	0·2281
	S_2	2	13	127·29	1·2729	22	23	6·96	0·9744
Z	Z	5	2	3·60	0·0068	—	—	—	—

* $P < .05$

References

Bettini, T. M. (1960). Acc. Econ. Agr. dei Georgofili, Firenze.

Rendel, J. (1959). VI. International blood-group Congress in Munich.

Salerno, A. and Gatti, L. (1963). Prod. Anim. 2 : 107.

A CONTRIBUTION TO THE PROBLEM OF BLOOD GROUPS IN DUCKS

V. DROBNÁ, J. HORT, P. IVÁNYI, J. MARDIAK

Poultry Research Institute, Ivánka pri Dunaji,
and Institute of Experimental Biology and Genetics, Czechoslovak Academy of Sciences, Prague

In connection with the experiments of Hašek and coworkers on the induction of immunological tolerance to homologous erythrocytes in adult ducks, we obtained specific antibodies against red cell antigens. We used Peking ducks in these experiments. 30 exsanguination transfusions in adult ducks were carried out. In six (i.e. 20%) tolerance was elicited, in 24 (i. e. 80%) antibody formation took place. Among these, 19 sera (i. e. 80%) had a titre lower than 32, mostly 2—8; these sera reacted weakly and gave badly reproducible results, their specificity was not yet studied in detail. Only 5 sera (i.e. 20%) had an antibody titre higher than 64, gave strong reactions and excellently reproducible results. Specificity of these sera was examined in detail and the results showed that they all reacted in the same way. We denoted them anti-A, and the ducks, according to the reaction to these sera, A+ and A−.

Table 1

Type of mating	Number of progeny	Frequencies of phenotypes A+		A−		X^2	$P_{(1)}$
		expected	observed	expected	observed		
Aa × Aa	425	318·7	306	106·3	119	2·023	0·10
Aa × aa	380	190	185	190	195	0·262	0·50
aa × aa	107	0	1	107	106	1·009	0·30
AA × Aa	35	35	35	0	0	—	—

Antibody titre was determined in physiological saline, further by the trypsin and Coombs test. The latter gave higher titres and the results could be better evaluated. Sera whose titre (with the trypsin test) was higher than 64 were used as reagents.

The purpose of this paper is to study the segregation ratio of phenotypes of ducklings and to find the relationship between blood group A and the productive characteristics in ducks. In this part of our study we investigated the relationship between blood group and body weight at the age of 56 days.

The ratio of phenotypes was examined by the chi-square test. Variability in body weight was studied in an extensive breeding experiment. The ducklings were hatched individually and weighed at the age of 56 days. At the same time, they were tested for their blood group. Body weight differences between single groups were evaluated by Student's t-test.

Table 2

Phenotype	Number	Average weight
A−	431	2333·0 g.
A+	544	2341·6 g.
A₁	224	2341·9 g.
A₂	320	2341·3 g.

The hypothesis, that A antigen is inherited as a dominant character has been verified by means of the segregation ratio of phenotypes (table 1). The only exception in the group of mating aa × aa is probably due to a technical error. In all cases, the P value (df 1) is higher than $P_{0.05}$; differences are insignificant and they prove the dominant mode of inheritance of the A antigen.

Comparison of body weights at 56 days of A+ and A− ducklings showed completely insignificant differences (table 2). The regression coefficient of body weight to blood group antigen is negligible: b_x 0·007. Having divided A+ ducklings into two subgroups by means of agglutination strength, no differences were found (table 2).

Summary

By means of isoimmune sera an antigen, designated A can be identified in Peking ducks. This antigen is inherited as a dominant character against its absence. No relationship was found between blood group A and the weight of ducklings at the age of 56 days.

INVESTIGATIONS OF THE BLOOD TRANSFUSION IN CATTLE

B. BUSCH

German Academy of Agricultural Sciences at Berlin, Institute of Animal Breeding Research, Dummerstorf

The practical use of animal blood group research is particularly in animal breeding. But it is also the task of blood group investigators to study the problems of general biology and veterinary medicine. Therefore we started investigations of blood transfusion in cattle.

In our country blood transfusions in cattle are performed very seldom. We suppose that this is due to technical problems, frequent incompatibility reactions of the recipients and the lack of an in vitro test to avoid them.

I will give a short report of our experiments without technical details.

In first transfusions, but more frequently in repeated transfusions of the same donor's blood we observed reactions of the recipient, which can be classified into 3 groups.

1. Muscle trembling, a little excitement. After the interruption of the transfusion the animal returned quickly to a normal state.
2. Excitement, higher respiratory and pulse frequency, salivation. After interruption of the transfusion the animal returned to a normal state after an hour.
3. Heavy excitement, abdominal respiration, high respiratory and pulse frequency, salivation, stupor, suffering appearance. After interruption of the transfusion these signs continued, food was refused. The animals returned to a normal state after about 5 hours.

The reactions of group 1 appeared in several cases and disappeared in the course of the transfusion. The reactions of group 2 were observed after having transfused 200·0 bis 500·0 ml. of blood and transfusion was stopped in most cases. If the transfusion was continued, the reactions became more serious, as in group 3.

The causes of these reactions may be:

1. the technique of transfusion
2. protein incompatibility
3. blood group incompability.

The technique was not involved, because the same technique led in some cases to reactions, in others not. And protein incompatibility must be excluded too, because the results of the Ouchterlony agar gel diffusion test were negative.

Transfusion of plasma to animals, which had reacted to whole blood, also gave no reactions.

Only one cause was possible, the blood group incompatibility. When we transfused washed erythrocytes, the animals showed the same reactions as to a whole blood transfusion.

We investigated, whether the haemolytic test can be used as an in vitro test to avoid transfusion reactions. But in most cases the positive test was followed by reactions and the negative test did not prevent these reactions. The haemolytic test was not suitable.

In a modified haemolytic test, the number of positive reactions was higher, but there was no correlation with the presence or absence of transfusion reactions.

Furthermore, we tried to determine incomplete antibodies by using a rabbit-anti-cattle-globulin serum (Coombs-test), but we did not succeed.

The consequences of our investigations are:

1. The reactions, we observed in the course of transfusions were specific and depended on blood group incompatibility.
2. The normal and a modified haemolytic test and the Coombs test were not suitable to detect antibodies, which are the cause of transfusion reactions.
3. We suppose that it will be possible to detect antibodies by other serologic tests and in this direction we will continue our experiments.

THE SEROLOGICAL ANALYSIS OF THE A BLOOD GROUP SUBSTANCE(S)

J. F. BOREL

Central Laboratory of the Blood Transfusion Service of the Swiss Red Cross in Bern

Methods for the diagnosis of the haemolytic disease of the newborn due to the ABO blood group incompatibility are still quite unsatisfactory (Damerow (1963)). In an attempt to find a new approach, our first step was to survey all the work previously done. This paper reviews briefly what is known about the serological analysis of the A antigen mosaicism.

The properties of the different subgroups of the A antigen in humans are described by Race & Sanger (1962). In their decreasing order of antigenicity these antigens are: A_1, A_2, A_3, A_x, and A_m.

Furuhata (1960) summarizes the analytic studies on the complex structure of the ABO group substances (GS) as follows: "The A-GS can be partitioned into A_I (man part), A_{II} (dog part), A_{III} (hog part), and A_{IV} (sheep part). Each of these partial antigens has been demonstrated to produce the corresponding partial antibody". The mosaic structure of the A antigen can be understood the following way:

Human	A-GS contains 4 partial antigens
Dog	A-GS contains 3 partial antigens
Swine	A-GS contains 2 partial antigens
Sheep and goat	A-GS contains 1 partial antigen.

Similarly, the B- and O-GS can also be subdivided into three partial antigens each. About the ontogenetic development of the GS, Furuhata states further that in the early stages of gestation, the maturation of the GS of the lower orders is more advanced than of the higher orders.

Ueno & al. (1959) have made experiments which indicate marked individual differences in the composition ratio of the partial A and B antigens in the red blood cells and in the saliva. They think that the composition ratio of the partial antigens are a factor inherited along with A and B genes, and that the quantity of the partial antigens is determined by the secretor locus.

Dahr (1938) has discussed the similarities of ape GS with the ABO antigens. The B antigen of the orangutan contains all three partial antigens present in the human B-GS. Furthermore he shows that normal human sera contain heteroagglutinins against primates in contrast to ape sera which lack heteroantibodies against human cells. No differences between naturally occurring isoantibodies of type anti-A and anti-B in man and ape could be found. Recently, Wiener & Moor (1963) have reviewed the results related in previous papers on this subject. Accordingly, the chimpanzees possess either A- or O-GS, the A antigen being more closely related to the A_2 subtype than to the A_1. The orangutans and gibbons display A and B, but no O properties. The gorillas having anti-A in their serum and their erythrocytes being only weakly agglutinated by anti-B reagents, they are classified as B-like. The blood groups of gorillas as well as of baboons are best determined by GS from secretions, the latter having the A and B, but no O antigens.

Schermer (1935) seems to have found an A-like GS in horse red blood cells. Two other authors cited by Dujarric de la Rivière & Eyquem (1953), are also said to have established the presence of A- and B-GS similar to human GS in the erythrocytes and in the saliva of horses. Quite opposite to this statement is a report of Eyquem & al. (1962) which says, that "there is no relation between the red cell antigens and the A- or B-GS which can be found in the stomach or organs of some horses".

Witebsky & Okabe (1927) discovered the existence of A-GS on the blood cells and in the serum of certain cattle. By using bovine erythrocytes, rabbit immune anti-A serum and complement, they observed haemolysis in about 25% of the animals tested. They also detected by absorption tests a naturally occurring specific anti-A in the serum of some individuals. These results have been independently confirmed much later by Neimann & al. (1954), who designated the A-GS of cattle as the J factor. They have also demonstrated that the human A erythrocytes removed all antibodies for J, but that anti-A was not absorbed by J positive cells. Thus, the A-GS of man contains all partial antigens of the J factor, but the reverse is not true. It is, however, not stated which of the four previously mentioned partial antigens are shared by both the J- and the A-GS (Stone (1962)). We can only suppose that both sheep R- and bovine J-GS have at least one partial antigen in common, because anti-R does not lyse J positive cells, but anti-J does react with R positive cells. The anti-R specificity would therefore be directed against another receptor of R than the one they have in common.

Andresen (1962) presents an up-to-date account of the blood group research in pigs. The long known A system has an A-GS which occurs primarily in the serum and which is sometimes also present on the erythrocytes in the same way as the J-GS of cattle and the R-GS of sheep. The A antigen of hog crossreacts strongly with bovine anti-J. Hojný & Hála (1964) were able to distinguish between two types of A antigens in pigs by their differing capacities in inhibiting anti-A serum from rabbits immunized with human A_1 cells. Winstanley & al. (1957) have studied the effect of natural and immune anti-A from humans against A positive pig blood cells. They observed that natural human anti-A isoantibodies would not lyse A positive pig cells, but A-GS from hog gastric mucin or human A secretor saliva would inhibit these antibodies. The so-called "immune" human anti-A, however, did lyse A pig cells and could be absorbed with these blood cells. The conclusion to be drawn is that A-GS from the saliva of A secretor pigs and the A-GS derived from gastric mucin possess one component more than the A-GS of the erythrocytes. Yokoyama & Fudenberg (1964) demonstrated that human natural (γ_{1M}) antibodies agglutinated strong A positive pig cells, while immune (γ_{ss}) antibodies produced lysis of those cells. We were able to produce a specific anti-A reagent for pig as well as for human blood cells by absorbing O-type normal pig serum containing anti-A with human O cells. Human A cells completely absorbed these natural swine isoantibodies (Borel (1964), unpubl. data).

Rasmusen (1962) has studied the R-O system of sheep, a system serologically closely related to the A and H properties of human GS, the A-O system of pig, and the J-Oc system of cattle. Rendel (1957) compares the close genetic relationship existing for the production of the R- and O-GS in sheep and the Le^b-, H-, and A-GS in man.

Hara (1930) detected an A-like GS in the serum of individual rabbits by means of a complement consumption test. These results were confirmed and extended by Terajima (1942). He could show that only O-type rabbits were able to produce anti-A antibodies when injected with A-type erythrocytes of goats, sheep, pigs, dogs, and human. His experiments indicate further that the A-GS in the saliva of rabbits contains all four human partial antigens. Bednekoff & al. (1963) have again confirmed the presence of a J-like GS in the serum, saliva, and urine of some rabbits, which was capable of inhibiting the reaction between cattle anti-J serum and J positive cells. J-type rabbits were unable to produce anti-J. The heredity of this A-like GS was disclosed and shown to be controlled by an autosomal dominant gene.

About two thirds of the minks tested also possess an A-like GS, both on their erythrocytes and in their serum, which is detectable by the indirect antiglobulin test with pig sera containing naturally occurring incomplete anti-A (Saison (1964)).

In an immuno-genetic study of amphibians and reptiles Hildemann (1962) states: "No haemagglutinogens similar to human A or B have been observed in reptiles, whereas antigens similar or identical to human B are definitely present on the red cells of certain amphibian species". Dujarric, Eyquem & Fine (1954) have consistently found specific heteroagglutinins against human A and B antigens in the normal serum of a viper species (Vipera aspis).

Active GS are also known to occur in simple organisms as bacteria and plants. Springer & al. (1962) showed a B-GS activity in E. coli O_{86}. Chemical analysis disclosed the presence of the same sugars known to compose the human

B mucoid. However, not all bacteria possessing these same sugars exhibit B-GS activity. From two plants (Sassafras albidum and Taxus cuspidata) they extracted two polysaccharides possessing a high and specific H(O) activity in the heterologous eel anti-H(O) serum and O erythrocyte system. The interesting point lies in the fact that these plants contain non-identical compounds, though related groupings, and yet show activities very similar to those of human blood group mucoids when tested with heterologous reagents. Mori (1951) observed that the A antigen of Shigella flexneri (R.) contains all four human A partial antigens. By transformation from S to R (induced by antiserum) the GS were made to appear or disappear.

Tomcsik (1945) and Doerr (1948) have discussed the serological cross-reactivity existing between human A cells and the Pneumococcus type XIV. The crossreactive polysaccharides analyzed by Morgan (1960), showed considerable chemical similarity.

Furthermore, Yokoyama & Fudenberg (1964) have demonstrated a variety of heterogenetic antibodies directed against A positive pig blood cells in sera from human subjects immunized with specific soluble A-GS, typhoid and paratyphoid vaccine, or tetanus antitoxin, and in sera from rabbits injected with polysaccharides from Pneumococcus type XIV. Five human anti-Le^a sera also showed anti-A pig activity.

When discussing the A-GS, the Forssman antigen (F_A) has also to be mentioned. The controversy about human A-GS containing F_A or not, is not yet settled. We agree with Stormont & Suzuki (1958), who suggest that the problem regarding the F_A classification resides largely in the various criteria used by the different investigators. Armangué (1945), in a review on heterogenetic antigens, gives the following definition of the F_A: "Any antigens present in either the tissues, cells, fluids or microorganisms, which, when injected into a rabbit, cause the formation of antisheep haemolysins" (translated). Furthermore, he defines the F_A-type haemolysins as: "all anti-sheep haemolysins, either naturally occurring or immune, which can be absorbed by guinea pig and horse kidneys, or their alcoholic extracts" (translated).

Schiff & Adelsberger (1924) demonstrated in normal serum of some guinea pigs the presence of haemolysins reacting with sheep and human A and AB red cells, but inactive against human O and B erythrocytes. Sheep red cells were not able to absorb all antibodies present in rabbit anti-human A sera. Immune sera of rabbits injected with guinea pig kidneys had their anti-F_A activity exhausted if absorbed with sheep blood cells, but not with human A cells. They consequently assumed the existence of a common receptor between sheep and human A erythrocytes, but not identical with F_A. Zeki (1929) demonstrated that anti-human A immune sera after reacting with alcoholic extracts from horse and guinea pig kidneys or from sheep and human A erythrocytes would specifically bind complement. The same happened with antihorse kidney immune sera, except that the alcoholic extract of human A cells did no longer fix complement. This again confirms non-identity between the A-GS and the F_A.

Thomsen (1936) distinguished between three different receptors: F_{Me}, a receptor common to all human and sheep erythrocytes; F_A, a receptor common to all human A-type and sheep red cells, called "Schafanteil"; and finally the specific human A receptor, designated A sensu strictiori. He showed the presence of a specific anti-A in immune sera of guinea pigs immunized with human A_1 blood cells and subsequently absorbed with human O cells. He assumed that this anti-A was different from the anti-F_A, because this very F_A is present in all guinea pigs, and that A_1 must be a component of the F_A, as no anti-A_1 was ever produced in these animals.

Andersen (1938a) could not increase the normal anti-R titre in sheep by immunizing them with A_1 blood cells. The antibody can nevertheless be completely exhausted after incubation with human A cells, but not with O erythrocytes. Andersen (1938b) also showed that anti-A haemolysins, but not specific anti-A_1, were formed in rabbits injected with human A_1 or A_2 blood cells. Rabbit anti-sheep cells immune sera absorbed with A_2 cells contained a specific anti-A_1 agglutinin, which could be absorbed by any kind of sheep cells. Absorptions with human A_1 or A_2 cells did not lower the sheep haemolysin content. From several experiments he concluded that the A_1- as well as the A_2-GS share the same partial antigens with sheep blood cells.

Terajima (1942) made a distinction between two types of anti-sheep haemolysins formed in rabbits, namely the anti-A being only produced in O-type animals and the true anti-F_A antibodies occurring in both O- and A-type rabbits. As these anti-F_A antibodies could not be absorbed either by human A cells or by human kidney, the human A-GS was obviously devoid of F_A-GS.

The anti-A of lectins from Vicia cracca, Dolichos biflorus and Phaseolus limensis do not possess any anti-F_A specificity (Krüpe (1956)).

Andersen (1938c) analyzed the occurrence of normal anti-sheep antibodies in human sera. In the serum of patients suffering from Mononucleosis infectiosa, a remarkable increase of sheep agglutinins can be found, which can be absorbed by human A red cells. He reported in three cases the presence of anti-R, which could also be absorbed with human A cells. Lee & al. (1963) have recently studied sera from Mononucleosis patients, showing the characteristic increase in sheep agglutinins as well as beef haemolysins. These antibodies, however, do not have a true anti-F_A activity, as they are not absorbed by guinea pig kidney (Landsteiner (1945)). Different physicochemical methods could not separate the two types of antibodies. The only differences being, that the beef haemolysins were incompletely absorbed by sheep blood cells and more sensitive to temperature changes than the sheep agglutinins. They assumed different combining sites on the same molecule.

Current studies by Rose (1964) may soon reveal precise knowledge of the chemical constitution of both the A and the F_A antigens.

The probable explanation for all these confusing results has been fore-shadowed by Landsteiner (1945), who wrote: "Heterogenetic reactions do not signify that the same substance is present in the serologically similar materials. It may suffice to mention differences in the GS A and B and in the paradigmatic F_A antigens. In absorption and immunization experiments, and also by comparing various immune sera, it has been demonstrated that the F_A reactions, which possibly depend upon the presence of relatively small characteristic groupings, are due to similarity rather than to identity of the respective substances."

References

Andersen, T. (1938a): Z. Rassenphysiol. 10 : 88—103.
Andersen, T. (1938b): Z. Rassenphysiol. 10 : 154—165.
Andersen, T. (1938c): Z. Rassenphysiol. 10 : 166—180.
Andresen, E. (1962): Ann. N. Y. Acad. Sci. 97 : 205—225.
Armangué, M. (1945): Path. Bakt. 8 : 360—422.
Bednekoff, A. G., Tolle, A., Datta, S. P., Friedman, J. & Stone, W. H. (1963): J. Immunol. 91 : 369—373.
Dahr, P. (1938): Z. Rassenphysiol. 10 : 78—87.
Damerow, R. (1963). Ergebn. inn. Med. Kinderheilk. 19 : 132—205.
Doerr, R. (1948). Immun. Forsch. 3 : 65—74.
Dujarric de la Rivière, R. & Eyquem, A. (1953): Les groupes sanguins chez les animaux. p. 153. (Ed. méd. Flammarion, Paris).
Dujarric de la Rivière, R., Eyquem, A., & Fine, J. (1954): Experientia 10 : 159—165.
Eyquem, A., Podliachouk, L. & Millot, P. (1962): Ann. N. Y. Acad. Sci. 97 : 320—328.
Furuhata, T. (1960): Special lecture. 8th Congr. int. Soc. Blood Transfusion, Tokyo 1960.
Hara, K. (1930): Z. Immun. Forsch. 67 : 125—136.
Hildemann, W. H. (1962): Ann. N. Y. Acad. Sci. 97 : 139—152.
Hojný, J. & Hála, K. (1965): Proc. IXth Congr. europ. Ass. Anim. Blood Group Res. p. 153.
Krüpe, M. (1956): Blutgruppenspezifische pflanzliche Eiweisskörper. (F. Enke Verlag, Stuttgart.)
Landsteiner, K. (1945): The specificity of serological reactions. Chapter III. Rev. ed. (Harvard Univ. Press, Cambridge, Mass.).
Lee, C. L., Takahashi, T. & Davidsohn, I. (1963): J. Immunol. 91 : 783—790.
Morgan, W. T. J. (1960): Proc. roy. Soc. (B) 151 : 308—347.
Mori, T. (1951): Jap. J. Legal Med. 5 : 155—165.
Neimann-Sörensen, A., Rendel, J. & Stone, W. H. (1954): J. Immunol. 73 : 407—414.
Race, R. R. & Sanger, R. (1962): Blood groups in man. p. 28. 4th ed. (Blackwell Sci. Publ., Oxford.)
Rasmusen, B. A. (1962): Ann. N. Y. Acad. Sci. 97 : 306—319.
Rendel, J. (1957): Acta Agric. Scand. 7 : 224—259.

Rose, J. K. (1964): Personal commun.

Saison, R. (1964): J. Immunol. 94 : 20—23.

Schermer, S. (1935): Z. Rassenphysiol. 7 : 33—42.

Schiff, F. & Adelsberger, L. (1924): Z. ImmunForsch. 40 : 335—367.

Springer, G. F., Williamson, P. & Readler, B. L. (1962): Ann. N. Y. Acad. Sci. 97 : 104 to 110.

Stone, W. H. (1962): Transfusion (Philad.) 2 : 172—177.

Stormont, C. & Suzuki, Y. (1958): J. Immunol. 81 : 276—284.

Terajima, M. (1942): Jap. J. med. Sci. 4 : 1—9.

Thomsen, O. (1936): Z. ImmunForsch. 87 : 335—365.

Tomcsik, J. (1945): Path. Bakt. 8 : 345—359.

Ueno, S., Matsuzawa, S., Kitamura, S. & Mishima, H. (1959): J. Immunol. 82 : 385—396.

Wiener, A. S. & Moor-Jankowski, J. (1963): Science 142 : 67—69.

Winstanley, D. P., Konugres, A. & Coombs, R. R. A. (1957): Brit. J. Haemat. 3 : 341—347.

Witebsky, E. & Okabe, K. (1927): Klin. Wschr. 6 : 1095.

Yokoyama, M. & Fudenberg, H. H. (1964): J. Immunol. 92 : 413—424.

Zeki, J. (1929): Z. ImmunForsch. 62 : 207—218.

DISCUSSION

B. Busch. The report of Mr. Millot was of great interest to me because we tried to detect incomplete antibodies in cattle, too as I reported in my report but we did not succeed. And now I have two questions to Mr. Millot.

1. Have you tried to produce an anti-cattle-globulin serum by immunizing rabbits?

2. You have used a modified Coombs test. Have you found that the results are better than in the original Coombs test? May be that the differences in our results are due to the different techniques.

P. Millot: Dr. Busch has tried the Coombs reaction with rabbit antiglobulin without success. He asks me why this reaction is successful with goat antiglobulin and not with rabbit antiglobulin. I answer that I have rabbits now injected but I have no results on this species up to now. I think an analysis of proteins of our goat sera perhaps indicates what is the active fraction in the Coombs reaction with bovine cells.

A. Eyquem: The use of a goat antiserum for the study of incomplete anti-cattle blood groups could be supported by the study of incomplete antibodies in human blood groups. These goat antisera are not giving a prozone which is usually observed with rabbit antisera. I would like to ask Prof. Stormont to comment on the results published by Coombs et al. (1951) relative to the coupling of antibodies in cattle blood-typing reagents to form long antibody-chains capable of bringing about direct saline agglutination of cattle red cells.

C. Stormont: Dr. Eyquem when commenting on Dr. Millot's paper asked me if I would comment on results published by Coombs et al. (1951).

Cattle red cells are notorious for their inability to agglutinate even when heavily sensitized with blood-typing antibodies. This, of course, is why we were forced to use haemolytic techniques after the original methods of Bordet and Ehrlich. But this was all to our advantage because of the great sensitivity of these techniques when used to explore blood groups in species of the family Bovidae.

When Dr. Linus Pauling visited my laboratory in 1950 he suggested that the reason for the failure of cattle red cells to agglutinate when sensitized with antibodies could be due to the possibility that the antigens are situated in rather deep valleys on the surface of the red cell membrane and that the bound antibodies are not long enough to extend beyond the ridges separating the valleys. Consequently, agglutination of the sensitized red cells would not take place because of these "mechanical" difficulties. He reasoned that it should be possible to produce agglutination of cattle red cells by using polymerized (coupled) antibodies of sufficient length to extend beyond the ridges. Accordingly, an experiment was planned in which I would provide the isoimmune antisera and Pauling would produce long-chained antibody molecules by coupling or polymerizing these molecules. Although we did send antisera to Pauling he apparently was too busy with other matters to perform the coupling experiments. But, as it turned out, it was hardly necessary for us to go ahead with the experiment because within a year, as I recall, Coombs and colleagues succeeded in producing long-chained complexes of cattle antibodies which were capable of agglutinating cattle red cells. Furthermore, they were entirely unaware that a similar experiment had been planned perhaps almost at the same time they conceived the same idea.

M. Braend: Mrs. Lindström, have you tested for the relationship between the recessive lethal gene and your reagents SF 3 and have you done any investigations to find out if the SF 3 belongs to any of the previously known systems.

G. Lindström. The SF 3 reagent was not included when the animals were tested.

C. Stormont: Have these prolonged hairless calves been subjected to careful autopsy?

G. Lindström: No autopsy was made.

C. Stormont: The reason I ask this question is that at our University we have discovered (Stormont, Kendrick and Kennedy, 1956) a "new" syndrome of inherited lethal defects associated with abnormal gestation in Guernsey cattle (Genetics 41 : 663; Kennedy, Kendrick and Stormont, 1957): adenohypophyseal aplasia, an inherited defect associated with abnormal gestation in Guernsey cattle. (Cornell Vet. 47 : 160—178) a lethal trait in Guernsey cattle which, superficially, would appear to be very similar to the one you have described here in a Finnish breed of cattle. On autopsy it was found that they all lacked the anterior lobe of the pituitary gland (some lacked the entire gland), hence the name adenohypophyseal aplasia.

Subsequently, this same trait was found in Australian Jersey cattle. I am very hopeful that you can arrange to have autopsies performed whenever any more of these calves are delivered. I should add that we also found no association between the gene for adenohypophyseal aplasia and any of the blood group loci.

J. Moustgaard (Chairman).: I want to ask Dr. Stormont to comment on J. F. Borel's paper relative to defining Forssman activity.

C. Stormont: There is thoughout the animal and plant kingdoms a wide variety of substances having Forssman antigenic activity, that is, as judged by their ability to cross react with antisera, produced in rabbits following immunization with homogenates of guinea pig visceral organs (liver, spleen and kidneys). This is the original criterion for Forssman activity as set forth by Dr. Forssman. When it is followed it is found, for example, that the red cells of domestic sheep and goats are Forssman-positive whereas those of man and cattle (like those of the rabbit) are Forssman-negative. There are also all degrees of cross reactivity. Using the lytic activity of Forssman antisera on sheep red cells as a standard for measuring Forssman cross reactions, we examined the cross reactions of some 33 species of Artiodactyls representing four different families (Bovidae, Cervidae, Antelocapridae, and Camelidae) and described three Forssman subtypes (Stormont and Suzuki, 1958: The distribution of Forssman blood factors in individuals of various Artiodactyl species. J. Immunol. 81 : 276—284). There are undoubtedly many, many more to be described. We also discussed various criteria other than the original Forssman criterion in assaying Forssman activity. Naturally I am pleased to note that Dr.

Borel agrees with the position we took in that paper. There is, nevertheless, much confusion existing concerning the classification of substances with respect to Forssman activity.

E. M. Tucker: We have been investigating the Forssman system in Cambridge and although the work is not yet complete I should like to point out that it has become very evident that it is most important to know whether the rabbits which are being immunized are of the blood type A or O; also to know whether A substance is present in the material which is being injected to produce anti-Forssman antisera. Some confusion has been caused in the past because neither the blood type of the sheep red cells nor that of the rabbits have been taken into account.

REPORT OF THE BUSINNES MEETING OF THE E.S.A.B.R.

CHAIRMAN: MIKAEL BRAEND,
PRESIDENT OF THE E. S. A. B. R.

GENERAL BUSINESS

The members were informed that the session would be divided into two main parts: One concerning general business of the Society and one connected with technical aspects of comparison and reference tests.

The first subjects to be discussed were the Constitution and By-laws which were sent around to the members. The president stated that the Committee was well aware of the fact that the Constitution and By-laws were as yet not perfect. The committee, however, prepared it to the best of its abilities. The president then called for remarks and questions on the presented draft. Since no remarks were made by the members, the president concluded that the Constitution and By-laws were accepted.

On the basis of this acceptance the president concluded that the European Society for Animal Blood Group Research was started officially.

The next item was the financial affairs of the Society.

The president stated that the Secretary has many duties to perform and that much labour has to be performed by him. He therefore proposed that the Society should present an allowance to the Secretary of £ 50. The Society accepted this proposition by acclamation.

The Secretary presented a short report of the finances. Until August 15th 1964 about half of the institutes and of the members had paid their fees, resulting in a total of receipts of £ 242·— and a sum of overdue contributions of £ 248·—.

The expenditures were mainly due to shipments of reference reagents for a total sum of £ 74·—, while the total of the expenditures amounted to £ 133·—.

After this report the president proposed to the Society:

1. That calls for fees for 1964 should be sent out as soon as possible after the conference. Participants of this conference who were not members were kindly invited to contact the Secretary. Laboratories and individuals having certain difficulties with payments were invited, if such was desirable, to pay for two years at one time.

2. The laboratories receiving reference reagents should pay themselves for the expenditures connected with the shipment.

Both proposals were accepted without discussion.

Elections of officers

The president informed the conference that due to Constitution two groups of officers had to be elected: One consisting of the President — the Secretary and one member for a four years' period and one consisting of three members for a two years' period.

The president invited Mr Gahne to take the chair for the time of the elections.

The new chairman asked for proposal for officers.

Mr. McDermid proposed to continue for the following four years with Dr. Braend as president and Dr. Bouw as secretary, since they shared the main responsibilities in creating the Society. Furthermore, Mr. McDermid proposed as a third officer Dr. Matoušek who had done such a perfect job in the organization of this conference.

Prof. Moustgaard seconded this proposal by stating that "horses should not be changed in the middle of the river".

The proposal was accepted by acclamation.

Concerning the three other officers a proposal was made by Dr. Kovacz to elect Prof. Moustgaard who had always been very helpful for all laboratories which were in need for technical or material help and who also demonstrated on many occasions to be very able in affairs of organization. Dr. Kovacz furthermore proposed Dr Rendel who also had much experience in this field.

The next proposal was made by Dr. Efremov who suggested to continue also with Dr Böhm who had done a perfect job in the organization of the Ljubljana conference.

Mr. Imlah proposed to elect also a member working on chicken blood groups. On request Mr Imlah did, however, not make a proposal for the person to be elected. The chairman asked for further proposals, but none were made.

The Conference then elected by acclamation Prof. Moustgaard, Dr. Rendel and Dr. Böhm as members of the Committee for the two years' period.

The president then took over the chair in thanking the Conference for its faith in him. He expressed his wishes that he would try to perform his duties to the best of his abilities.

The next item was connected with place and time of the next Conference.

The president informed the members that the Committee had contacted representatives of the French laboratories and studied the possibilities for a meeting at Paris. The Committee found Paris to be an excellent city for the next meeting. The proposal to ask the French people to organize the next conference was accepted by acclamation.

On request Dr. Eyquem informed the meeting that the people from France would be happy to organize the meeting for 1966 as well as this was done in Prague.

In connection with this the president informed the members that the Committee considered to form a so-called "selecting Committee" which should take care of the selection of reports and other preparations for the Conference in Paris. The proposed committee would consist of: Dr. Eyquem, The Ex-officio member of commitee of the E. S. A. B. R., The President of the E. S. A. B. R., The Secretary of the E. S. A. B. R., Prof. Moustgaard.

After the settlement of this item president stated that the Committee had considered the selection of the next ex-officio member and agreed upon Dr. F. Grosclaude from Jouy-en-Josas (France). No objections were made. On request Dr. Grosclaude informed the meeting that he would accept this task.

With regard to communications about the activities of the committee the president stated that the committee was planning to send out letters of information.

For the time being the members were informed that there had been contacts with the "European Association for Animal Production" (E. A. A. P.). From the discussion about the relations with E. A. A. P. was concluded that it is not possible to affiliate closely with this organisation, since the aims of the Society differ too much from those of the E. A. A. P.

Interesting and succesful contacts were made with the Food and Agricultural Organization of the United Nations (F. A. O.) which organization instituted a panel of blood group scientists.

In Japan the workers on blood groups in aminals are collaborating closely with each other. The committee will exchange information with this group of scientists.

After these communications the president proceeded with the technical part of the Agenda.

Technical aspects of the business meeting

Reports on comparison tests for blood groups in cattle and pigs were presented by Dr. J. Matoušek, Miss Gunvor Lindström and Mr. J. Hojný.

Reports on tests for the selection of reference reagents for blood typing of cattle as performed in 1963 and 1964 were presented by Dr. J. Bouw.

31*

Furthermore it was decided that the next reference test should be performed by the laboratory at Wageningen — the Netherlands with the machine methods.

In this test attempts will be made for the selection of references for the factors:

$$A_1, B, R', Y_1, E'_3, J'_1, \text{ and } E.$$

In a written request the participating laboratories will be invited to inform the Secretary which reagents they have available for the purpose of these reference tests.

For the next comparison test of cattle blood groups the Hungarian laboratories were invited as duty laboratories, and for the comparison test for blood groups of pigs the laboratory at Copenhagen.

After having finished the technicalities the president of the E. S. A. B. R. closed the 9th European Conference of Animal Blood Groups.

CLOSING OF THE CONFERENCE

M. BRAEND
President of the E. S. A. B. R.

Dear friends,

We have now finished with the scientific part of the 9th European Conference of Animal Blood Groups. Personally I must say that it has been a very good conference in all respects. There are many people to thank for this. Thus we must thank the chairmen, those who contributed with papers, those who went into discussions and those who only listened. Those to be thanked most however, are those who arranged this conference. They have done an excellent job. I am not able to express my gratitude in words perfectly but believe me it comes from my heart and I am quite certain that you all agree with me. In trying to show our gratitude our committee on behalf of our Society wants to give the two representatives from the Czechoslovak Academy of Sciences something that for ever will remind about our thankfulness. Would you please Mr. Hašek and Mr. Matoušek come forward. It is a great pleasure for me to present these presents to you. And again our very best thanks.

There are also many more people to be thanked, people about whom we very well know have been doing lots of work in the background. As representatives for these we have five ladies here. We want to express our gratitude through these flowers.

And after having gone through the Business programme the official part of the 9th European Conference of Animal Blood Group Research is closed.

Adalsteinsson 42
Adams 229, 234
Adelsberger 474, 477
Alexander 421
Alexandrowicz 123
Allen 173, 175, 177, 178, 276
Allison 272, 276
Alter 273, 276
Amos 405, 406, 413
Andersen 71, 155, 475, 476
Andresen 59, 61, 71, 73, 88, 91, 110, 121, 130, 159, 161, 163, 167, 168, 472, 476
Aptekman 197, 203, 205
Armangué 474, 476
Armstrong 413
Aschaffenburg 290, 292, 293
Ashton 87, 91, 112, 121, 245, 251, 288, 293, 301, 304, 305, 306, 311, 312, 313, 316, 319, 321, 322, 325, 328, 329, 333, 336, 345

Baborovská 203
Baker 71, 73, 88, 91, 163, 167, 168
Balbierz 337
Baldwin 292, 293, 396
Ballantyne 403
Bandrowsky 283
Bangham 87, 91, 257, 259, 295, 296, 299, 309, 310
Barnum 99
Baxi 299
Bearn 271, 272, 276, 277, 283, 293
Bednekoff 155, 161, 476
Bell 147
Bellis 143, 148
Bence 274
Benditt 293
Benedict 117
Benoit 187, 191

Berg 273, 274, 276
Berliner 242, 243
Bettini 466
Bhatia 299
Bialy 381, 386
Biel 271, 272, 277
Billingham 205, 395, 405, 406, 411, 413
Bílková 406
Blackshaw 381, 386
Blumberg 272, 273, 276
Blunt 345
Bogart 381, 386
Bogden 197, 198, 201, 202, 203, 205, 206
Booth 395
Bordet 99, 103, 479
Borel 11, 37, 471, 473, 480, 481
Bouquet 11, 27, 33, 37
Bouw 11, 25, 27, 34, 37, 39, 42, 69, 72, 73, 74, 89, 91, 105, 262, 458, 486, 487
Bowley 215, 395
Bowman 174, 395
Boyd 401, 403
Böhm 11, 105, 169, 486
Braend 11, 20, 21, 27, 37, 39, 42, 63, 66, 67, 72, 74, 223, 228, 242, 243, 245, 251, 253, 258, 259, 262, 288, 290, 293, 313, 314, 319, 343, 344, 345, 423, 480, 483, 486, 489
Braun 203
Bräuner-Nielsen 161, 163, 167, 168, 306
Brdička 11, 197, 198, 203
Brent 395, 413
Briles 71, 74, 173, 177, 178, 179, 181, 185, 186
Broman 283, 293
Brooks 268, 277
Brucks 84, 85
Brummerstedt-Hansen 49, 61, 272, 276, 279, 293, 301, 306
Brunet 337

Burhoe   197, 201, 203, 205
Buruianu   369, 379
Busch   11, 445, 469, 479
Buschmann   11, 127, 169, 262, 301, 305, 306
Buys   37, 74

Cabannes   253, 259, 295, 299, 313, 319
Carr   302, 306
Castle   194
Celano   242, 243
Chai   195, 196
Chapman   299
Charles   293
Chepov   403
Chown   393
Chutná   402, 403, 405, 411, 413
Cleve   271, 272, 276, 277
Cloudman   414
Cohen   193, 194, 195, 196, 275, 277, 319
Cole   405, 412, 414
Connell   268, 277
Coombs   99, 100, 103, 124, 155, 156, 163, 164, 167, 207, 217, 424, 449, 467, 470, 477, 479, 480
Cotterman   214, 215
Craig   397, 399
Crockett   297, 299, 344
Czaja   93, 97
Czambelová   403
Černý   396
Čumlivský   396, 421
Čuta   11, 85,

Dabczewski   229, 245
Dahr   194, 471, 476
Damerow   471, 476
Davidsohn   476
Datta   49, 55, 61, 72, 74, 161, 301, 304, 306, 476
Deckart   333, 336
De Ligny   11
Derlogea   431
De San Martin   11
Deutsch   231, 234
Dikov   11, 447
Dixon   268, 277, 448
Doerr   474, 476

Dola   11, 39, 42
Doria   405, 413
Dornetzhuber   403
Dostál   11
Dörner   333, 336
Dray   272, 276
Drobná   12, 467
Dubiski   97
Duggleby   243
Dujarric de la Riviére   203, 472, 473, 476
Duncan   299, 310
Dunsford   214, 215, 395

Ebertus   12, 63, 105, 439
Edelman   274, 277
Efremov   12, 253, 259, 262, 313, 314, 318, 319, 344, 345, 424, 486
Ehrlich   28, 37, 394, 395, 448, 479
Elliot   63, 67, 174
Erhard   12, 30, 43, 47, 235, 243, 457
Erlandson   273, 276
Esslová   396
Evans   313, 318, 319, 345
Eyquem   12, 156, 159, 161, 197, 203, 205, 217, 229, 262, 263, 387, 401, 403, 472, 473, 476, 479, 487

Fabián   12, 193, 196, 402, 403
Fahey   274, 277
Falconer   133
Fallon   301, 306
Feldman   401, 403
Felton   423
Fenton   237, 243
Ferguson   25, 26, 27, 28, 37, 39, 42, 63, 67, 69, 74
Fésüs   12
Fine   473, 476
Finney   119
Fisher   59, 71, 194, 401, 403
Fishman   287, 293
Flammarion   203
Fleischer   12
Fleischman   274, 277
Foord   396
Franklin   275, 277, 278
Franks   55, 61, 229, 234, 237, 243
Fudenberg   275, 277, 278, 473, 474, 477

Frenzl   12, 197, 198, 199, 203, 205, 399
Freund   206, 276, 402, 449
Friedberger   197, 203
Friedman   161, 476
Friedmann   26, 37
Fulka   12, 381, 382, 386
Furuhata   471, 476

Gahne   12, 42, 54, 61, 87, 91, 217, 225, 228, 254, 259, 301, 304, 306, 307, 311, 312, 329, 486
Gajos   251, 338, 341
Galatius-Jensen   120
Gasparska   12
Gasparski   12, 39, 40, 42, 93, 97, 106
Gatti   466
Gavrilet   431
Gerner-Nowakowa   97
Giblett   268, 269, 270, 277, 311, 312
Gilman   230, 231, 234, 242, 243
Gilmour   173, 174, 177, 178, 179, 180, 185, 186
Gippert   12
Golders   122
Goodwin   122
Gorer   405, 406, 412, 413
Gowe   173, 178
Grabar   351, 358, 360, 368
Graetzer   279
Granciu   12, 106, 431
Greenwood   421
Gregory   74
Grimes   295, 299, 309, 310
Grodecka   194
Grosclaude   12, 30, 31, 32, 37, 79, 85, 105, 106, 458, 487
Grosdanovič   402, 403
Grubb   274, 276, 277

Hafs   384, 386
Hála   12, 155, 156, 158, 161, 163, 164, 167, 168, 217, 472, 476
Hall   12, 106, 261
Hancock   402, 403
Handler   293
Handscombe   147
Hara   473, 476
Harboe   13, 106, 169, 170, 267, 274, 275, 276, 277, 345, 424

Hardy   163, 195
Harris   112, 121, 295, 299, 313, 319, 320, 448
Hašek   13, 19, 20, 21, 391, 392, 393, 395, 396, 399, 421, 423, 448, 467, 489
Hašková   396, 397, 405, 411, 413
Hauge   122
Haupt   271, 272, 277
Havskov Sørensen   307
Hedal   194, 401, 403
Heide   271, 272, 277
Helle   319
Helmbold   194
Hernandez   13
Hess   13
Hesselholt   13, 49, 56, 57, 61, 112, 114, 119, 221, 228, 229, 230, 231, 235, 237, 243, 261, 262, 279, 281, 293, 306, 343, 344, 345
Hibino   197
Hickman   87, 91, 269, 277, 301, 307, 311, 312
Hildemann   473, 476
Hilgert   406
Hiller   109, 122
Hirschfeld   270, 271, 272, 273, 276, 277
Hložánek   13
Hoecker   413, 414
Hojný   13, 155, 156, 158, 161, 163, 164, 166, 167, 168, 321, 327, 329, 472, 476, 487
Holmberg   283, 293
Holz   47
Horsfall   272, 277
Hort   13, 396, 399, 467
Horváth   13
Hosoda   235
Højgaard   161, 301, 307
Hraba   392, 395, 396, 421
Hradecký   13
Hughes   275, 278
Huisman   295, 299, 318, 320, 424
Hunter   384, 386
Hutchinson   215, 395

Icha   386
Ikin   395
Imlah   13, 106, 109, 112, 114, 116, 119, 279, 283, 293, 343, 424, 486

Ingram 13, 99, 100, 103, 106, 155, 159, 161, 169
Irwin 25, 26, 37, 38, 39, 42, 44, 47, 61, 63, 67, 72, 74, 95, 179, 180, 186, 359, 368
Iványi 13, 193, 194, 196, 200, 393, 396, 401, 402, 403, 424, 467
Iványiová 402, 403
Iya 433, 434

Jackson 293
Jaemeri 197
Jaffe 397, 399
Jakóbiec 39, 42, 211
James 395
Jamieson 63, 67, 106
Jenkins 396
Jensen 197
Johansson 54, 61
Jones 274
Johns 302, 307
Jonsson 270, 277
Johnston 293
Josipovič 13
Jovanovič 13, 105, 149, 453
Jover 13, 459
Joysey 194, 401, 403
Jylling 161

Kabat 276
Kaczmarek 14, 123, 229, 235, 242, 243
Kaempffer 155, 161
Kaminski 14, 245, 251, 337, 338, 340, 341, 345
Kapitchnikov 401, 402, 403
Keeler 194
Kellner 194, 401, 403
Kendrick 480
Kennedy 480
Kenning 448
Kevek 14
Khanolkar 295, 299
Kiddy 291, 293
King 112, 114, 118, 122, 147, 148, 279, 285, 293, 319
Kindler 401, 403
Kinský 399
Kirk 272, 277
Kitamura 477

Klein 14, 405, 411
Kleinman 272, 277
Knížetová 14, 397, 399
Knopfmacher 194
Koger 299
Končar 14, 105, 453
Konugres 477
Kopečný 14
Kořínek 358
Koubek 14
Koutková 14
Kovácz 14, 345, 486
Kownacki 229, 234, 235
Kraay 14, 74, 87, 89, 91, 106
Kracht 243
Král 203
Krasińska 97
Kraus 203
Kristjansson 87, 91, 112, 115, 122, 223, 228, 254, 259, 279, 293, 360, 368
Krummen 14, 436
Krüpe 242, 243, 475, 476
Křen 14, 197, 199, 203, 205, 399
Křenová-Peclová 200, 203
Kunkel 275, 276, 277

Lackman 102, 103
Lampkin 395, 396, 421
Landsteiner 26, 32, 33, 37, 242, 394, 396, 475, 476
Lane 215, 396, 421
Lang 14
Lange 242, 243
Larsen 14, 49, 54, 61, 161, 293, 306
Lasley 381, 386
Lassiter 299, 310
Latter 173, 178
Laurell 274, 277, 283, 293
Laza 14
Lederer 358
Lee 475, 476
Lehmann 257, 259, 295, 299
Lehnert 55, 61
Leiva 14
Lengerová 396
Levi 395, 396
Levine 213, 215, 242, 243
Li 91
Lie 15, 72, 74

Lille-Szyckowicz   235, 242, 243
Lindström   15, 461, 480, 487
Linnet-Jepson   109, 120
Lipecka   15
Little   26, 37
Lundsden   197, 203
Lutwak-Mann   381, 386

Mácha   15
Madden   140, 147
Madeyska-Lewandowska   15
Maijala   461
Malandkar   298
Mančič   43, 47, 105
Mann   371, 381, 386
March   396
Marcussen   194
Mardiak   467
Martensson   276, 277
Martin   293
Martínek   15
Martínková   406
Mason   51, 61
Mather   88, 91
Matoušek   15, 19, 20, 21, 22, 80, 85, 167,
   168, 331, 333, 336, 349, 359, 360, 369,
   382, 385, 386, 387, 486, 487, 489
Matsumoto   194, 230, 235
Matsuzawa   477
McAllister   283, 293
McDermid   15, 173, 174, 178, 217, 397,
   399, 424, 459, 486
McDougall   301, 306, 313, 319
McGibbon   74, 179, 186
McIndoe   319, 320
Medawar   395, 401, 403, 405, 413
Meltzer   275, 277
Mendel   112, 177, 188
Mervartová   15
Michalec   358
Michl   406
Míček   15
Mikešová   15
Mikulska   406, 413
Millar   383, 386
Miller   38, 61, 79, 85
Millot   15, 37, 75, 79, 85, 217, 449, 476, 479
Milovanov   381, 386
Mishima   477

Mitchison   393, 396, 399, 405, 412, 413
Mitscherlich   28, 37
Moor   472
Moor-Jankowski   477
Moore   415, 421
Moores   395
Morell   283, 293
Morgan   366, 368, 474, 476
Morgenroth   28, 37, 394, 395
Mori   474, 476
Morris   38, 74, 425
Morton   49, 61, 231, 234
Moustgaard   15, 49, 61, 72, 74, 147, 161,
   217, 279, 293, 304, 306, 307, 343, 433,
   480, 486, 487
Möller   405, 411, 412, 413
Møller   49, 61, 161, 293, 306, 307
Mullan   402, 403
Munk-Andresen   159, 161
Musil   358
Müller   435

Naik   15, 295, 299, 344, 425
Nair   433, 434
Nance   268, 277
Nasrat   34, 37, 69, 74, 89, 91
Neimann-Sörensen   27, 28, 33, 37, 39, 40,
   42, 67, 72, 74, 89, 91, 161, 441, 472, 476
Nelken   194
Newton   137
Nicholas   395, 396
Nielsen   49, 59, 61, 155, 293
Nikolajczuk   15, 262
Nordskog   182, 186, 397, 399
Novotný   15

Ogden   313, 320
Okabe   472, 477
Okerman   15, 424
Oprescu   16
Osterhoff   63, 67, 301, 302, 309, 310, 311,
   312, 344, 416
Osterland   275, 276, 277
Osterlee   16, 344, 345
Otte   26, 38
Ottenberg   26, 37
Ottensooser   243
Owen   25, 38, 39, 42, 43, 47, 74, 197, 198,
   201, 203, 205, 214, 215, 391, 395, 396

Pain 274, 277
Palludan 49, 293, 306
Palm 198, 202, 203, 205, 206
Papp 16
Parker 272, 277, 406
Pauling 479, 480
Pavlok 16, 386
Pavlu 369, 379
Payne 397, 399
Perramon 16, 179
Petrovská 369
Petrovský 16, 349, 358, 368, 369, 384, 386
Pilz 16, 429
Pizarro 405, 412, 414
Plaut 395
Podliachouk 16, 55, 56, 57, 61, 156, 159,
  161, 187, 221, 224, 228, 229, 476
Pokorná 403
Pollitzer 243
Popovici 229, 235
Porter 274, 277
Poulik 254, 259, 274, 277, 283, 287, 288,
  293, 360, 368
Poviet 450
Powell 174
Prunier 272, 277
Puza 393, 396

Queval 229, 235
Quittet 82, 85

Race 71, 150, 215, 395, 471, 476
Ramos 414
Rapacz 16, 39, 42, 74, 211, 215, 217
Rasmusen 71, 74, 155, 159, 161, 473, 476
Rasmuson 270, 277
Ray 434
Readler 477
Reckel 38
Reinskou 270, 271, 272, 277
Rendel 28, 37, 39, 42, 54, 61, 63, 67, 72,
  74, 97, 221, 225, 228, 259, 307, 441, 465,
  466, 473, 476, 486
Renkonen 37
Rhode 61, 221, 223, 228, 235
Richards 102, 103
Riddel 273, 276
Rivat 275, 277
Roberts 421

Robertson 89, 91, 173, 178
Robinson 273, 276
Rodero 16, 459
Ropartz 275, 277
Rose 476, 477
Rosenberg 197
Rous 241, 436
Rousseau 275, 277
Rowson 415, 416, 420, 421
Rubinstein 414
Russell 405, 412, 414
Rusu 431
Růžička 358
Rychlíková 403

Saison 155, 159, 161, 211, 215, 473, 477
Salerno 16, 229, 463, 466
Salisbury 295, 299, 310, 381, 386
Sanger 150, 215, 395, 471, 476
Sanghvi 295, 299
Santner 47
Sas — Kortsak 283, 293
Schacht 272, 277
Schechtman 396
Scheinberg 38, 283, 293
Schermer 26, 38, 472, 477
Scheuch 43, 47
Schierman 182, 186, 397, 399
Schiff 474, 477
Schindler 435
Schmid 16, 43, 45, 47, 105, 229, 230, 235,
  237, 242, 243, 245, 251, 261, 262, 288,
  293, 301, 305, 306, 307, 309, 310, 457
Schott 155, 161
Schröffel 16, 85, 168, 321, 327, 329, 331,
  333, 336, 360, 368
Schultz 16
Schultze 271, 272, 277
Schwarz 234, 242, 243
Schwerdtner 16
Sebens 320
Sedláková 16
Serain 253, 259, 295, 299, 313, 319
Shackelford 211, 215
Shaw 72, 74
Shreffler 295, 299, 309, 310
Sieblitz 47
Silvers 405
Simmons 405, 412, 414

Sirbu   229, 235, 431
Slee   415, 416, 418, 420, 421
Smetana   203
Smith   147, 148, 293, 381, 386, 393, 396
Smithies   87, 91, 109, 112, 122, 254, 259,
    267, 268, 269, 271, 277, 293, 301, 304,
    307, 311, 312, 321, 322, 329, 360, 368
Snell   412, 414
Socean   431
Soos   17
Sorokovoj   17
Spiteri   205
Sprague   38, 61
Springer   473, 477
Spryszak   39, 40, 42
Stanworth   275
Steinberg   274, 276, 278
Stetkiewicz   155, 161
Stetson   403
Stevens   384, 386
Stojanovič   17, 149, 169
Stone   28, 37, 38, 49, 55, 61, 63, 67, 72,
    73, 74, 85, 97, 161, 180, 186, 301, 306,
    359, 368, 472, 476, 477
Stormont   17, 25, 26, 27, 28, 31, 32, 34, 37,
    38, 39, 42, 44, 47, 49, 52, 55, 61, 63, 67,
    69, 72, 74, 75, 79, 80, 83, 85, 87, 91, 169,
    214, 215, 221, 222, 223, 224, 228, 230,
    235, 242, 243, 251, 253, 258, 259, 261,
    288, 290, 293, 343, 395, 396, 419, 421,
    423, 424, 425, 474, 477, 479, 480
Stratil   349, 368, 386
Student   468
Stukovsky   17
Sturgeon   396
Suzuki   38, 61, 72, 74, 85, 221, 222, 223,
    224, 228, 230, 235, 253, 425, 474, 477, 480
Szabo   122
Szczekin-Krotow   39, 42
Széky   196, 403
Szeniawska   17
Szent-Ivanyi   124
Sziszkowicz   401, 403
Szymanowski   155, 161
Szynkiewicz   97
Šereda   17
Šiler   17
Šmerha   17
Štark   17, 197, 198, 203, 205, 397, 399
Šulcová   381, 386

Takahashi   476
Taslokwa   197, 203
Terajima   473, 475, 477
Thompson   215, 291, 292, 293, 395
Thomsen   475, 477
Thymann   279
Tichonov   17, 424, 437
Tobies   396
Tolar   193
Tolarová-Koutková   193, 401
Tolle   26, 28, 37, 38, 161, 439, 441, 476
Tomášková   193, 194, 196
Tomcsik   474, 477
Townsend   275, 278
Tucker   17, 28, 38, 345, 415, 423, 424, 425,
    481
Tyan   405, 412, 414
Tyler   74

Ueno   471, 477
Uhlenbruck   242, 243
Ujhélyiová   403
Urbaschek   28, 37
Uriel   283, 288, 293, 351, 358

Valenta   17, 349, 358, 359, 360, 368, 381,
    382, 385, 386
van der Helm   299
van Furth   272, 277
Van Heerden   301, 309, 310, 311, 312
Van Vliet   299, 320
Varga   17, 193, 401
Vasilev   17
Venge   54, 61, 259, 307
Veselý   203
Visser   299
Vojenčiak   17
Vojtíšková   396
von Jettmar   99, 103

Wachler   155, 161
Wadowski   229
Wake   292, 293
Waler   275, 277, 278
Walker   267
Wallerstein   234, 242, 243

Warren 295, 299, 313, 319, 320
Watkins 366, 368
Weir 215, 396, 421
White 287
Wiatroszak 97, 123
Widdowson 17, 137
Wiener 71, 74, 472, 477
Williams 421
Williamson 477
Wilson 278
Winstanley 477
Witebsky 472, 477
Wojciechowska 229, 235, 242, 243
Wood 396
Woodworth 414
Wright 194

Wróblewski 123, 167, 168, 229, 235
Wunderly 349, 358

Yamaguchi 230, 235
Ycas 28, 38
Yokoyama 473, 474, 477

Zavřel 17
Zeki 474, 477
Zilber 395, 396
Zotikov 402, 403
Zumpft 413
Zwolinski 229, 235, 242, 243
Želev 17
Žurkowski 17

Blood Groups of Animals

E R R A T A

page	line	for	read
12	3	Artificia	Artificial
59	13	facort	factor
89	13	the questio ben	the question of
89	14	can of	can be
150	26	ol	of
193	19	form	from
201	Table 4	Hydrid group	Hybrid group
231	42	weak	week
273	13	pricipitin	precipitin
334	29	gonadotrophinsalone	gonadotrophins alone
351	Fig. 3.	Fig. 3. Phenotypes aBC etc.	Fig. 2. Phenotypes Ab etc.
365	Table 3	31	21
381	26	from in cauda	from cauda
384	Fig. 1.	Fluids unsed	fluids used
385	1	penhomenon	phenomenon
434	1	indivated	indicated
487	30	aminals	animals